Advances in Robot Control

Sadao Kawamura   Mikhail Svinin   (Eds.)

Advances in Robot Control

From Everyday Physics
to Human-Like Movements

Springer

Professor Sadao Kawamura
Robotics Department
Faculty of Engineering
Ritsumeikan University
Kusatsu, Shiga, 525-8577
Japan

Doctor Mikhail Svinin
Bio-Mimetic Control Research Center
RIKEN
Anagahora, Shimoshidam
Moriyama-Ku, Nagoya 463-0003
Japan

Library of Congress Control Number: 2006930409

ISBN-10 3-540-37346-2 Springer Berlin Heidelberg New York
ISBN-13 978-3-540-37346-9 Springer Berlin Heidelberg New York

This work is subject to copyright. All rights are reserved, whether the whole or part of the material is concerned, specifically the rights of translation, reprinting, reuse of illustrations, recitation, broadcasting, reproduction on microfilm or in any other way, and storage in data banks. Duplication of this publication or parts thereof is permitted only under the provisions of the German Copyright Law of September 9, 1965, in its current version, and permission for use must always be obtained from Springer. Violations are liable for prosecution under the German Copyright Law.

Springer is a part of Springer Science+Business Media
springer.com
© Springer-Verlag Berlin Heidelberg 2006

The use of general descriptive names, registered names, trademarks, etc. in this publication does not imply, even in the absence of a specific statement, that such names are exempt from the relevant protective laws and regulations and therefore free for general use.

Typesetting: by the authors and techbooks using a Springer LaTeX macro package
Cover design: *design & production* GmbH, Heidelberg

Printed on acid-free paper   SPIN: 11588429   89/techbooks   5 4 3 2 1 0

This book has been written by friends, former students and close collaborators of Suguru Arimoto on the occasion of his 70th birthday. It was an unforgettable memory and a great pleasure to cooperate with him and, for some of us, work under his supervision. In greatest respect to his strong and inquiring mind, his enthusiasm in the pursuit of the science, and his passion in education we dedicate our book to him.

Preface

Robotics is still a young science, but we can already identify the people who defined its primary course of development. Suguru Arimoto is one of them. His early works laid the foundations of what nowadays is called modern robot control, and we believe it is both appropriate and necessary to write a book on recent advances in this field in the context of his scientific interests.

While presenting recent advances in robot control is the main intention of this book, we also think it is appropriate to highlight Suguru Arimoto's research career, main scientific achievements, and his personality, too. This can be very inspiring and instructive, especially for young researchers.

What are the most remarkable features of Suguru Arimoto? On the personal side, his vitality is striking. He is always focused on a research target, and it is always a fun and a pleasure to discuss with him scientific problems and to learn from him. His passion to explain things that might not appear obvious is endless. It is very encouraging to younger researchers that, at this stage of his career, he is still a very active, approachable, and influential researcher, and a person who leads by example. On the scientific side, we should stress his research philosophy. He believes that the final result should be simple and have a clear physical (or physiological, in his recent research) interpretation. This simplicity is always supported by rigorous mathematical proofs. You can see this in all of his papers. He advocated and articulated this approach throughout his influential career, and we have tried to adopt it for this book.

The book begins with a welcome message—a short essay of Suguru Arimoto on his perception of human robotics. It clearly outlines two important research problems: the physics-based robot control for coping with the so-called everyday physics problems on one hand, and the challenge of reproducing beautiful, human-like movements on the other hand. Attacking these problems defined much of his research career, at least as far as robotics is concerned, and not coincidentally we split the book into two corresponding parts. Part I, stretching in the keyword space from everyday physics to robot control, deals with the physics-based principles and control algorithms, while

Part II links robot control with the control of biological machines. In particular, it is directed to the understanding of the control of the best manipulator yet developed, the human arm, and the most versatile tool, the human hand.

Part I starts with a study on a natural motion of robotic manipulators. Many great scientists tried to define natural motion in the realm of the analytical mechanics. This topic has a very rich history but still attracts the attention due to its irresistible appeal which is only increased with the advent of control. Surely, the pursuit to find out a definition of the natural motion will continue for a long time. In this connection, Chapter 1 introduces a new viewpoint by linking the definition to the singularity of the inverse kinematic solutions. The problem is addressed here via a non-linear, differential-geometric approach. A natural motion component is first identified at the kinematic level, and then a dynamic analysis reveals the essence of this component as a nondissipative motion along the prescribed end-effector path.

Chapters 2, 3, and 4 are conceptually closely connected. They have grown up from the original research of Suguru Arimoto on quasi-natural potential feedback and model-based adaptive control. Chapter 2 introduces and analyzes several adaptive control laws for robot manipulators with uncertainties in kinematics and dynamics. The main feature of the controllers proposed is the use of approximate manipulator's Jacobian and the analysis how the approximation may affect the stability property. Chapter 3 extends the formulation of the control task by addressing problems of visual servoing. Here, the adaptive controllers utilize feedback signals from the camera image plane. The key idea is the use of a depth-independent image Jacobian matrix, which enables the unknown camera parameters to enter linearly the closed-loop dynamics and resolves the bottleneck problem of the conventional visual servoing control techniques. This is supported by a rigorous proof of asymptotic convergence of the image errors when the nonlinear dynamics of the robot are fully taken into account, and confirmed by experiments. Chapter 4 addresses the constrained manipulation and extends Arimoto's orthogonalization control principle to the case of visual servoing. In the extended principle it is possible to fuse the image coordinates into orthogonal complements of the joint velocities and the contact forces. Based on this principle new adaptive controllers, featuring exponential convergence for the position and force errors, can be constructed.

Chapters 5 and 6 can be traced back to Arimoto's notion of potential energy shaping that was the precursor for the passivity-based control. Chapter 5 treats the multi-agent networked coordination and control problem from an input-output passivity perspective. This novel formulation unifies many results available in the literature, and offers constructive tools for solving a wide variety of problems such as the open problem of exponential synchronization of Kuramoto oscillators, the output synchronization of Lagrangian systems, and the multi-robot coordination problem. Chapter 6 brings the idea of an artificial potential into the realm of nonholonomically constrained mechanical systems. This is done through the introduction of special navigation functions.

Using these functions, one can address the control of nonholonomic systems in dynamic settings by lifting conventional 1st order (gradient-based) hybrid feedback controllers to their 2nd order dynamical counterparts.

Chapter 7 deals with the iterative learning control, another mainstream of Arimoto's research. The novel idea introduced in this chapter is the acquisition of the input torque patterns by iterative learning with the use of non-linear time scale transformations. This can be done without estimating the unknown parameters of the robot and the environment. The effectiveness of this idea is verified under experiments on control of a geometrically constrained robot and an underwater manipulator.

Part II of the book starts with the expository Chapter 8 on what robotics may learn from the brain. It considers functional modeling of the central nervous system, stressing on its modular, feedback-based architecture. Using nonlinear contraction theory, a comparatively recent analysis tool, this chapter analyzes synchronization as a model of computations at different scales in the brain and discuss its applications in robotics and system neuroscience.

Chapter 9 and 10 deal with the control of muscle-activated systems. Chapter 9 establishes the mathematics of the force and stiffness analysis and comment them in the context of human motor control. The key issues here are the force-dependent stiffness and the stabilizing effect of muscle impedance. This chapter presents many ideas about force production in the mammalian biomechanical system that can be inspiring in designing new types of actuators and control schemes. Chapter 10 analyzes dynamic control mechanisms of human musculo-skeletal system using a Hill-like type muscle model. The sensory-motor control strategy used here is an extension of the original Arimoto's research on modeling of reaching movements and resolving Bernstein's problem of redundant degrees of freedom. The strategy is tested for a reaching task, where it is shown it produces human-like reaching movements featured by the bell-shaped velocity profiles, and for the task of pinching an object by two fingers.

The following Chapter 11 shows that the principle of superposition, introduced by Suguru Arimoto for the control of multi-fingered robotic hands, is applicable to the control of prehensile actions by humans. This is a remarkable example when progress in robotics proves to be fruitful for analysis of human hand actions. As reported in this chapter, the principle of superposition holds with respect to reactions to expected and unexpected mechanical perturbations applied to a hand-held object.

Chapter 12 deals with modeling of human-like reaching movements in the manipulation of a multi-mass flexible object (underactuated system) with the elimination of residual vibrations. This a complex, sport-like movement task where the shape of the hand velocity profiles can be quite different from the classical bell shape and may feature multiple phases. This chapter develops a minimum hand jerk model that takes into account the dynamics of the flexible object, and shows that it gives a satisfactory prediction of human movements.

The concluding Chapters 13 and 14 are on the applied side of human robotics. Chapter 13 directly connects the everyday physics with the human movements. It reports the development of a robot force control application in order to augment human haptic feedback during human-robot co-manipulation tasks. The application features a push-poke theory of tool-environment interaction and an adaptive force scaling algorithm that estimates the environment compliance on-line and provides asymptotically exact force scaling/reflection for smooth time-varying user-applied forces. Shifting the focus from cooperation to competition, Chapter 14 presents a table tennis robot that rallies with a human being. Here, the challenge of hybrid control problems is met by learning input-output maps, used for motion planning, by means of locally weighted regression. The design also features a feed-forward control scheme based on iterative learning.

The editors would like to express their sincere gratitude to everyone who contributed to this book. We are also very grateful to Dr. Thomas Ditzinger and Ms. Heather King of Springer-Verlag for their guidance and invaluable help. It was a great pleasure to work with them and learn from them. We would also like to thank Prof. Bruno Siciliano for his support and encouragement in undertaking this wonderful project.

Kusatsu, Nagoya, *Sadao Kawamura*
August 2006 *Mikhail Svinin*

Curriculum Vitae: Suguru Arimoto

Suguru Arimoto was born on August 3, 1936 in Hiroshima, Japan. He received the B.S. degree in mathematics from Kyoto University, Kyoto, Japan, in 1959, and the Dr. Eng. degree in control engineering from the University of Tokyo, Tokyo, Japan, in 1967. From 1959 to 1961, he was with Oki Electric Industry Co. Ltd., Tokyo, Japan, as an Engineer in the Electric Computer Department. From 1962 to 1967, he was Research Assistant, and from 1967 to 1968, Lecturer, in the Department of Mathematical Engineering and Information Physics, University of Tokyo. In 1968, he joined the Faculty of Engineering Science, Osaka University, Osaka, Japan, as Associate Professor, and in 1973, he was promoted to Professor of Systems Engineering. In 1988, he was invited to join the University of Tokyo as Professor of the Department of Mathematical Engineering and Information Physics. In 1997, he retired from the University of Tokyo and moved to Ritsumeikan University, Shiga, Japan, where he contributed to the establishment of a new department. Since 1997, he has been a Professor in the Department of Robotics. His research interests are in information theory, control theory, cybernetics, robotics, and machine intelligence.

Professional Activities

During his professional career Dr. Arimoto served in the Robotics Society of Japan (AdCom Member, 1989-1991; Vice President, 1993-1995; President, 1995-1997), the Society of Instrument and Control Engineers (AdCom Member, 1991-1993), the Institute of Electronics, Information and Communication Engineers (Chairman of the Society for Fundamentals of Electronics, Communications and Computer Sciences, 1989-1991), the IEEE (Member of the Board of Givernors of the IT Society, 1985-1989; Chairman of the Tokyo Chapter for the IT Society, 1988-1990; Chairman of the Tokyo Chapter for the Robotics and Automation Society, 1990-1992), the Institute of Systems, Control, and

Information Engineers (AdCom Member, 1978-1985), and the Society for Information Theory and Its Applications (Vice President, 1992-1993; President 1994-1995).

He also served in the Science Council of Japan (the Professional Committee of Mechatronics; AdCom member, 1994-1997; Chairman, 1997-2000), and in the Japanese University Accreditation Association (Member of the Accreditation Committee, 1999-2005).

He was Program Chairman of the IEEE International Conference on Information Theory, 1988, Kobe, Japan, the Japan-USA Symposium in Flexible Automation, 1988, Minneapolis, Minnesota, and the IEEE International Conference on Robotics and Automation, 1995, Nagoya, Japan.

He was on the editorial boards of the Robotica Journal (Deputy Editor, 1980-2005; Associate Editor, 2005-present time), the International Journal of Robotics Research (Associate Editor, 1980-2002), Journal of Robotic Systems (Associate Editor, 1984-present time), IEEE Transactions on Robotics and Automation (Associate Editor, 1996-1999), Autonomous Robots (Associate Editor, 1994-present time), the International Journal of Intelligent Control and Systems (Associate Editor, 1997-present time), Journal of Circuits, Systems and Computers (Associate Editor, 1991-2003), Systems and Control Letters (Associate Editor, 1996-present time), Journal of Chinese Institute of Engineers (Associate Editor, 1997-present time), Annals of the Institute of Statistical Mathematics (Associate Editor, 1993-2000), International Journal of System Science (Associate Editor, 1973-1997), International Journal of Adaptive Control and Signal Processing (Subject Editor, 2004-present time), IEICE Transactions on Fundamentals (Editor-in-Chief, 1994-1996), and SICE Transactions (Deputy Editor, 1991-1992; Editor-in-Chief, 1992-1993).

Awards and Honors

1967 The Yonezawa Promotion Award from the Institute of Electronics, Information and Communications Engineers (IEICE)
1968 The Best Paper Award from the Society for Instrument and Control Engineers (SICE)
1974 The Best Paper Award from the IEEE Information Theory Society
1976 The Best Paper Award from the SICE
1983 The IEEE Fellowship
1987 The Best Paper Award from the SICE
1987 The Sawaragi Memorial Award from the Institute of Systems, Control, and Information Engineers (ISCIE)
1994 The Best Paper Award from the Robotics Society of Japan (RSJ)
1997 The Meritorious Award from the Japan Society of Mechanical Engineers (JSME) on the 100th Anniversary of the JSME
1998 The Meritorious Award from the Robotics & Mechatronics Society of the JSME

2000 The IEICE Fellowship
2000 The IEEE Third Millennium Medal
2000 The Royal Medal with a Purple Ribbon from the Japanese Government
2002 The Russel Springer Professorship, University of California at Berkeley, California, USA
2003 The Best Book Publication Award from the SICE
2003 The Best Paper Award, the Japan-USA Symposium on Flexible Automation
2003 The RSJ Fellowship
2005 The JSME Fellowship
2005 The Honorary Membership from the Society of Information Theory and Its Applications (SITA)
2006 The Pioneer in Robotics and Automation Award from the IEEE Robotics and Automation Society (RAS) for *"his work on PD and PID control, iterative learning control, and passivity-based control of nonlinear mechanical systems, that represents a source of reference for virtually any scientists dealing with complex robotic systems."*

List of Ph.D. Students

1. Youssef Gaafar, "Nonlinear control theory," Osaka University, 1977.
2. Hikaru Shimizu, "Modeling of the total amount of water inflow to Lake Biwa by nonlinear Kalman filtering," Osaka University, 1977.
3. Takanori Minami, "Vibration suppression for automobiles," Osaka University, 1979.
4. Yoshimi Monden, "Fast algorithms in signal processing," Osaka University, 1980.
5. Takeshi Hashimoto, "Channel coding theorems for convolution coding," Osaka University, 1981.
6. Toshihiko Inoue, "Measurement and prediction of two-phase flow," Osaka University, 1981.
7. Morikazu Takegaki, "Nonlinear control theory of robot manipulators," Osaka University, 1981.
8. Fumio Miyazaki, "Hierarchical control for biped robots," Osaka University, 1982.
9. Yasuo Tsukamoto, 'Numerical methods in inverse problems," Osaka University, 1983.
10. Hiroyoshi Morita, "Image data compression based on Shannon's entropy," Osaka University, 1983.
11. Akira Kikuchi, "Design of digital filters," Osaka University, 1984.
12. Sadao Kawamura, "Iterative learning control," Osaka University, 1986,
13. Hiroshi Noborio, "Octrees approach for robot path planning," Osaka University, 1987.

14. Kouji Nagaoka, "Linear stochastic systems theory," Osaka University, 1987.
15. Hisashi Suzuki, "Learning through self-organized and tree-structured data base," Osaka University, 1988.
16. Shiro Tamaki, "Optimal control algorithms for a class of linear discrete time systems," Osaka University, 1988.
17. Ho-Gil Lee, "Control of flexible robots," Osaka University, 1989.
18. Takahiro Masuda, "Modeling of complex robot dynamics," Osaka University, 1989.
19. Nobuo Ishimura, "Ship positioning by simulated rader analysis through marine charts," Osaka University, 1989.
20. Hiroshi Sugiyama, "Extension of Shannon's sampling theorem," The University of Tokyo, 1991.
21. Yun-Hui Liu, "Algorithmic study on path and motion planning of robots," The University of Tokyo, 1992.
22. Hajime Sato, "Zero-error problems in information theory," The University of Tokyo, 1993.
23. Tsutomu Kawabata, "Universal source coding," The University of Tokyo, 1993.
24. Tomohide Naniwa, "Coordinated control of dual robot arms," The University of Tokyo, 1994.
25. Vicente Parra-Vega, "Adaptive sliding mode control for robot manipulators," The University of Tokyo, 1994.
26. Hiroshi Koga, "Source coding with fidelity criterion," The University of Tokyo, 1995.
27. Hiroshi Kameyama, "Pattern recognition of hand-written letters based on knowledge base," The University of Tokyo, 1996.
28. Tetsuya Yuasa, "Image reconstruction for medical CT based on a novel modality," The University of Tokyo, 1997.
29. Kouzou Kawada, "Matched filters for pattern recognition," The University of Tokyo, 1997.
30. Yoshifumi Kawaguchi, "Development of a climbing robot for detection of failures," The University of Tokyo, 1997.
31. Soon-Yong Yang, "Development of stroke detection cylinders for construction machines," The University of Tokyo, 1997.
32. Hiroyuki Ohnishi, "Pattern matching based on detection of rotation and translation by using Hough and Fourier transforms," Ritsumeikan University, 1999.
33. Anh Nguyen, "Dexterous manipulation of objects by single fingers and pairs of fingers with soft-tips," Ritsumeikan University, 2002.
34. Kenji Tahara, "Sensory feedback for dynamic stable pinching and orientation control of an object by means of a pair of robot fingers," Ritsumeikan University, 2003.
35. Ji-Hun Bae, "Object manipulation by a pair of robot fingers from the bio-mimetic viewpoint," Ritsumeikan University, 2004.

36. Hiroe Hashiguchi, "A dynamic control method based on stability on a manifold for a family of redundant robots and under-actuated robots," Ritsumeikan University, 2006.
37. Morio Yoshida, "2-D and 3-D object grasping and manipulation under nonholonomic constraints," Ritsumeikan University, in progress.
38. Masahiro Sekimoto, "The generation of human-like multi-joint reaching movements under DOF redundancy," Ritsumeikan University, in progress.

Main Scientific Contributions

1 Information Theory

People in the robotics research community are perhaps unfamiliar with the fact that in the beginning of his research career Professor Arimoto was very active in information theory. Here, one should mention the paper [1], where a type of error-correcting code similar to the so-called Reed-Solomon code was established, and an efficient decoding algorithm equivalent to Peterson's algorithm was found. The code is nowadays popular in IT technology. It is implemented in VLSI-chips in compact disks and digital video disks. A summarized paper was published lately in [2]. An efficient algorithm for computing a channel capacity of any type of memoryless channels was proposed in [3]. This algorithm was subsequently discovered also by R. Blahut in the same year, so nowadays it is called the Arimoto-Blahut algorithm. The contribution [3] received the Best Paper Award from the IEEE Information Theory Society. Another paper [4] concerning the strong converse to the coding theorem became fundamental and stimulated the derivation of new proofs of the channel coding theorems. Also related to this direction are the papers [5, 6, 7, 8, 9, 10].

Selected Publications

1. Arimoto S (1961) Encoding and decoding of p-nary group codes and the correction systems. Information Processing 2(6):320–325
2. Arimoto S (1962) On a non-binary error-correcting code. Information Processing in Japan 2:22–23
3. Arimoto S (1972) An algorithm for computing the capacity of arbitrary discrete memoryless channels. IEEE Trans. on Information Theory 18(1):14–20
4. Arimoto S (1973) On the converse to the coding theorem for discrete memoryless channels. IEEE Trans. on Information Theory 19(3):357–359
5. Arimoto S (1976) Computation of random coding exponent functions. IEEE Trans. on Information Theory 22(6):665–671

6. Hashimoto T, Arimoto S (1979) Computational moments for sequential decoding of convolutional codes. IEEE Trans. on Information Theory 25(5):584–591
7. Hashimoto T, Arimoto S (1980) Universally optimum block codes and convolutional codes with maximum likelihood decoding. IEEE Trans. on Information Theory 26(3):272–277
8. Hashimoto T, Arimoto S (1980) On the rate-distortion function for the nonstationary gaussian autoregressive process. IEEE Trans. on Information Theory 26(4):478–480
9. Hashimoto T, Arimoto S (1981) A hierarchy of codes for memoryless channels. IEEE Trans. on Information Theory 27(3):348–350
10. Morita H, Arimoto S (1983) SECT—a coding technique for black/white graphics. IEEE Trans. on Information Theory 29(4):559–570

2 Signal Processing

A matrix extension of Rouche's theorem to investigate the location of zeros of polynomial matrices and its application to to multivariate autoregressions was proposed in [1]. A fast algorithm for fitting ARX (m, n) models and determining their orders from the covariance and cross-covariance information of input and output processes was presented in [2]. Compared to the usual Cholesky decomposition method, which requires a number of operations proportional to $O[(m+n)4]$, this algorithm reduces the computational complexity to $[O(m + n)2]$. Next, a new method for statistical design of approximately linear-phase autoregressive-moving average (ARMA) digital filters was introduced in [3]. The key idea of this method is that a time-delayed ARMA filter is used to approximate a high-order FIR filter that meets the prescribed amplitude spectrum sufficiently well.

Selected Publications

1. Monden Y, Arimoto S (1980) Generalized Rouche's theorem and its application to multivariate autoregressions. IEEE Trans. on Acoustics, Speech, and Signal Processing 28(6):733–738
2. Monden Y, Yamada M, Arimoto S (1982) Fast algorithm for identification of an ARX model and its order determination. IEEE Trans. on Acoustics, Speech, and Signal Processing 30(3):390–399
3. Monden Y, Komatsu T, Arimoto S (1984) Statistical design of nearly linear-phase ARMA filters. IEEE Trans. on Acoustics, Speech, and Signal Processing 32(5):1097–1100

3 General Control Theory

In 1964, prior to Potter's publication in 1966, an efficient algorithm for computing the unique positive definite matrix solution for a stationary matrix

Riccati equation was established in [1]. The Arimoto-Potter algorithm is based on the spectrum factorization of an extended Hamilton's matrix. By using this algorithm, optimal regulators as well as the Kalman filters, which are a class of linear dynamical systems with gain matrices in the state feedback defined as solutions to Riccati's matrix equations, can be computed [2]. Note that if the states are not accessible, it is normally recommended to use state observers for the state estimation. However, when calculating the optimal gain matrix, the overall quadratic performance index deteriorates owing to the incorporation of the state observer. This problem of performance deterioration was solved completely by the exact evaluation of the deterioration amount for the case the Kalman filter as a state estimator [3] and for the case of a minimum-dimension Luenberger's observer [4].

Selected Publications

1. Arimoto S (1964) An analytical design method for multi-variable control systems. In: Proc. Annual SICE Conf. Volume 1. 207–220
2. Arimoto S (1966) Optimal feedback control minimizing the effects of noise disturbance. Trans. of SICE 2(1):1–7
3. Arimoto S, Porter B (1973) Performance deterioration of optimal regulators incorporating Kalman filters. International Journal of Systems Science 4(2):179–184
4. Arimoto S, Hino H (1975) Performance deterioration of optimal regulators incorporating state estimators. International Journal of Control 19(6):1133–1142

4 Theory of Robot Control

4.1 PD Feedback for Robot Control by Means of Artificial Potentials

This is perhaps the most well-known contribution of Professor S. Arimoto. It was incubated in joint discussions with then Ph.D. student M. Takegaki and Assistant Professor F. Miyazaki. The original idea of both joint-space and task-space PD feedback schemes with damping shaping and gravity compensation first appeared in [1]. The global asymptotic stability for set-point position control by using PD feedback schemes was proved on the basis of Lyapunov's stability analysis. The use of Jacobian transpose associated with the task-space feedback was also initiated in this paper. In early 1980s this was truly a pioneering work in robot control. Nowadays, PD feedback with damping shaping becomes standard and is implemented in many experimental and industrial robotic systems. This control scheme is based upon the idea of introduction of an artificial potential function [2]. The PD feedback scheme was subsequently extended to a PID feedback control without compensating the gravity term [3], which assures asymptotic stability of the target equilibrium state in a local sense. The global asymptotic stability of such a PID feedback scheme for robotic arms was in the sequel established by introducing a saturated position error term, which was called the SP-ID feedback scheme [4]. The robustness of the task-space PD feedback control was formally established in [5, 6, 7].

Selected Publications

1. Takegaki M, Arimoto S (1981) A new feedback method for dynamic control of manipulators. Trans. ASME, Journal of Dynamic Systems, Measurement, and Control 103(2):119–125
2. Miyazaki F, Arimoto S, Takegaki M, Maeda Y (1984) Sensory feedback based on the artificial potential for robot manipulators. Proc. 9th IFAC Congress. Volume 8. 27–32
3. Arimoto S, Miyazaki F (1984) Stability and robustness of PID feedback control for robot manipulators of sensory capability. In Brady M, Paul R, eds.: Robotics Research: First International Symposium. MIT Press 783–799
4. Arimoto S (1995) Fundamental problems of robot control: Part I. Innovations in the realm of robot servo-loops. Robotica 13(1):19–27
5. Cheah C, Hirano M, Kawamura S, Arimoto S (2003) Approximate Jacobian control for robots with uncertain kinematics and dynamics. IEEE Trans. on Robotics and Automation 19(4):692–702
6. Cheah C, Kawamura S, Arimoto S, Lee K (2001) H-∞ tuning for task-space feedback control of robot with uncertain Jacobian matrix. IEEE Trans. on Automatic Control 46(8):1313–1318
7. Cheah C, Hirano M, Kawamura S, Arimoto S (2004) Approximate Jacobian control with task-space damping for robot manipulators. IEEE Trans. on Automatic Control 49(5):752–757

4.2 Passivity-Based Control

It was observed in [1] that not only a conjugate input-output pair composed of the torque control input vector and the joint angular velocity vector satisfies passivity but also a pair of the torque input and a linear sum of a saturated position error vector and the joint angular velocity vector satisfies passivity. This idea played a key role in establishing a new control-theoretic approach called "passivity-based control", which is now admitted to be a very effective tool for the control of nonlinear mechanical systems [2]. A nonlinear position-dependent circuit theory for the mechanical systems was outlined in [3]. The passivity-based approach was applied to the tracking control of robotic arms [4], force control in geometrically constrained manipulations [5, 6], and to the cooperative control of multiple robots carrying a common object [7, 8]. The basic ideas of this approach in the applications to robot control were highlighted in the keynote talk [9] given at a special session, dedicated to the passivity-based control, of the IEEE ICRA 2000 Conference.

Selected Publications

1. Arimoto S (1994) A class of quasi-natural potentials and hyper-stable PID servo-loops for nonlinear robotic systems. Trans. of SICE 30(9):1005–1012
2. Arimoto S (1996) Control Theory of Nonlinear Mechanical Systems: A Passivity-Based and Circuit-Theoretic Approach. Oxford University Press, U.K.
3. Arimoto S, Nakayama T (1996) Another language for describing motions of mechatronics systems: a nonlinear position-dependent circuit theory. IEEE/ASME Trans. on Mechatronics 1(2):168–180
4. Parra-Vega V, Arimoto S, Liu YH, Hirzinger G, Akella P (2003) Dynamic sliding PID control for tracking of robot manipulators: theory and experiments. IEEE Trans. on Robotics and Automation 19(6):967–976
5. Whitcomb L, Arimoto S, Naniwa T, Ozaki F (1997) Adaptive model-based hybrid control of geometrically constrained robot arms. IEEE Trans. on Robotics and Automation 13(1):105–116
6. Whitcomb L, Arimoto S, Naniwa T, Ozaki F (1996) Experiments in adaptive model-based force control. IEEE Control Systems Magazine 16(1):49–57
7. Liu YH, Kitagaki K, Ogasawara T, Arimoto S (1999) Model-based adaptive hybrid control for manipulators under multiple geometric constraints. IEEE Trans. on Control Systems Technology 7(1):97–109
8. Liu YH, Arimoto S (1996) Distributively controlling two robots handling an object in the task space without any communication. IEEE Trans. on Automatic Control 41(8):1193–1198
9. Arimoto S (2000) Passivity-based control. Proc. IEEE Int. Conf. on Robotics and Automation, San Francisco, CA 227–232

4.3 Iterative Learning Control

In 1976 a simple but original idea of the effectiveness of learning of robotic motion through repeated exercises was presented by M. Uchiyama, a Professor of Tohoku University. Professor S. Arimoto and his colleagues reformulated this idea into an axiomatic framework of iterative learning control (ILC) [1], which gave rise to a new field of control theory. Initially, the framework was based on a simple and effective sufficient condition for convergence of the iterative control scheme [1, 2]. A more efficient iterative learning control scheme, named P-type ILC, was subsequently proposed in [3]. Later on, a unified approach for not only the ILC but also for the repetitive (or periodic) control was presented in [4, 5]. It is shown there that, as far as linear dynamical systems are concerned, the ability of learning corresponds to the system characteristics such as the output-dissipativity or the strict positive realness with an extra condition. The application of the ILC to the control of geometrically constrained robot arms was outlined in [6].

Selected Publications

1. Arimoto S, Kawamura S, Miyazaki F (1984) Bettering operation of robots by learning. Journal of Robotic Systems 1(2):123–140

2. Kawamura S, Miyazaki F, Arimoto S (1988) Realization of robot motion based on a learning method. IEEE Trans. on Systems, Man and Cybernetics 18(1):126–134
3. Arimoto S (1990) Learning control theory for robotic motion. International Journal of Adaptive Control and Signal Processing 4(6):543–564
4. Arimoto S, Naniwa T (2000) Equivalence relations between learnability, output-dissipativity, and strict positive realness. International Journal of Control 73(10):824–831
5. Arimoto S, Naniwa T (2001) Corrections and further comments to "equivalence relations between learnability, output-dissipativity and strict positive realness". International Journal of Control 74(14):1481–1482
6. Naniwa T, Arimoto S (1995) Learning control for robot tasks under geometric endpoint constraints. IEEE Trans. on Robotics and Automation 11(3):432–441

4.4 Navigation of Autonomous Robot Vehicles

A new data-structure, named tangent graph, that can be effectively used in navigation of autonomous robot vehicles was introduced in [1]. This data-structure is computationally superior to the conventional data-structure based on the visibility graphs because the total number of nodes can be drastically reduced in comparison with the conventional data-structure [2]. The tangent graph data-structure can be implemented in autonomous robotic vehicles equipped with range sensors. Various path planning techniques exploiting this data-structure were proposed in [3, 4].

Selected Publications

1. Liu YH, Arimoto S (1991) Proposal of tangent graph for path planning of mobile robots. Proc. IEEE Int. Conf. on Robotics and Automation, Sacramento, CA 312–317
2. Liu YH, Arimoto S (1992) Path planning using a tangent graph for mobile robots among polygonal and curved obstacles. The International Journal of Robotics Research 11(4):376–382
3. Liu YH, Arimoto S (1994) Computation of the tangent graph of polygonal obstacles by moving-line processing. IEEE Trans. on Robotics and Automation 10(6):823–830
4. Liu YH, Arimoto S (1995) Finding the shortest path of a disc among polygonal obstacles using a radius-independent graph. IEEE Trans. on Robotics and Automation 11(5):682–691

4.5 Dynamic Bipedal Walking

In 1979 the research group at Osaka University, led by Professor S. Arimoto, developed a biped robot, named "Idaten", that could walk stably in a dynamic sense with a regular human walking speed [1, 2]. The development was largely based on the theoretical analysis of the dynamics of the biped walking in the sagittal plane. To stabilize the robot motion, a hierarchical control system based on the singular-perturbation technique was proposed [1]. The use of an early prototype of the iterative learning control scheme also contributed to the success in the experiments.

Selected Publications

1. Miyazaki F, Arimoto S (1980) A control theoretic study on dynamical biped locomotion. Trans. ASME, Journal of Dynamic Systems, Measurement, and Control 102:233–239
2. Arimoto S, Miyazaki F (1984) Biped locomotion robots. Japan Annual Review in Electronics, Computers & Telecommunications 12:194–205

4.6 Intelligent Control of Multi-Fingered Robotic Hands

This research area was in the scope of the scientific interests of Professor S. Arimoto starting from the early paper [1], where the basic modeling was analyzed. The role of the sensory-motor coordination in the control of multi-fingered hands was explored in [2, 3, 4, 5]. In [2, 3] it was shown that rolling contacts produce constraint forces tangent to object surfaces, by which stable pinching in a dynamic sense can be realized. Concepts of the stability on a manifold and the transferability to a submanifold were introduced in [6] and shown to be crucial in dealing with robot dynamics that are are nonlinear, redundant and under-actuated. A single sensory-motor control signal for stable pinching that needs neither parameters of the object kinematics nor the external sensing was established in [7, 8]. This control scheme was extended for the objects with non-parallel surfaces, and it was shown that a blind grasping can be implemented in robotic hands in the same way as humans grasp an object stably when they close eyes [9]. Note that the papers [7, 9, 11] were among the three finalists for the IEEE ICRA Best Manipulation Paper Award in, respectively, 2004, 2005, and 2006. Recently, it was shown that 3D-object grasping and manipulation by two fingers are possible in a blind manner even when the instantaneous axis of object rotation is changeable and yields non-holonomic constraints [10]. In addition, a physically faithful modeling of 3-D object manipulation, taking into account the spinning motion around the opposition axis, was established in [12, 13].

Selected Publications

1. Arimoto S, Miyazaki F, Kawamura S (1987) Cooperative motion control of multiple robot arms or fingers. Proc. IEEE Int. Conf. on Robotics and Automation, Raleigh, North Carolina 1407–1412
2. Arimoto S, Nguyen P, Han HY, Doulgeri Z (2000) Dynamics and control of a set of dual fingers with soft tips. Robotica 18(1):71–80

3. Arimoto S, Tahara K, Yamaguchi M, Nguyen P, Han HY (2001) Principle of superposition for controlling pinch motions by means of robot fingers with soft tips. Robotica 19(1):21–28
4. Arimoto S, Tahara K, Bae JH, Yoshida M (2003) A stability theory on a manifold: concurrent realization of grasp and orientation control of an object by a pair of robot fingers. Robotica 21(2):163–178
5. Arimoto S, Yoshida M, Bae JH, Tahara K (2003) Dynamic force/torque balance of 2D polygonal objects by a pair of rolling contacts and sensory-motor coordination. Journal of Robotic Systems 20(9):517–537
6. Arimoto S (2004) Intelligent control of multi-fingered hands. Annual Review in Control 28(1):75–85
7. Ozawa R, Arimoto S, Yoshida M, Nakamura S (2004) Stable grasping and relative angle control of an object by dual finger robots without object sensing. Proc. IEEE Int. Conf. on Robotics and Automation, New Orleans, USA 1694–1699
8. Ozawa R, Arimoto S, Nakamura S, Bae JH (2005) Control of an object with parallel surfaces by a pair of finger robots without object sensing. IEEE Trans. on Robotics 21(5):965–976
9. Arimoto S, Ozawa R, Yoshida M (2005) Two-dimensional stable blind grasping under the gravity effect. Proc. IEEE Int. Conf. on Robotics and Automation, Barcelona, Spain 1208–1214
10. Arimoto S, Yoshida M, Bae JH (2006) Stable "blind grasping" of a 3D object under non-holonomic constraints. Proc. IEEE Int. Conf. on Robotics and Automation, Orlando, USA 2124–2130
11. Tahara K, Luo ZW, Ozawa R, Bae JH, Arimoto S (2006) Bio-mimetic study on pinching motions of a dual-finger model with synergetic actuation of antagonist muscles. Proc. IEEE Int. Conf. on Robotics and Automation, Orlando, USA 994–999
12. Arimoto S, Yoshida M, Bae JH (2006) Stability of 3-d object grasping under the gravity and non-holonomic constraints. Proc. 17th Int. Symp. on Math. Theory of Networks and Systems, Kyoto, Japan
13. Arimoto S, Yoshida M, Bae JH (2006) Modeling of 3-d object manipulation by multi-joint robot fingers under non-holonomic constraints and stable blind grasping. Proc. Asian Congress of Multi-body Dynamics, Tokyo, Japan

4.7 Human-like Reaching Movements

Dealing with famous Bernstein's problem of the coordination of multiple degrees of freedom in human movements, it was suggested [1] that a typical problem of ill-posedness of the inverse kinematics for redundant robotic systems can be resolved in a natural way without introducing any artificial cost function to determine the inverse or without calculating any pseudo inverse of the Jacobian matrices. It was shown [2] that in the case of redundant multi-joint reaching movements a task-space position feedback with a single stiffness parameter together with damping shaping in the joint space can generate human-like skilled reaching motions (the endpoint trajectory is quasi-linear, the velocity profile is bell-shaped and the acceleration signals have double peaks). Based on the mathematical analysis and physiological interpretations of this fact, a *Virtual Spring/Damper Hypothesis* was introduced in [3, 4]. The hypothesis can successfully compete with various (and sometimes controversial) reasonings proposed in physiology. A mathematical verification of the effectiveness of the *Virtual Spring/Damper Hypothesis* on the basis of differential geometry is given in [5], and its neurophysiological meaning is discussed in [6].

Selected Publications

1. Arimoto S, Sekimoto M, Hashiguchi H, Ozawa R (2005) Natural resolution of ill-posedness of inverse kinematics for redundant robots: A challenge to Bernstein's degrees-of-freedom problem. Advanced Robotics 19(4):401–434
2. Arimoto S, Sekimoto M, Hashiguchi H, Ozawa R (2005) Physiologically inspired robot control: A challenge to Bernstein's degrees-of-freedom problem. Proc. IEEE Int. Conf. on Robotics and Automation, Barcelona, Spain 4511–4518
3. Arimoto S, Hashiguchi H, Sekimoto M, Ozawa R (2005) Generation of natural motions for redundant multi-joint systems: A differential-geometric approach based upon the principle of least actions. Journal of Robotic Systems 22(11):583–605
4. Arimoto S, Sekimoto M (2006) Human-like movements of robotic arms with redundant dofs: Virtual spring-damper hypothesis to tackle the Bernstein problem. Proc. IEEE Int. Conf. on Robotics and Automation, Orlando, USA 1860–1866
5. Arimoto S (2006) A differential-geometric approach for Bernstein's degrees-of-freedom problems. Proc. 17th International Symposium on Mathematical Theory of Networks and Systems, Kyoto, Japan
6. Arimoto S (2006) What is a breakthrough toward human robotics? Proc. IEEE/RSJ Int. Conf. on Intelligent Robots and Systems, Beijin, China (an extended version of this paper is to appear in Advanced Robotics, VSP International Science Publishers).

List of Contributors

Suguru Arimoto
Department of Robotics, Ritsumeikan University, Kusatsu, Shiga 525-8577, Japan
arimoto@se.ritsumei.ac.jp

Chien Chern Cheah
School of Electrical and Electronic Engineering, Nanyang Technological University, 50 Nanyang Avenue, 639798, Republic of Singapore
ecccheah@ntu.edu.sg

Nikhil Chopra
Coordinated Science Laboratory, University of Illinois at Urbana-Champaign, 1308 West Main Street, Urbana, IL 61801, USA
nchopra@uiuc.edu

Emmanuel Dean-Leon
Mechatronics Division, Research Center for Advanced Studies, IPN Av. 2508, San Pedro Zacatenco, Mexico City, 07300, Mexico
edean@cinvestav.mx

Igor Goncharenko
3D Incorporated, 1-1 Sakaecho, Kanagawa-ku, Yokohama, Kanagawa 221-0052, Japan
igor@ddd.co.jp

Neville Hogan
Department of Mechanical Engineering & Department of Brain and Cognitive Sciences, Massachusetts Institute of Technology, 77 Massachusetts Avenue, Cambridge, MA 02139, USA
neville@mit.edu

Shigeyuki Hosoe
Bio-Mimetic Control Research Center, RIKEN, Anagahora, Shimoshidami, Moriyama-ku, Nagoya, Aichi 463-0003, Japan
hosoe@bmc.riken.jp

Sadao Kawamura
Department of Robotics, Ritsumeikan University, Kusatsu, Shiga 525-8577, Japan
kawamura@se.ritsumei.ac.jp

Daniel E. Koditschek
Department of Electrical & Systems Engineering, University of Pennsylvania, 200 33rd Street, Philadelphia, PA 19103, USA
kod@seas.upenn.edu

Mark L. Latash
Department of Kinesiology, Rec.
Hall-268N, The Pennsylvania State
University, University Park, PA
16802, USA
ml111@psu.edu

Yun-Hui Liu
Department of Automation and
Computer Aided Engineering, The
Chinese University of Hong Kong,
Shatin, NT, Hong Kong
yhliu@acae.cuhk.edu.hk

Gabriel Lopes
Electrical Engineering and Computer
Science Department, University of
Michigan, Ann Arbor, MI, USA
glopes@umich.edu

Zhi-Wei Luo
Department of Computer and Systems Engineering, Kobe University,
1-1 Rokkodai, Nada-ku, Kobe,
Hyogo, 657-8501 Japan
luo@gold.kobe-u.ac.jp

Michiya Matsushima
Graduate School of Engineering,
Osaka University, Osaka 565-0871,
Japan
matsushima@mapse.eng.
osaka-u.ac.jp

Fumio Miyazaki
Graduate School of Engineering
Science, Osaka University, Osaka
560-8531, Japan
miyazaki@me.es.osaka-u.ac.jp

Dragomir Nenchev
Musashi Institute of Technology,
Tamazutsumi 1-28-1, Setagaya-ku,
Tokyo, 158-8557 Japan
nenchev@sc.musashi-tech.ac.jp

Vicente Parra-Vega
Mechatronics Division, Research
Center for Advanced Studies, IPN
Av. 2508, San Pedro Zacatenco,
Mexico City, 07300, Mexico
vparra@cinvestav.mx

Daniel L. Rothbaum
Atlantic Ear, Nose, and Throat, P.A.
2705 Rebecca Lane, Suite A, Orange
City, FL 32763, USA
d_rothbaum@yahoo.com

Jaydeep Roy
GE Global Research, Materials
Systems Technology, One Research
Circle, K1-3A20, Niskayuna, NY
12309, USA
jaydeep.roy@ge.com

Norimitsu Sakagami
Tokai University, Department of
Naval Architecture and Ocean Engineering, 3-20-1 Orido, Shimizu-ku,
Shizuoka-shi, 424-8610, Japan
sakagami@scc.u-tokai.ac.jp

Jean-Jacques Slotine
Department of Mechanical Engineering & Department of Brain and
Cognitive Sciences, Massachusetts
Institute of Technology, 77 Massachusetts Avenue, Cambridge, MA
01239, USA
jjs@mit.edu

Mark W. Spong
Coordinated Science Laboratory,
University of Illinois at Urbana-
Champaign, 1308 West Main Street,
Urbana, IL 61801, USA
spong@uiuc.edu

Mikhail Svinin
Bio-Mimetic Control Research
Center, RIKEN, Anagahora, Shimoshidami, Moriyama-ku, Nagoya,
Aichi 463-0003, Japan
svinin@bmc.riken.jp

Kenji Tahara
Bio-Mimetic Control Research
Center, RIKEN, Anagahora, Shimoshidami, Moriyama-ku, Nagoya,
Aichi 463-0003, Japan
tahara@bmc.riken.jp

Masahiro Takeuchi
Akashi National College of Technology, Hyogo 674-8501, Japan
takeuchi@akashi.ac.jp

Hesheng Wang
Department of Automation and
Computer Aided Engineering, The
Chinese University of Hong Kong,
Shatin, NT, Hong Kong
yhliu@acae.cuhk.edu.hk

Louis L. Whitcomb
Johns Hopkins University, Department of Mechanical Engineering,
3400 N. Charles Street, 123 Latrobe
Hall, Baltimore, MD 21218, USA
llw@jhu.edu

Vladimir M. Zatsiorsky
Department of Kinesiology, 039
Recreation Building, The Pennsylvania State University, University
Park, PA 16802, USA
vxz1@psu.edu

Contents

Human Robotics: A Vision and A Dream
Suguru Arimoto 1

Part I From Everyday Physics to Robot Control

Natural Motion and Singularity-Consistent Inversion of Robot Manipulators
Dragomir N. Nenchev 9

Approximate Jacobian Control for Robot Manipulators
Chien Chern Cheah 35

Adaptive Visual Servoing of Robot Manipulators
Yun-Hui Liu, Hesheng Wang 55

Orthogonalization Principle for Dynamic Visual Servoing of Constrained Robot Manipulators
Vicente Parra-Vega, Emmanuel Dean-Leon 83

Passivity-Based Control of Multi-Agent Systems
Nikhil Chopra, Mark W. Spong 107

Navigation Functions for Dynamical, Nonholonomically Constrained Mechanical Systems
Gabriel A. D. Lopes, Daniel E. Koditschek 135

Planning and Control of Robot Motion Based on Time-Scale Transformation
Sadao Kawamura, Norimitsu Sakagami 157

Part II From Robot Control to Human-Like Movements

Modularity, Synchronization, and What Robotics May Yet Learn from the Brain
Jean-Jacques Slotine 181

Force Control with A Muscle-Activated Endoskeleton
Neville Hogan 201

On Dynamic Control Mechanisms of Redundant Human Musculo-Skeletal System
Kenji Tahara, Zhi-Wei Luo 217

Principle of Superposition in Human Prehension
Mark L. Latash, Vladimir M. Zatsiorsky 249

Motion Planning of Human-Like Movements in the Manipulation of Flexible Objects
Mikhail Svinin, Igor Goncharenko, Shigeyuki Hosoe 263

Haptic Feedback Enhancement Through Adaptive Force Scaling: Theory and Experiment
Jaydeep Roy, Daniel L. Rothbaum, Louis L. Whitcomb 293

Learning to Dynamically Manipulate: A Table Tennis Robot Controls a Ball and Rallies with a Human Being
Fumio Miyazaki, Michiya Matsushima, Masahiro Takeuchi 317

Human Robotics: A Vision and A Dream

Suguru Arimoto[1]

Department of Robotics, Ritsumeikan University, Kusatsu, Shiga, 525-8577 Japan
arimoto@se.ritsumei.ac.jp

Summary. On the celebration of my seventieth birthday, I would like to ask you to allow me to introduce my present dream: human robotics. Before explicating what is human robotics, it is important for me to spell out why such dream has been incubated in my mind during the past decade.

First, let me recollect the very beginning of my engagement in research. It was in the spring of 1960, but I remember quite vividly the occasion when the leader of our small group engaged in the development of an electric computer requested me to quickly learn error-correcting codes and devise an operating system that could protect stored programs in magnetic core memories from bit errors. I graduated from the Department of Mathematics at Kyoto University in 1959, but I can not say that I learned very much about mathematics. Nevertheless, I was fortunately able to quickly discover a special class of algebraic codes together with a powerful decoding algorithm and succeeded in implementing it in an error-protecting operation system. From then on I began to be confident of having some sort of talent necessary for engagement in research and development even in the academic society. Since then I have been always concerned with discovering something new and useful in both the scientific and engineering areas of computing, information processing, and control. During the first two decades of my academic career, Claude Elwood Shannon and Alan Mathison Turing were my admiration.

When I was appointed a Professor of the 10th Chair of the Department of Mechanical Engineering at Osaka University, I felt of some reluctance because I was not well qualified in the technical fields of mechanical engineering. Notwithstanding it, I accepted this appointment in the sequel. The reason was simple—Shannon was also interested in chess-playing robots, juggling robots, and designed and made by himself a maze-solving machine (micro-mouse). Thus, I have been infatuated with robotics for over three decades, beginning from around 1976 and till now. A remarkable and memorable day happened in November 1984 when Shannon was awarded the first Kyoto Prize in the area of fundamental science. I was asked to organize a workshop on information theory celebrating his award and had a rare chance of watching Shannon's 8mm film on the screen. I was amused at how swiftly his chess-playing robot hand with three fingers picked and placed the chess pieces. The years centering 1984 were perhaps the most productive and fruitful time in the robotics research in my laboratory at Osaka University. Many ideas for the the physics-based robot control were conceived, and the foundations of a control theory of nonlinear mechanical systems [1] were laid and contoured during that time.

In 1988 when I came back to the University of Tokyo, I was implicated in an artificial intelligence debate. Professors Hubert Dreyfus and Stuart Dreyfus claimed in the book [2] that the progress of creating "artificial intelligence" would be blocked by the commonsense knowledge problem including everyday physics or commonsense physics. They expatiated on: "Can there be a theory of the everyday world as rationalist philosophers have always held? Or is the commonsense background rather a combination of skills, practices, discrimination, and so on, which are not intentional states and so, a fortiori do not have any representational content to be explicated in terms of elements and rules?" I thought that, differently from the commonsense reasoning that may be crucial in AI, the commonsense physics can be spelt out in a set of facts and rules. It should be a scientific domain called "everyday physics".

In 1999, I published an article in the third millennium commemorative issue of the International Journal of Robotics Research [3], where the term "everyday physics" was used as a scientific domain related to accountability of dexterous accomplishment of ordinary tasks we encounter in our everyday life. In the article, three stages of research for the exploration of the everyday physics were proposed. They are shown in Fig. 1.

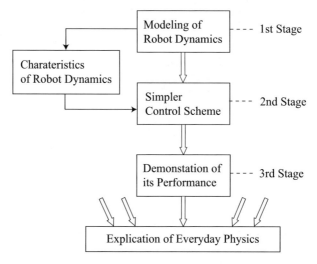

Fig. 1. Robotics research should evolve from precise modeling of a robot task to explication of everyday physics.

In preparation of the article I just started a new research project on control of multi-fingered hands together with master-course students in Ritsumeikan University which I had joined in the spring of 1997 after retirement from the University of Tokyo in compliance with the age limit of 60 years old. The memory of Shannon's chess-playing robot with three fingers has always encouraged me to tackle this challenging research subject through which I believe we would be able to explore certain physical principles that may govern the secret of dexterity of our hands. With doctor-course students we discovered a crucial role of tangential forces arising from rolling constraints (Fig. 2) and a principle of linear superposition of the feedback and feedforward signals for the coordination control based upon the passivity relation. In 2006 we at last succeeded in the derivation of a mathematical model, faithfully expressing the dynamics of 3-D object grasping and manipulation under the gravity effect and nonholonomic constraints. Now, we see that even a pair of 3-D robot fingers can grasp a 3-D object stably without exactly knowing the object kinematics and without using any external sensing. A robot hand can grasp an object in a blind manner, just like a human does with his or her eyes closed.

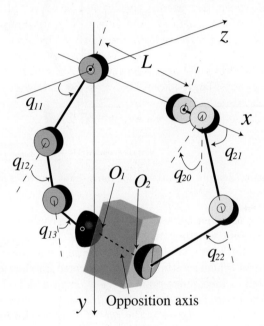

Fig. 2. Three-dimensional object grasping and manipulation.

During the process of exploring the dexterity of multi-fingered hands, I recognized around 2003 that we were already solving the so-called Bernstein problems[4], because the total number of degrees of freedom (DOF) of the overall fingers-object system possibly becomes redundant in comparison with the number of independent physical variables necessary for describing imposed tasks. In the same year a Japanese translation of Bernstein's book [5] was published; its original manuscript was written in Russian more than a half century ago. Also in 2003, we theoretically proved, without considering the DOF redundancy, the convergence of the grasping motion toward its state of force/torque balance. This research introduced a new concept of the stability on a constraint manifold and showed how it can be used in the verification of the grasping stability understood in a dynamic sense.

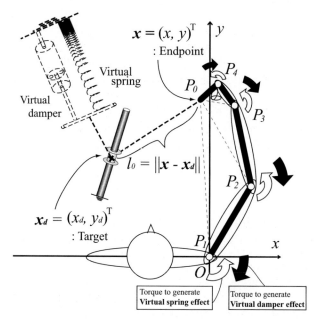

Fig. 3. Spring/damper hypothesis for multi-joint reaching movements.

All these taken together coincidentally led me to challenging the Bernstein problem of multi-joint reaching movements under redundant DOFs. Subsequently we found that a skilled human-like reaching movement can be realized by a simple control scheme without solving inverse kinematics or dynamics. From the Newtonian mechanics point of view, such a skilled motion can be generated as if the endpoint of the whole arm is drawn by parallel connection of a spring with a damper placed at the target point as shown in Fig. 3. This means that the control is of the form: $u = -C_0 \dot{q} - J^T (c\dot{x} + k\Delta x)$, where q denotes the joint vector, $\Delta x = x - x_d$, J stands for the Jacobian matrix of x in q, and C_0 does for the joint damping matrix.

Having sketched the structure of the feedback control signal, one can proceed further and ask, what kind of neuro-motor signals does the central nervous system (CNS) emanate in a feedforward manner and how do they excite a set of muscles involved in the joint rotations leading to quasi-straight reaching movements? If $-C_0\dot{q}$ can be interpreted as passive damping at the joints, then the signal $-kJ^T\Delta x$ corresponds to the generation of spring-like forces induced from the muscle contractions. Then, where and how does the remaining part, $-J^T(c\dot{x})$, come from in the case of human movements. My prediction is that it may follow from co-contraction of antagonist muscles excited through a spinal-reflex loop supervised by the cerebellum provided that it behaves as a velocity observer for $\dot{x}$. As any prediction, it remains to be tested and hopefully verified on the difficult road toward the creation of truly humanoid robots.

In conclusion, I am still dreaming of "human robotics" that directs us to the design of an artificial CNS that would cope with the everyday physics just beautifully. Human robotics should be a core of robotics research that attempts to unveil the mysteries of dexterity and versatility in human intelligent behaviours seen in everyday life. To develop and systematize it, many breakthroughs awaiting your exploration are indispensable. Sooner or later I will step down and leave the playing list of active researchers, but I hope I will be able to enjoy your challenges from the gallery.

References

1. Arimoto S (1996) Control Theory of Nonlinear Mechanical Systems: A Passivity-Based and Circuit-Theoretic Approach. Oxford University Press, U.K.
2. Graubard S (ed.) (1988) The Artificial Intelligence Debate: False Starts, Real Foundations. MIT Press, Boston
3. Arimoto S (1999) Robotics research toward explication of everyday physics. The International Journal of Robotics Research 18(11):1056–1063
4. Latash M (1996) The Bernstein problem: how does the nervous system make its choices? In Latash M, Turvey M, ed.: Dexterity and Its Development. Lawrence Erlbaum Associates, Mahwah, New Jersey, 277–303
5. Bernstein N (1967) The Co-ordination and Regulation of Movements. Pergamon Press, Oxford, U.K.

Part I

From Everyday Physics to Robot Control

Natural Motion and Singularity-Consistent Inversion of Robot Manipulators

Dragomir N. Nenchev

Musashi Institute of Technology, Tamazutsumi 1-28-1, Setagaya-ku, Tokyo, 158-8557 Japan, nenchev@sc.musashi-tech.ac.jp

Summary. Numerous robotic tasks require the solution of the inverse problem, known to be ill-conditioned in the neighborhood of kinematic singularities. The problem is addressed here via a non-linear, differential-geometric approach. A "natural motion" component is identified thereby at the velocity level. Dynamic analysis reveals that the essence of this component is nondissipative motion along the prescribed end-effector path, with nonstationary initial condition. A kinematic feedback controller and a dynamic feedback controller are introduced and shown to ensure stable motion initialization from kinematic singularities, as well as tracking of prescribed paths that pass through such singularities.

1 Introduction

Using visual feedback, humans are able to position their upper/lower limbs appropriately to meet various task requirements. Robot application tasks quite often also require the solution of the so-called inverse problem, in terms of inverse kinematics or inverse dynamics in task coordinates [1]. The formal solution to the problem is not as straightforward as might be expected.

For nonredundant structures, the inverse problem has been most often tackled by employing the inverse of the Jacobian matrix. Though, as recognized at an early stage [2], the ill-conditioning of the inverse problem at and around kinematic singularities can destabilize the controls. Trying to avoid such singularities, as proposed by some authors, is an oversimplified solution, inconsistent with our everyday praxis. Just imagine what would it be if we were not able to fully stretch an arm or a leg along a specific path.

One possible formalization of the inverse problem has been based on an approximate "inverse" solution, obtained via the transpose of the Jacobian. Satisfactory performance can be achieved then, provided simple position feedback control is present, as noted almost three decades ago in [3] (p.246). Jacobian transpose based control schemes have been since widely discussed in literature, see e.g. [4]–[6]. Very recently, attention is drawn again to the Jacobian transpose in the hope that it may provide a "natural" solution to the inverse kinematic problem, for kinematically redundant structures inclusively [7].

Another formalization, based on a different type of approximate inverse kinematics solution, is the so-called "damped least-squares method" (DLS method) [8], [9]. The essential mapping is again the Jacobian transpose, scaled, however, by the inverse of a quadratic form containing a variable "damping factor." The latter is used to balance between the error and the infeasible velocity norm resulting from the ill-conditioning of the inverse at and around a kinematic singularity. The DLS method has been successfully tested experimentally [10], [11]. It must be pointed out, though, that the process of determining the damping factor is not straightforward and may require the evaluation of the singular values of the manipulator Jacobian [12]. Also, since the error introduced by the method is along the *singular direction*, it is not always possible to reach a kinematically singular configuration [13].

On the other hand, efforts have been made to approach the problem of ill-conditioning of the inverse task from a nonlinear point of view. The possibility to initialize a feasible motion along the singular direction at a kinematic singularity has been pointed out in [14], see also [15]–[17] among others.

The nonlinear approach to the inverse kinematics problem has also proven to be quite useful according to our own experience, while developing the *singularity-consistent method* (SC method) [18]. The method has been tested in practice with various mechanical structures, e.g. a serial six degree-of-freedom (DOF) slave arm of a teleoperation system [19], [20], a six-DOF parallel robot HEXA [21], [22], a seven-DOF serial arm [23] and very recently, with a humanoid robot [24].

The aim of this work is to summarize results based on the singularity-consistent method, with special emphasis on *natural motion* — a type of motion obtained in a straightforward manner within the notation [25], [26]. By extending analysis to the level of dynamics and control law design, we show also that our treatment of the inverse problem meets Koditschek's expectation that dynamical models "...which in some way match the internal dynamics of the robot will afford more accurate performance with less effort" [27].

2 Singularity-Consistent Formulation of Inverse Kinematics

Let us consider a serial-link robot with n joint variables. Each manipulator configuration is defined as an n-tuple $\boldsymbol{q} = (q_1, q_2, ..., q_n)$. The set of all possible manipulator configurations will be referred to as the *configuration space* $\mathcal{C}$. The end-effector position and orientation, related to a given manipulator configuration via the *direct kinematics*, is expressed by a set of m independent variables $\boldsymbol{x} = (x_1, x_2, ..., x_m)$. In the general case, $m = 6$.

The velocity $\boldsymbol{v}$ of a characteristic point on the end-effector, and the angular velocity $\boldsymbol{\omega}$ of the end-effector are obtained from the well known relation:

$$\begin{bmatrix} v \\ \omega \end{bmatrix} = J(q)\dot{q}, \tag{1}$$

where $J(q)$ is the manipulator Jacobian. The pair $(v, \omega) \equiv \nu \in \Re^6$ denotes the coordinates of the *end-effector twist*.

Consider now the following expansion of the Jacobian:

$$J(q) = \begin{bmatrix} t_1 & t_2 & ... & t_n \end{bmatrix}, \tag{2}$$

where $t_i(q)$ is the normalized end-effector twist obtained with unit-speed motion in joint i only: $\dot{q}_i = 1$ rad/s. For simplicity, we will assume that all joint variables represent angles. The configuration space $\mathcal{C}$, a subspace of $\Re^n$, will be then uniform in terms of physical units[1]. The usual vector norm $\|\circ\|$ can be employed then.

The composite end-effector twist, due to motion in all joints, is then:

$$\nu(q, \dot{q}) = \sum_{k=1}^{n} t_k \dot{q}_k. \tag{3}$$

This equation solves the direct kinematics problem for the velocities.

Next, we focus on the *inverse kinematics* problem which is frequently used in analysis and control. Assume the end-effector moves along a smooth parameterized path $x(q_*)$, q_* denoting the path parameter. The end-effector twist, at a given point q_*, can be represented as

$$\nu(q_*, \dot{q}_*) = t_*(q_*)\dot{q}_* \tag{4}$$

where $t_*(q_*)$ is the normalized end-effector twist, obtained when the speed of the parameter is one: $\dot{q}_* = 1$ rad/s. Equations (3) and (4) can be combined as:

$$t_* \dot{q}_* = \sum_{k=1}^{n} t_k \dot{q}_k \tag{5}$$

which is just another representation of linear system (1).

Let us now rewrite the last equation as follows:

$$\sum_{k=1}^{n} t_k \dot{q}_k - t_* \dot{q}_* = 0. \tag{6}$$

This is a closure equation for the kinematic chain. Its form gives us a hint: it would be convenient to augment the configuration space with the path parameter q_*. Doing so, we obtain the *augmented configuration space* $\mathcal{C}^*$, a subspace of $\Re^{n+1}$, with elements $\bar{q} \equiv (q, q_*)$. Equation (6) can be rewritten in compact form, as:

---

[1] In case of a mixed-joint structure (with rotational and translational joints), uniformity can be ensured with proper scaling.

$$\overline{J}(\bar{q})\dot{\bar{q}} = 0 \tag{7}$$

where

$$\overline{J}(\bar{q}) \equiv \begin{bmatrix} J(q) & -t_*(q_*) \end{bmatrix}$$

is called *the column-augmented Jacobian*. To ensure uniformity in terms of physical units within $\mathcal{C}^*$ space, the path parameter q_* should be defined as an angular variable. This implies also that the usual vector norm can be employed.

It should be apparent that Eq. (7) is underdetermined, and hence, the number of solutions is infinite. But this is also a homogeneous equation. Therefore, all the solutions must be in the kernel of matrix $\overline{J}$. Let us assume now that the manipulator is *nonredundant*: $m = n$. For a full-rank column-augmented Jacobian, the kernel contains a single nonzero element then. The set of solutions of Eq. (7) is:

$$\dot{\bar{q}} = b\bar{n}(\bar{q}) \tag{8}$$

where b is an arbitrary scalar, and $\bar{n}(\bar{q}) \in \ker \overline{J}(\bar{q}) \subset \Re^{n+1}$. A vector from the kernel can be derived in analytical form with the help of the cofactors of matrix $\overline{J}$ [28], [29]:

$$\bar{n}(\bar{q}) = \begin{bmatrix} C_1 & C_2 & ... & C_{n+1} \end{bmatrix}^T \tag{9}$$

where $C_i = (-1)^{i+1} \det \overline{J}_i$ is the i-th cofactor, and $\overline{J}_i$ is obtained from matrix $\overline{J}$ by removing the i-th column. The last expression can be rewritten as:

$$\bar{n}(\bar{q}) \equiv \begin{bmatrix} n^T(\bar{q}) & \det J(q) \end{bmatrix}^T \tag{10}$$

where $n(\bar{q}) \equiv [\mathrm{adj} J(q)] t_*(q_*)$, $\mathrm{adj}(\circ)$ denoting the adjoint matrix [20]. Note that $n(\bar{q})$ maps vectors from the augmented configuration space to the tangent space of $\mathcal{C}$ at q: $\{n(\bar{q}) : \mathcal{C}^* \subset \Re^{n+1} \to T\mathcal{C}_q \subset \Re^n\}$.

Equation (8) can be split into two parts to obtain the joint motion differential

$$\dot{q} = b\, n(\bar{q}) \tag{11}$$

and the path parameter differential

$$\dot{q}_* = b \det J(q). \tag{12}$$

Usually, path parameter q_* is considered an independent variable. Then, one determines the scalar b from Eq. (12), and substitutes it back into Eq. (11) to obtain the joint velocity. Note, however, that in the vicinity of kinematic singularities this may lead to instability since the determinant of the Jacobian is close to zero.

In fact, the above formulation of the inverse kinematics has been developed to deal with motion control around and at kinematic singularities [18]. Note, when the Jacobian is singular, and in addition, linear system (1) or (5) is inconsistent, a nontrivial solution can be obtained only when the end-effector is (instantly) at rest, i.e. the twist is set to zero. This condition must be satisfied in order to comply with the existing physical motion constraint at the singularity [30]. This can be easily achieved with the above formulation, by exchanging the roles of q_* and b, such that b becomes the new independent variable, while q_* is regarded as the dependent one. Then, from Eq. (12) it is evident that when b is constant, the end-effector twist magnitude $\dot{q}_*$ will be zero at a kinematic singularity, since the determinant is zero.

Note that the above procedure does not lead to *path tracking* error. Indeed, we have

$$\boldsymbol{\nu} - \boldsymbol{J}\dot{\boldsymbol{q}} = \dot{q}_*\boldsymbol{t}_* - b\boldsymbol{J}(\mathrm{adj}\boldsymbol{J})\boldsymbol{t}_* \qquad (13)$$
$$= (\dot{q}_* - b\det\boldsymbol{J})\boldsymbol{t}_*.$$

Thus, the error is always along the normalized end-effector twist $\boldsymbol{t}_*$ that represents the path tangent. In other words, via b, we can effectively reparameterize the end-effector path $\boldsymbol{x}(q_*)$ in a way that any inconsistency with the singularity motion constraint, mentioned above, will be avoided. Because of this special property of the formulation, Eq. (11) has been called the *singularity-consistent inverse kinematic solution*.

3 Natural Motion of a Manipulator

The trivial reparameterization, b being a nonzero *constant*, plays an important role throughout this work. Note that when b is constant, Eq. (8) can be regarded as an autonomous dynamical system. The flow of vector field $\bar{\boldsymbol{n}}(\bar{\boldsymbol{q}})$ is represented then by a set of spatial curves in augmented configuration space $\mathcal{C}^*$. Motion along a specific curve (determined from the initial condition) can be associated with the so called *self-motion* of a kinematically redundant manipulator obtained from the original manipulator by adding a virtual joint with q_* as the joint variable. The curve will be referred to as the *self-motion manifold* [31].

We are interested in a positive definite metric on the self-motion manifold. The candidate metric is $\langle \bar{\boldsymbol{n}}, \bar{\boldsymbol{n}} \rangle$ which is positive semidefinite, though. Hence, stationary point analysis for the autonomous system is needed, to obtain first the set where $\bar{\boldsymbol{n}}(\bar{\boldsymbol{q}}) = \boldsymbol{0}$. It is not difficult to recognize that this set is composed of (i) kinematic singularities where the codimension is higher than one (all $n-1 \times n-1$ minors are zero then), and (ii) kinematic singularities of codimension one where the linear system (5) is consistent[2]. Henceforth, we will refer to such singularities as *stationary point singularities*. We emphasize the fact that those codimension-one kinematic singularities for which the linear system is <u>inconsistent</u>, are regular points of the dynamical system[3]. Hence, it would be straightforward to apply existing control laws in the neighborhood

---

[2] A consistent linear system implies that $\boldsymbol{t}_*$ is in the column-space of $\boldsymbol{J}$, which mapped by $\mathrm{adj}\boldsymbol{J}$, becomes zero.

[3] Such singularities have been named "ordinary singularities" by Kieffer [15].

of, or even at such kinematic singularities. These singularities will be referred to as *regular point singularities*.

Henceforth, the subdomain of regular points will be referred to as the *domain of complete integrability*. The self-motion manifold can then be characterized by the invariant arc lenght λ, called also natural parameter. λ is determined uniquely up to an additive constant, via:

$$\dot{\lambda} = \|\bar{n}\|. \tag{14}$$

$\dot{\lambda}$ will be referred to as the *natural speed* along the self-motion manifold.

From Eq. (8) we obtain:

$$\dot{\bar{q}} = \dot{\lambda}\hat{\bar{n}}(\bar{q}) \tag{15}$$

where $\hat{\bar{n}}$ is the tangent vector of unit length at $\bar{q}$, and the constant b has been set to one, without loss of generality.

Definition (Natural motion)
Manipulator motion with generalized velocities in proportion to the natural speed $\dot{\lambda}$ is called *natural motion*.

Corollary 1
Under natural motion, the magnitude of the end-effector twist is in proportion to $|\det \boldsymbol{J}|$.
Proof: Follows directly from Eqs. (12) and (15).

Corollary 2
Natural motion requires nonzero initial conditions.
Proof: Since $\dot{\lambda} \neq 0$ everywhere within the domain of complete integrability.

Corollary 3
In the vicinity of stationary points, natural motion decays exponentially.
Proof: Follows from the solution type of the autonomous dynamical system.

4 Vector-Parameterization of Configuration Space

We are interested in the projection of vector field $\bar{n}(\bar{q})$ onto n-dimensional $\mathcal{C}$ space. The projection, $\boldsymbol{n}(\bar{q}) = [\mathrm{adj}\boldsymbol{J}(\boldsymbol{q})]\boldsymbol{t}_*(q_*)$, can be interpreted as a parameterized vector field, $\boldsymbol{t}_*(q_*)$ playing the role of an n-parameter. Consequently, Eq. (11) can be regarded as a *parameterized autonomous dynamical system* [32]. Its domain will be referred to as the *parameterized $\mathcal{C}$-space*. Any point from the parameterized $\mathcal{C}$-space is determined by the couple $(\boldsymbol{q}, \boldsymbol{t}_*)$. Note that from definition (10) it follows that, within the domain of complete integrability, there is a one-to-one mapping between points $(\boldsymbol{q}, \boldsymbol{t}_*)$ and $\boldsymbol{n}(\bar{q})$.

The vector-parameterization just introduced can be quite useful in such applications as telerobotics [19], [20], or "hand-in-eye" systems [33] where no a priori information about the path $\boldsymbol{x}(q_*)$ is available, and only the velocity of the end-effector (i.e. the twist $\dot{q}_*\boldsymbol{t}_*$) is known.

5 Second-Order Singularity-Consistent Inverse Kinematics

Second-order kinematic relations play an important role for the dynamic analysis to be presented in the following section. Two representations of joint accelerations will be introduced. First, we refer to the underdetermined linear system (7). Differentiating w.r.t. time, we obtain:

$$\overline{\boldsymbol{J}}(\bar{\boldsymbol{q}})\ddot{\bar{\boldsymbol{q}}} + \dot{\overline{\boldsymbol{J}}}(\bar{\boldsymbol{q}})\dot{\bar{\boldsymbol{q}}} = \boldsymbol{0}. \tag{16}$$

This equation admits the following solution:

$$\ddot{\bar{\boldsymbol{q}}} = \beta \bar{\boldsymbol{n}}(\bar{\boldsymbol{q}}) - \overline{\boldsymbol{J}}^{+}(\bar{\boldsymbol{q}})\dot{\overline{\boldsymbol{J}}}(\bar{\boldsymbol{q}})\dot{\bar{\boldsymbol{q}}} \tag{17}$$

where β denotes an arbitrary small value, and $(\circ)^+$ stands for the pseudoinverse. The representation decomposes the joint acceleration into two orthogonal components. This is evident from the fact that the pseudoinverse component is known to be orthogonal to the null space component.

The last equation can be split into two parts, to obtain second differentials of joint motion and path motion, respectively:

$$\ddot{\boldsymbol{q}} = \beta \boldsymbol{n}(\bar{\boldsymbol{q}}) - \boldsymbol{J}^{T}(\boldsymbol{q})\boldsymbol{J}_*^{-1}(\bar{\boldsymbol{q}})\dot{\overline{\boldsymbol{J}}}(\bar{\boldsymbol{q}})\dot{\bar{\boldsymbol{q}}} \tag{18}$$

and

$$\ddot{q}_* = \beta \det \boldsymbol{J}(\boldsymbol{q}) + \boldsymbol{t}_*^T(q_*)\boldsymbol{J}_*^{-1}(\bar{\boldsymbol{q}})\dot{\overline{\boldsymbol{J}}}(\bar{\boldsymbol{q}})\dot{\bar{\boldsymbol{q}}}, \tag{19}$$

where $\boldsymbol{J}_*(\bar{\boldsymbol{q}}) \equiv \boldsymbol{J}(\boldsymbol{q})\boldsymbol{J}^T(\boldsymbol{q}) + \boldsymbol{t}_*(q_*)\boldsymbol{t}_*^T(q_*)$.

We pay special attention to matrix

$$\boldsymbol{J}^{\dagger}(\bar{\boldsymbol{q}}) \equiv \boldsymbol{J}^T(\boldsymbol{q})\boldsymbol{J}_*^{-1}(\bar{\boldsymbol{q}}),$$

a generalized inverse of $\boldsymbol{J}(\boldsymbol{q})$. Note that the inverse in the above expression exists for all nonsingular manipulator configurations and all regular point singularities. Thus, the joint accelerations in Eq. (18) are valid within the domain of complete integrablity. We will refer to matrix $\boldsymbol{J}^{\dagger}(\bar{\boldsymbol{q}})$ as the *singularity-consistent generalized inverse*. The variable β can be determined from Eq. (19), or, at a singularity where the determinant is zero, β will be properly modified.

Another representation of joint acceleration is obtained by differentiating the joint velocity in Eq. (8):

$$\ddot{\bar{\boldsymbol{q}}} = \dot{b}\bar{\boldsymbol{n}}(\bar{\boldsymbol{q}}) + b^2 \frac{\partial \bar{\boldsymbol{n}}(\bar{\boldsymbol{q}})}{\partial \bar{\boldsymbol{q}}} \bar{\boldsymbol{n}}(\bar{\boldsymbol{q}}). \tag{20}$$

This representation is helpful in analysis related to natural motion. When $b = \text{const}$, the first term on the r.h.s obviously vanishes. The remaining acceleration term $b^2 \frac{\partial \bar{\boldsymbol{n}}(\bar{\boldsymbol{q}})}{\partial \bar{\boldsymbol{q}}} \bar{\boldsymbol{n}}(\bar{\boldsymbol{q}})$ is an exact differential then. Henceforth, we will refer to it as the *integrable joint acceleration component*. Note that the covector mapping $\frac{\partial \bar{\boldsymbol{n}}(\bar{\boldsymbol{q}})}{\partial \boldsymbol{q}}$ contains information about the curvatures.

On the other hand, the first component on the r.h.s. of Eq. (20), $\dot{b}\bar{\boldsymbol{n}}(\bar{\boldsymbol{q}})$, exists only when $b \neq$ const. This component represents joint acceleration along the vector field; it may never be an exact differential due to the $\dot{b}$. We will call it therefore the *nonintegrable joint acceleration component*. The component necessarily vanishes at stationary points of the dynamical system (see Corollary 3 above). Note, however, that the integral curves of the vector field (to be interpreted as velocity curves) are well defined in the neighborhood of stationary points. Therefore, we can conclude that the nonintegrable joint acceleration plays the important role of ensuring motion via stationary points.

The projection of Eq. (20) onto $\mathcal{C}$-space is given by the following three components:

$$\ddot{\boldsymbol{q}} = \dot{b}\boldsymbol{n}(\bar{\boldsymbol{q}}) + b^2 \frac{\partial \boldsymbol{n}(\bar{\boldsymbol{q}})}{\partial \boldsymbol{q}} \boldsymbol{n}(\bar{\boldsymbol{q}}) + b^2 \frac{\partial \boldsymbol{n}(\bar{\boldsymbol{q}})}{\partial q_*} \det \boldsymbol{J}(\boldsymbol{q}). \tag{21}$$

It is seen that the integrable joint acceleration consists of two terms. These terms, $\frac{\partial \boldsymbol{n}(\bar{\boldsymbol{q}})}{\partial \boldsymbol{q}} \boldsymbol{n}(\bar{\boldsymbol{q}})$ and $\frac{\partial \boldsymbol{n}(\bar{\boldsymbol{q}})}{\partial q_*} \det \boldsymbol{J}(\boldsymbol{q})$, represent curvature-related joint acceleration due to the nonlinear manipulator kinematics and the nonlinearity of the imposed end-effector path $\boldsymbol{x}(q_*)$, respectively. Note also that while the former term vanishes at stationary points, the latter one does so at any kinematic singularity.

6 Dynamic Analysis

In this section, we first derive the singularity-consistent parameterized form of the equation of motion, and thereafter discuss the dynamics of natural motion.

6.1 Singularity-Consistent Parameterization of the Equation of Motion

The equation of motion is:

$$\boldsymbol{M}(\boldsymbol{q})\ddot{\boldsymbol{q}} + \boldsymbol{C}(\boldsymbol{q},\dot{\boldsymbol{q}})\dot{\boldsymbol{q}} + \boldsymbol{g}(\boldsymbol{q}) = \boldsymbol{\tau} + \boldsymbol{\tau}_e, \tag{22}$$

where $\boldsymbol{M}(\boldsymbol{q})$ is the symmetric positive definite manipulator inertia matrix, $\boldsymbol{C}(\boldsymbol{q},\dot{\boldsymbol{q}})\dot{\boldsymbol{q}}$ and $\boldsymbol{g}(\boldsymbol{q})$ denote Coriolis and centrifugal, and gravity forces, respectively, $\boldsymbol{\tau}$ stands for the driving joint torque and $\boldsymbol{\tau}_e$ denotes "disturbance" and/or external forces, e.g. friction in the joints etc. Making use of the singularity-consistent notation (21), the equation of motion can be rewritten as

$$\left(\dot{b}\boldsymbol{M}(\boldsymbol{q}) + b\boldsymbol{A}(\bar{\boldsymbol{q}})\right)\boldsymbol{n}(\bar{\boldsymbol{q}}) + \boldsymbol{g}(\boldsymbol{q}) = \boldsymbol{\tau} + \boldsymbol{\tau}_e, \tag{23}$$

where
$$A(\bar{q}) \equiv bM(q)\frac{\partial n(\bar{q})}{\partial \bar{q}} + C(q, \dot{q}).$$

Two remarks are due. First, the above form of the equation of motion incorporates the imposed end-effector constraint. When compared to similar formulations, Eq. (23) has the advantage that it can be applied in the vicinity of any kinematic singularity and at all regular point singularities. Second, note that force components due to the two terms of matrix A are usually treated separately: $M(q)\frac{\partial n(\bar{q})}{\partial \bar{q}}$ maps joint acceleration to inertial forces, while the $C(q, \dot{q})$ map generates centrifugal and Coriolis forces. The presence of the curvature-related covector map $\frac{\partial n(\bar{q})}{\partial \bar{q}}$ in the former term, however, shows that this term can be regarded as a nonlinear force generating term.

We are referring to Eq. (23) as the *singularity-consistent parameterization of the equation of motion* [25].

Next, note that Eq. (23) represents a set of $n = 6$ equations. The dimension can be reduced, if the dynamics are rewritten in terms of *input power*:

$$\dot{q}^T \tau = bn^T \left(\dot{b}M + bA\right) n + bn^T g - bn^T \tau_e. \tag{24}$$

For compactness, the functional dependence has been omitted in this notation. Further on, note that matrix $\dot{M} - 2C$ is anti-symmetric [27], [34]:

$$n^T(\bar{q})C(q, \dot{q})n(\bar{q}) = \frac{1}{2}n^T(\bar{q})\left[b\frac{\partial M(q)}{\partial q}n(\bar{q})\right]n(\bar{q}). \tag{25}$$

Hence, we have
$$n^T(\bar{q})A(\bar{q})n(\bar{q}) = bn^T(\bar{q})D(\bar{q})n(\bar{q}) \tag{26}$$

where matrix $D(\bar{q})$ is defined as

$$D(\bar{q}) \equiv M(q)\frac{\partial n(\bar{q})}{\partial \bar{q}} + \frac{1}{2}\frac{\partial M(q)}{\partial q}n(\bar{q}).$$

The input power (24) can be then rewritten as:

$$\dot{q}^T \tau = b\dot{b}n^T Mn + b^3 n^T Dn + bn^T g - bn^T \tau_e. \tag{27}$$

Now, note that the kinetic energy is

$$\mathcal{T} = \frac{1}{2}b^2 n^T(\bar{q})M(q)n(\bar{q}), \tag{28}$$

and therefore,
$$\frac{d}{dt}\mathcal{T} = b\dot{b}n^T Mn + b^3 n^T Dn. \tag{29}$$

With this, the input power is represented as:

$$\dot{q}^T \tau = \frac{d}{dt}\mathcal{T} - \frac{d}{dt}\mathcal{U} - bn^T \tau_e, \tag{30}$$

where $\mathcal{U} = \mathcal{U}(q)$ is the gravity potential at q, such that $g^T(q) = -\frac{\partial \mathcal{U}(q)}{\partial q}$.

6.2 Dynamic Analysis of Natural Motion

We consider again the special case of natural motion $\underline{b} = \text{const}^4$. Then, the force component $\dot{b}M(q)n(\bar{q})$ becomes zero, and the equation of natural motion can be written as:

$$\underline{b}A(\bar{q})n(\bar{q}) + g(q) = \tau + \tau_e. \tag{31}$$

The force component $\dot{b}M(q)n(\bar{q})$ just removed is in the direction of generalized momentum. Note also that this force component results from the nonintegrable joint acceleration.

The input power of natural motion is derived from Eq. (27) as:

$$\dot{q}^T\tau = \underline{b}^3 n^T Dn + \underline{b}n^T g - \underline{b}n^T \tau_e \tag{32}$$

$$= \frac{d}{dt}\mathcal{T} - \frac{d}{dt}\mathcal{U} - \underline{b}n^T \tau_e, \tag{33}$$

where the kinetic energy time differential

$$\frac{d}{dt}\mathcal{T} = \underline{b}^3 n^T Dn \tag{34}$$

is now configuration-dependent only: $\mathcal{P}(\bar{q}) \equiv \frac{d}{dt}\mathcal{T}$, $\mathcal{P}(\bar{q})$ denoting (instantaneous) mechanical power. But then, in the absence of external forces ($\tau_e = 0$), we obtain a pure potential system:

$$\dot{q}^T\tau = \frac{d}{dt}\mathcal{T}(\bar{q}) - \frac{d}{dt}\mathcal{U}(q). \tag{35}$$

This means that the joint driving forces can be derived from a configuration-dependent potential. Hence:

$$\tau_i = -\frac{d}{dq_i}\mathcal{V}(\bar{q}),\ (i = 1, ..., n), \tag{36}$$

where $\mathcal{V}(q)$ is the configuration-dependent potential. Plugging τ into Eq. (35) and integrating, we obtain

$$-\mathcal{V} = \mathcal{T} - \mathcal{U} + \underline{\mathcal{E}} \tag{37}$$

where $-\underline{\mathcal{E}}$ is a nonzero integration constant. We will refer to $\mathcal{V}$ as the *internal potential*. The internal potential exists if and only if the motion is natural. Thus, we have proven the following

Theorem: Natural motion conserves the energy of internal motion.

---

[4] Here and henceforth we will use underline to denote a constant value.

As noted, natural motion requires b to be a constant. Hence, the kinetic energy at configuration $\bar{q}_0$ considered to be the initial configuration for natural motion, will be

$$\mathcal{T}(\bar{q}_0) = \frac{1}{2}b^2 n^T(\bar{q}_0) M(q_0) n(\bar{q}_0) \neq 0.$$

In addition, we may assume that $\mathcal{V}(\bar{q}_0) = \mathcal{U}(q_0)$. Then, the integration constant is uniquely determined from Eq. (37) as:

$$\mathcal{E} = -\mathcal{T}(\bar{q}_0). \tag{38}$$

It should be apparent now that natural motion of a manipulator with a prescribed end-effector path is nothing else than nondissipative motion with prespecified initial energy. Note also that energy conserving systems are called *natural* [35]. Thus, the term *natural motion*, which was introduced in Section 3 to describe a pure geometrical phenomenon, is reaffirmed in a convincing way also from the viewpoint of energy conservation. But this is not surprising: conservation of energy is related to Jacobi's principle, which, in turn is instructive in how to represent motion as a pure geometrical problem [36].

7 An Example

A simple planar 2R manipulator with its end-tip moving along a circular arc will be used to demonstrate the relations derived above. The current point on the arc is determined by the path parameter angle q_* (see Fig. 1). The end-tip position vector is then:

$$\boldsymbol{x} = \begin{bmatrix} x_c \\ y_c \end{bmatrix} - r \begin{bmatrix} \cos q_* \\ \sin q_* \end{bmatrix} \tag{39}$$

where x_c, y_c are the arc center coordinates, r is the arc radius. The end-tip twist is:

$$\boldsymbol{t}_* = r \begin{bmatrix} \sin q_* \\ -\cos q_* \end{bmatrix}. \tag{40}$$

The vector field over the augmented configuration space is derived from the null-space of the column-augmented Jacobian $\overline{\boldsymbol{J}}(\bar{q})$ as:

$$\bar{\boldsymbol{n}} = \begin{bmatrix} -r l_2 \sin(q_1 + q_2 - q_*) \\ r[l_2 \sin(q_1 + q_2 - q_*) + l_1 \sin(q_1 - q_*)] \\ l_1 l_2 \sin q_2 \end{bmatrix} \tag{41}$$

where l_1 and l_2 are the link lengths. Kinematic singularities are defined by the set $\{\sin q_2 = 0\}$. The stationary points of the vector field are those kinematic singularities where $q_* = q_1 \pm 2k\pi$, $k = 0, 1, 2, \ldots$. An example of a stationary point singularity is the point where the circular arc is tangent to the workspace

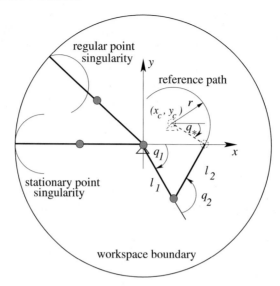

Fig. 1. Example of notations with a 2R manipulator.

boundary. An example of a regular point singularity is the point where the circular arc intersects the workspace boundary.

Numerical simulation of motion based on velocity-level kinematics, as in Eq. (11), will be demonstrated. Both link lengths are set to 1 m.

First, natural motion will be simulated to highlight behavior around a regular point singularity and around a stationary point singularity. The initial manipulator configuration is set to $(-60, 120)$ deg, the respective initial end-tip coordinates are $(1.0, 0.0)$ m. Also, the initial path parameter value is set to 180 deg. Natural motion is obtained with $\underline{b} = 1.0$ rad/m²s. The arc radius is set to 1.6 m, so that the arc becomes transversal to the outer workspace boundary. The kinematic singularity at the intersection is a regular point singularity. Simulation data are shown in Fig. 2. The $x - y$ graph in the lower left part of the figure shows the workspace boundary (the full circle) and the path described by the end-tip (the circular arc). Note that the q_* speed graph is actually the graph of determinant det $\boldsymbol{J}$. After about 3 s, change of its sign is observed which indicates motion *through* the kinematic singularity at the intersection point between workspace boundary and end-tip arc. Thereby, the end-tip is reflected from the workspace boundary back along the arc. Prolonging the simulation time to 12 s generates a circular arc that intersects the workspace boundary at two points. A repeatable motion cycle is then obtained. The respective selfmotion manifold in augmented configuration space $\mathcal{C}^*$ is shown in the lower right part of the figure.

Next, the arc radius is decreased to $r = 1.5$ m, so that the arc center coordinates become $(-0.5, 0.0)$ m. Note that the arc is then tangent to the

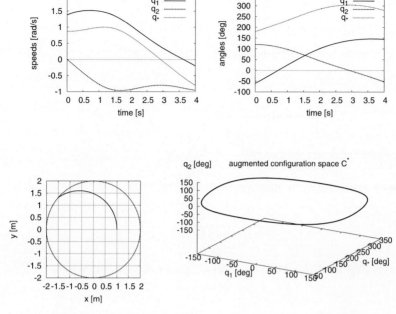

Fig. 2. Natural motion through a regular point singularity.

outer workspace boundary at the west pole with coordinates $(-2.0, 0.0)$ m — a stationary point of the dynamical system. Simulation data are shown in Fig. 3. The exponential decay of all vector field components is apparent from the speed graphs.

We repeat the same simulation, but instead of natural motion, constant twist magnitude is required ($\dot{q}_* = 1$ rad/s). The results are shown in Fig. 4. It is seen that motion through the stationary point has been achieved. The lower left plot shows the graphs of the three vector field components. The stationary point, occurring when all graphs intersect the zero axis, is clearly seen. Despite the (instantaneously) vanishing vector field, end-tip motion can be maintained, as required. This is possible, because constant twist magnitude implies a determinant $\det \boldsymbol{J}$ (vector field component n_3) appearing as denominator in the scalar b (cf. Eq. (12)).

The next two simulations demonstrate the possibility to *initialize* motion at a singularity. A simple way to achieve this is to choose input b as a linear function of time: $b = k_1 t$, where k_1 is an arbitrary constant and t denotes time. Figure 5 shows the result of a simulation when motion was initialized at a regular point singularity. The initial manipulator configuration and the initial path parameter values have been set to $(150, 0)$ deg and zero, respectively, whereas $k_1 = 0.8$ rad/(ms)^2. From the graphs it is evident that motion can be initialized without any instability.

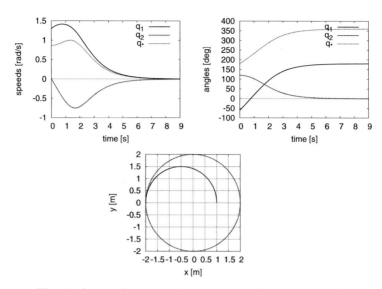

Fig. 3. Approach to a stationary point with natural motion.

On the other hand, it should be apparent that it is impossible to initialize motion from a stationary point since no data for the direction of motion is available. Motion can be initialized, though, from within an infinitesimal vicinity of the stationary point. This is demonstrated with the following sim-

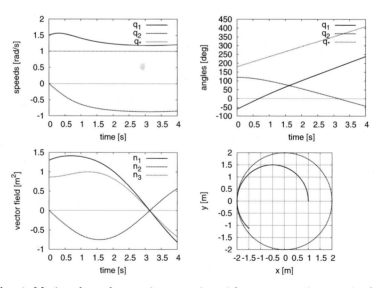

Fig. 4. Motion through a stationary point with constant twist magnitude $\dot{q}_*$.

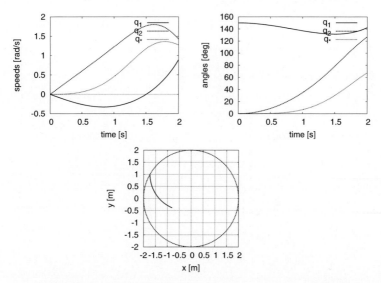

Fig. 5. Initializing motion at a regular point singularity.

ulation. The initial manipulator configuration has been set to (180, 0.01) deg, b was chosen again to be a linear function of time, as above. The result of the simulation is shown in Fig. 6. It is seen that motion can be initialized without any instability. Note, however, that due to the weakness of the vector field in the vicinity of the stationary point, with b being a linear function of time, some time delay is observed before joint velocity builds up.

8 Singularity-Consistent Controllers

Two types of closed-loop controllers will be presented. First, a "kinematic" closed-loop controller similar to the velocity-command generator-type controller based on the Jacobian inverse [4], [5] or the Jacobian transpose [6] will be introduced. Second, a dynamic closed-loop controller will be introduced.

8.1 Singularity-Consistent Kinematic Controller

In the presence of an unknown disturbance, the end-effector deviates from the desired path. Denote the end-effector position/orientation error as e, see e.g. [37]. With some abuse in the notation, widely done in the robotics literature, we will use $e(t) = x_d(t) - x(t)$ where $x_d \equiv x(q_*(t))$ and $x(t)$ is the end-effector position/orientation obtained from joint angle sensor data via the direct kinematics. The error appears then in the kinematic closure equation (7), which takes the form

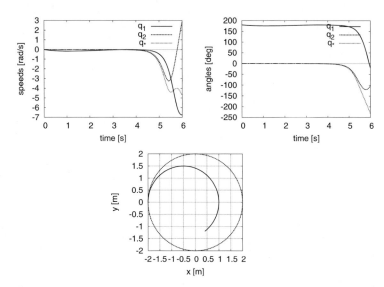

Fig. 6. Initializing motion very close to a stationary point singularity.

$$\overline{\boldsymbol{J}}(\bar{\boldsymbol{q}})\dot{\bar{\boldsymbol{q}}} = \boldsymbol{K}\boldsymbol{e}(t), \tag{42}$$

$\boldsymbol{K}$ denoting a constant feedback gain matrix. This equation is still underdetermined, but it is not anymore a homogeneous one. Hence, the set of solutions is:

$$\dot{\bar{\boldsymbol{q}}} = b\bar{\boldsymbol{n}}(\bar{\boldsymbol{q}}) + \overline{\boldsymbol{J}}^+(\bar{\boldsymbol{q}})\boldsymbol{K}\boldsymbol{e}(t). \tag{43}$$

Projecting it onto configuration space, we obtain the following kinematic feedback control law:

$$\dot{\boldsymbol{q}}_{sc} = b\boldsymbol{n}(\bar{\boldsymbol{q}}) + \boldsymbol{J}^\dagger(\bar{\boldsymbol{q}})\boldsymbol{K}\boldsymbol{e}(t) \tag{44}$$

together with the relation

$$\dot{q}_* = b\det\boldsymbol{J}(\boldsymbol{q}) - \boldsymbol{t}_*^T \boldsymbol{J}_*^{-1}(\bar{\boldsymbol{q}})\boldsymbol{K}\boldsymbol{e}(t). \tag{45}$$

As already explained, the roles of $\dot{q}_*$ and b are interchangeable as control inputs. It is easy to show that the above control law ensures the error dynamics $\dot{\boldsymbol{e}} + \boldsymbol{K}\boldsymbol{e} = \boldsymbol{0}$, with b derived from Eq. (45). Or, alternatively, for any b, the same error dynamics is obtained with $\dot{\boldsymbol{x}}_d = \dot{q}_* \boldsymbol{t}_*$ where $\dot{q}_*$ is derived from Eq. (45).

The block scheme of the controller is shown in Fig. 7. Note that the null space component $b\boldsymbol{n}(\bar{\boldsymbol{q}})$ and the orthogonal component $\boldsymbol{J}^\dagger(\bar{\boldsymbol{q}})\boldsymbol{K}\boldsymbol{e}(t)$ of the solution yield the feedforward and feedback controller actions, $\dot{\boldsymbol{q}}_{ff}$ and $\dot{\boldsymbol{q}}_{fb}$, respectively. Note also that the feedforward action does not depend upon a generalized inverse, as in conventional schemes. Thus, instabilities related to ill-conditioning can be alleviated. The controller will be referred to as the *singularity-consistent kinematic controller*.

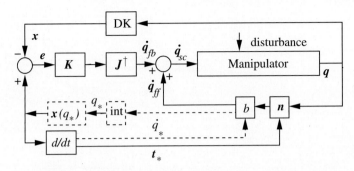

Fig. 7. Singularity-consistent kinematic controller.

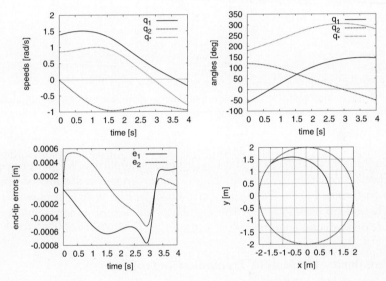

Fig. 8. Natural motion through a regular point singularity, with joint friction and kinematic closed-loop control.

Controller performance will be examined via simulations with the 2R manipulator introduced previously. The feedback gain matrix has been set to $K = \text{diag}\{100, 100\}$ (m rad)/s. The disturbance has the form of constant joint friction $-0.01\dot{q}_i$, $i = 1, 2$. The first simulation demonstrates natural motion via a regular point singularity with initial data as those in Fig. 2. The results are shown in Fig. 8. It becomes apparent that the singularity-consistent kinematic controller performs as expected, without instability. In addition, we also confirmed that motion initialization at the same type of singularity can be achieved, without any instability, too.

In fact, stable performance of the singularity-consistent kinematic controller is guaranteed within the domain of complete integrability. Stability is

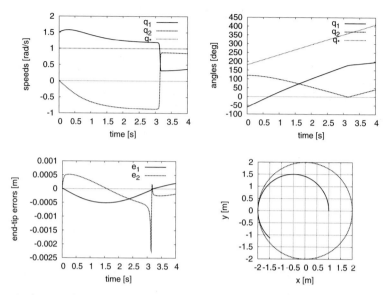

Fig. 9. Approaching a stationary point singularity, with constant twist magnitude, joint friction and kinematic closed-loop control.

not guaranteed, though, at a stationary point singularity. We performed a simulation of constant twist magnitude motion through a stationary point, similar to that in Fig. 4. The results are shown in Fig. 9. It is apparent that when the end-tip passes close by the singularity, an abrupt, inadmissible jump in the velocity occurs.

8.2 Singularity-Consistent Dynamic Controller

Dynamic controller design is based on the singularity consistent parameterization of the equation of motion (23). The *structure* of the proposed controller, though, has not to be a completely new one. We will make use of the controller structure proposed by Slotine and Li [38] which was found to be suitable for velocity vector field tracking.

Define first the *field tracking* error as

$$\boldsymbol{e}_f(t) = \dot{\boldsymbol{q}}_d(t) - \dot{\boldsymbol{q}}(t), \tag{46}$$

where the desired joint velocity is obtained from the vector field projection:

$$\dot{\boldsymbol{q}}_d(t) = b_d(t)\boldsymbol{n}(\bar{\boldsymbol{q}}_d). \tag{47}$$

Note that the above error is defined for the general case, i.e. b being time varying. It is easy to show then that, under the following feedback control law:

$$\boldsymbol{\tau}_{sc} = (\dot{b}\boldsymbol{M} + b\boldsymbol{A})\boldsymbol{n} + \boldsymbol{g} + \boldsymbol{K}_D\boldsymbol{e}_f, \qquad (48)$$

the function

$$V(t) = \frac{1}{2}\boldsymbol{e}_f^T\boldsymbol{M}\boldsymbol{e}_f \qquad (49)$$

is a Lyapunov function, and the above control law ensures that $\boldsymbol{e}_f(t) \to 0$ as $t \to \infty$ within the domain of complete integrability, and in the absence of disturbances. $\boldsymbol{K}_D$ in the above control law is a positive definite matrix that may be chosen to be time varying. To ensure zero steady state position error, $\boldsymbol{e}_f$ is replaced by the following sliding surface:

$$\boldsymbol{s} = \boldsymbol{e}_f + \boldsymbol{\Lambda}(\boldsymbol{q}_d - \boldsymbol{q}), \qquad (50)$$

$\boldsymbol{\Lambda}$ denoting a constant positive definite matrix [38].

Henceforth we refer to control law (48) as the *singularity-consistent dynamic* controller. The block scheme is shown in Fig. 10. The singularity-consistent dynamic controller combines feedforward action with nonlinear compensator within a single block: $\boldsymbol{\tau}_{ff\&nc} = (\dot{b}\boldsymbol{M} + b\boldsymbol{A})\boldsymbol{n} + \boldsymbol{g}$. Occasionally, gravity compensation will be switched off to model external disturbance. Note that the vector field projection appears also in the feedback component $\boldsymbol{\tau}_{fb} = \boldsymbol{K}_D\boldsymbol{s}$ through the desired quantities.

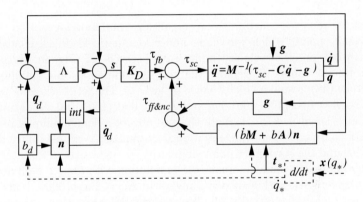

Fig. 10. Singularity-consistent dynamic controller.

Controller performance will be examined with the help of numerical simulations. As a base for comparison, we will employ the resolved acceleration controller [37]:

$$\boldsymbol{\tau}_{racc} = \boldsymbol{M}\boldsymbol{J}^{-1}\left(\dot{\boldsymbol{\nu}}_d + \boldsymbol{K}_d(\boldsymbol{\nu}_d - \boldsymbol{\nu}) + \boldsymbol{K}_p\boldsymbol{e} - \dot{\boldsymbol{J}}\dot{\boldsymbol{q}}\right) + \boldsymbol{C}\dot{\boldsymbol{q}} + \boldsymbol{g}, \qquad (51)$$

where $\boldsymbol{e}$ denotes end-effector position/orientation error, $\boldsymbol{\nu}_d$ and $\boldsymbol{\nu}$ stand for desired and current end-effector twist, respectively, and $\boldsymbol{K}_p$ and $\boldsymbol{K}_d$ are controller gain matrices.

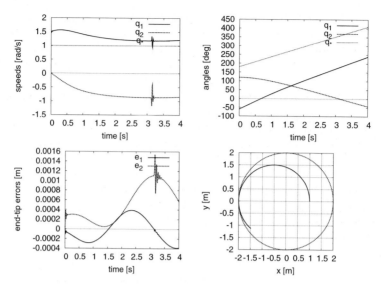

Fig. 11. Motion through a stationary point, with constant twist magnitude, gravity disturbance and dynamic closed-loop control.

First, we redo the constant twist-magnitude simulation for motion through a stationary point, which was done under kinematic feedback control, and where an inadmissible velocity jump was observed (cf. Fig. 9). Initial conditions are the same. We remove the gravity compensation capability from control law (48), to model disturbance by gravity. The feedback gain matrices are set to $\boldsymbol{\Lambda} = \text{diag}\{400, 400\}$ 1/s and $\boldsymbol{K}_D = \text{diag}\{100, 100\}$ Nm s/rad. The results are shown in Fig. 11. Note that no abrupt velocity jump is observed now, and continuous joint angle motion is obtained. *Local* instability has appeared around the stationary point singularity, though, as seen from the speed and error plots. We made an analogous simulation with the resolved acceleration controller (51), with feedback gains set to $\boldsymbol{K}_p = \text{diag}\{10000, 10000\}$ 1/s^2 and $\boldsymbol{K}_d = \text{diag}\{100, 100\}$ 1/s. The controller failed abruptly.

To cope with the local instability problem in the last simulation, we introduce a simple filter: set $\boldsymbol{\tau}_{ff\&nc} = \boldsymbol{0}$ when $|\sin q_2| < k_2$, where k_2 is a small positive constant. Simulation data is shown in Fig. 12. The constant was chosen as $k_2 = 0.1$; note that this choice is not a critical one. Comparing the resultant speed graphs and joint error graphs in Fig. 12 with those in Fig. 11, it is seen that the local instability at the singularity has been successfully removed. The right part of Fig. 12 shows total joint torques (upper plot) and two joint power components resulting from the integrable and nonintegrable joint acceleration components (lower plot). The action of the filter is evident from the latter.

Next, we demonstrate motion initialization from within an infinitesimal neighborhood of a stationary point singularity, similar to the simulation in Fig. 6. The result of the simulation is shown in Fig. 13. The errors are due to the gravity disturbance. It is confirmed that even in the presence of such disturbance, motion can be initialized without instability.

The final simulation demonstrates how motion initialization from within an infinitesimal neighborhood of a stationary point singularity can result in a relatively complex motion. The singularity is achieved through an almost folded-arm initial configuration of (90, 180.01) deg, and a reference circular arc, tangent to the first link (see Fig. 14, lower-right part). b is determined again as a linear function of time: $b = 0.8t$. The results of the simulation are shown in Fig. 14. First, only joint one rotates, keeping the arm configuration close to the singular one for about 2.7 s (stage 1 in the lower right figure). Thereafter, the arm unfolds, the end-tip tracking thereby the reference circular arc, as desired (stage 2, ibid.).

9 Conclusions

Until recently, not much attention has been paid to the useful role of kinematically singular configurations in motion tasks. The ill-conditioning of the inverse problem around such configurations has been tackled mostly within linear model frameworks. It has been shown here that a non-linear, differential-geometric approach to kinematic inversion can lead to improved models. With

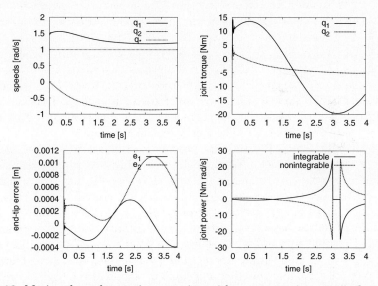

Fig. 12. Motion through a stationary point, with constant twist magnitude, gravity disturbance and dynamic closed-loop control with filter.

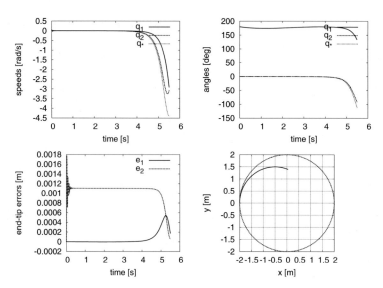

Fig. 13. Motion initialization close to a stationary point under gravity disturbance.

these models, certain kinematic singularities (called regular point singularities), can be treated as regular points of the nonlinear system. Already at the

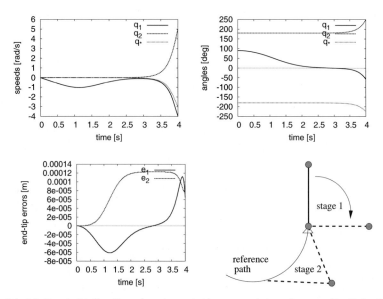

Fig. 14. Motion initialization close to a stationary point under gravity disturbance, resulting in two-stage motion.

velocity level, it was possible to identify the so-called natural motion, which, as shown by dynamic analysis, turns out to be nondissipative motion in a closed system, without external forces, leading also to an (internal) potential-type equation of motion. No ill-conditioning of the inverse problem occurs when regular point singularities are approached with natural motion. Motion initialization at and motion through such singularities is also possible. Another type of singularity, identified as stationary point singularity, requires more attention due to the vanishing vector field under natural motion.

Further on, a kinematic feedback controller and a dynamic feedback controller have been introduced and their performance examined via simulations. Stable motion control at and around regular point singularities has been confirmed, even in the presence of an external disturbance. A local instability problem, that occurred while moving through a stationary point singularity, has been alleviated by introducing a simple filtering approach.

References

1. Khatib O (1987) A unified approach to motion and force control of robot manipulators: The operational space formulation. IEEE Journal of Robotics and Automation 3(1):43–53
2. Uchiyama M (1979) A study of computer control of motion of a mechanical arm (1st report: Calculation of coordinative motion considering singular points.) Transactions of the JSME C-45(392):314–322
3. Popov E, Vereshchagin A, Zenkevich S (1978) Manipulation Robots: Dynamics and Algorithms. Nauka Publishers, Moscow (in Russian)
4. Wolovich W, Elliot H (1984) A computational technique for inverse kinematics. In: Proc. 23rd Conf. on Decision and Control, Las Vegas 1359–1363
5. Balestrino A, De Maria G, Sciavicco L (1984) Robust control of robotic manipulators. In: Preprints of the 9th IFAC World Congress, Budapest, Hungary 80–85
6. Chiaccio P, Siciliano B (1989) A closed-loop Jacobian transpose scheme for solving the inverse kinematics of nonredundant and redundant wrists. Journal of Robotic Systems 6(5):601–630
7. Arimoto S, Sekimoto M, Hashiguchi H, Ozawa R (2005) Natural resolution of ill-posedness of inverse kinematics for redundant robots: A challenge to Bernstein's degrees-of-freedom problem. Advanced Robotics 19(4):401–434
8. Nakamura Y, Hanafusa H (1986) Inverse kinematic solutions with singularity robustness for robot manipulator control. ASME Journal of Dynamic Systems, Measurment and Control 108:163–171
9. Wampler C (1986) Manipulator inverse kinematic solutions based on vector formulations and damped least-squares methods. IEEE Transactions on Systems, Man, and Cybernetics 16(1):93–101
10. Chiaverini S, Siciliano B, Egeland O (1994) Review of damped least-squares inverse kinematics with experiments on an industrial robot manipulator," IEEE Transactions on Control Systems Technology 2(2):123–134

11. Kirćanski M et al. (1994) An experimental study of resolved acceleration control in singularities: damped least-squares approach. In: Proc. IEEE Int. Conf. on Robotics and Automation, San Diego, California 2686–2691
12. Maciejewski A, Klein C (1988) Numerical filtering for the operation of robotic manipulators through kinematically singular configurations. Journal of Robotic Systems 5:527–552
13. Nenchev D, Tsumaki Y, Uchiyama M (2000) Singularity-consistent parameterization of robot motion and control. The International Journal of Robotics Research 19(2):159–182
14. Nielsen L, de Wit C, Hagander P (1991) Controllability issues of robots in singular configurations. In: Proc. IEEE Int. Conf. Robotics and Automation, Sacramento, California 2210–2215
15. Kieffer J (1994) Differential analysis of bifurcations and isolated singularities for robots and mechanisms. IEEE Transactions on Robotics and Automation 10(1):1–10
16. Tchoń K, Dulęba I (1993) On inverting singular kinematics and geodesic trajectory generation for robot manipulators. Journal of Intelligent and Robotic Systems 8:325–359
17. O'Neil K, Chen Y, Seng S (1996) Desingularization of resolved motion rate control of mechanisms. In: Proc. IEEE Int. Conf. on Robotics and Automation, Minneapolis, Minnesota 3147–3154
18. Nenchev D (1995) Tracking manipulator trajectories with ordinary singularities: a null space based approach. The International Journal of Robotics Research 14(4):399–404
19. Nenchev D, Tsumaki Y, Uchiyama M (1994) A telerobotic system with singularity consistent behaviour. In: Proc. Fifth Int. Conf. on Adaptive Structures, Sendai, Japan 501–510
20. Tsumaki Y, Nenchev D, Kotera S, Uchiyama M (1997) Teleoperation based on the adjoint Jacobian approach. IEEE Control Sytems Magazine 17(1):53–62
21. Nenchev D, et al. (1996) Singularity-consistent control of nonredundant robots In: Video Proceedings of the IEEE Int. Conf. Robotics and Automation, Minneapolis, Minnesota
22. Nenchev D, Uchiyama M (1997) Singularity-consistent path planning and motion control through instantaneous self-motion singularities of parallel-link manipulators. Journal of Robotic Systems 14(1):27–36
23. Nenchev D, Tsumaki Y, Takahashi M (2004) Singularity-consistent kinematic redundancy resolution for the S-R-S manipulator. In: Proc. IEEE/RSJ Int. Conf. on Intelligent Robots and Systems, Sendai, Japan 3607–3612
24. Kameta K, et al. (2005) Walking control using the SC approach for humanoid robot. In: Proc. 5th IEEE-RAS Int. Conf. on Humanoid Robots Tsukuba, Japan 289–294
25. Nenchev D, Uchiyama M (1997) Natural motion analysis based on the singularity-consistent parameterization. In: Proc. IEEE Int. Conf. on Robotics and Automation, Albuquerque, New Mexico 2683–2688
26. Nenchev D, Tsumaki Y, Uchiyama M (1998) Singularity consistency and the natural motion of robot manipulators. In: Proc. 37th IEEE Conf. Decision and Control, Tampa, Florida 407–412
27. Koditschek D (1984) Natural motion of robot arms. In: Proc. 23rd IEEE Conf. on Decision and Control, Las Vegas, Nevada 733–735

28. Shamir T (1990) The singularities of redundant robot arms. The International Journal of Robotics Research 9(1):113–121
29. Bedrossian N, Flueckiger K (1991) Characterizing spatial redundant manipulator singularities. In: Proc. IEEE Int. Conf. on Robotics and Automation, Sacramento, California 714–719
30. Singh S (1993) Motion planning and control of non-redundant manipulators at singularities. In: Proc. IEEE Int. Conf. on Robotics and Automation, Atlanta, Georgia 487–492
31. Burdick J (1991) A classification of 3R regional manipulator singularities and geometries. In: Proc. IEEE Int. Conf. on Robotics and Automation, Sacramento, California 2670–2675
32. Guckenheimer J, Holmes P (1986) Nonlinear Oscillations, Dynamical Systems, and Bifurcation of Vector Fields. New York, NY: Springer-Verlag New York Inc.
33. Nelson B, Khosla P (1995) Strategies for increasing the tracking region of an eye-in-hand system by singularity and joint limit avoidance. The International Journal of Robotics Research 14(3):255–269
34. Arimoto S, Miyazaki F (1984) Stability and robustness of PID feedback control for robot manipulators of sensory capability. In Brady M, Paul R, eds.: Robotics Research: First International Symposium. MIT Press 783–799
35. Arnold V (1989) Mathematical Methods of Classical Mechanics. Springer-Verlag, New York Inc.
36. Lanczos C (1970) The Variational Principles of Mechanics. Fourth Edition. Dover Pub. Inc., New York
37. Luh J, Walker M, Paul R (1980) Resolved acceleration control of mechanical manipulators. IEEE Transactions on Automatic Control 25(3):468–474
38. Slotine J-JE, Li W (1987) On the adaptive control of robot manipulators. The International Journal of Robotics Research 6(3):49–59

Approximate Jacobian Control for Robot Manipulators

Chien Chern Cheah

School of Electrical and Electronic Engineering
Nanyang Technological University
Block S1, Nanyang Avenue, S(639798)
Republic of Singapore
ecccheah@ntu.edu.sg

Summary. Most research so far in robot control has assumed either kinematics or Jacobian matrix of the robots from joint space to task space is known exactly. Unfortunately, no physical parameters can be derived exactly. In addition, when the robot picks up objects of uncertain lengths, orientations or gripping points, the kinematics and dynamics become uncertain and change according to different tasks. This paper presents several approximate Jacobian control laws for robots with uncertainties in kinematics and dynamics. Lyapunov functions are presented for stability analysis of feedback control problems with uncertain kinematics. We shall show that the end-effector's position converges to a desired position even when the kinematics and Jacobian matrix are uncertain.

1 Introduction

It is interesting to observe from human reaching movements that we do not need an accurate knowledge of kinematics and dynamics of our arms. We are also able to pick up a new tool or object and manipulate it skillfully to accomplish a task, using only the approximate knowledge of the length, mass, orientation and gripping point of the tool. Such basic ability of sensing and responding to changes without an accurate knowledge of sensory-to-motor transformation gives us a high degree of flexibility in dealing with unforseen changes in the real world.

The kinematics and dynamics of robot manipulators are highly nonlinear. By exploring physical properties of the robot system and using Lyapunov method, Takegaki and Arimoto [1], Arimoto and Miyazaki [2] showed that simple controllers such as the PD and PID feedback are effective for setpoint control despite the nonlinearity and uncertainty of the robot dynamics. To deal with trajectory-tracking control, Slotine and Li [3, 4] proposed an adaptive control law for robotic manipulator using Lyapunov method. After more than two decades of research, much progress has been made in control of robots with dynamic uncertainty [1]-[19].

However, most research on robot control has assumed that the exact kinematics and Jacobian matrix of the manipulator from joint space to Cartesian space are known. In the presence of uncertainty in kinematics, it is impossible to derive the desired joint angle from the desired end effector path by solving the inverse kinematics problem. In addition, the Jacobian matrix of the mapping from joint space to task space could not be exactly derived. This assumption leads us to several problems in the development of robot control laws today. In free motion [20], this implies that the exact lengths of the links, joint offsets and the object which the robot is holding, must be known. Unfortunately, no physical parameters could be derived exactly. In addition, when the robot picks up objects or tools of different lengths, unknown orientations and gripping points, the overall kinematics are changing and therefore difficult to derive exactly. Therefore, the robot is not able to manipulate the tool to a desired position if the length or gripping point of the tool is uncertain. Similarly, in hybrid position force control [21], the kinematics and constraint Jacobian matrix are also uncertain in many applications. When the control problem is extended to the control of multi-fingered robot hands [22], such assumption also limits its potential applications because the kinematics is usually uncertain in many applications of robot hands. For example, the contact points of the robot fingers are usually uncertain and changing during manipulation.

This paper presents the recent advances in control theory of robots with uncertain kinematics and dynamics. Several approximate Jacobian controllers, that do not require exact knowledge of either kinematics or dynamics of robots, are presented. The control problems are formulated based on Lyapunov analysis. By using sensory feedback of the robot end effector position, it is shown that the end effector is able to follow a desired motion with uncertainties in kinematics and dynamics. This gives the robot a high degree of flexibility in dealing with unforseen changes and uncertainties in its kinematics and dynamics, which is similar to human reaching movements and tool manipulation.

This paper is organized as follows. Section 2 formulates the robot dynamic equations and kinematics; Section 3 presents the approximate Jacobian setpoint controllers; Section 4 presents the adaptive Jacobian tracking controller; Section 5 offers brief concluding remarks.

2 Robot Kinematics and Dynamics

The equations of motion of robot with n degrees of freedom can be expressed in joint coordinates $q = [q_1, \cdots, q_n]^T \in R^n$ as [19, 23, 24]:

$$M(q)\ddot{q} + (\frac{1}{2}\dot{M}(q) + S(q,\dot{q}))\dot{q} + g(q) = \tau \qquad (1)$$

where $M(q) \in R^{n \times n}$ is the inertia matrix, $\tau \in R^n$ is the applied joint torque to the robot,

$$S(q,\dot{q})\dot{q} = \frac{1}{2}\dot{M}(q)\dot{q} - \frac{1}{2}\{\frac{\partial}{\partial q}\dot{q}^T M(q)\dot{q}\}^T$$

and $g(q) \in R^n$ is the gravitational force. Several important properties of the dynamic equation described by equation (1) are given as follows [19, 3, 23]:

Property 1 The inertia matrix $M(q)$ is symmetric and uniformly positive definite for all $q \in R^n$.

Property 2 The matrix $S(q,\dot{q})$ is skew-symmetric so that

$$\nu^T S(q,\dot{q})\nu = 0,$$

for all $\nu \in R^n$.

Property 3 The dynamic model as described by equation (1) is linear in a set of physical parameters $\theta_d = (\theta_{d1}, \cdots, \theta_{dp})^T$ as

$$M(q)\ddot{q} + (\frac{1}{2}\dot{M}(q) + S(q,\dot{q}))\dot{q} + g(q) = Y_d(q,\dot{q},\dot{q},\ddot{q})\theta_d$$

where $Y_d(\cdot) \in R^{n \times p}$ is called the dynamic regressor matrix. ◇

In most applications of robot manipulators, a desired path for the end-effector is specified in task space, such as visual space or Cartesian space. Let $x \in R^n$ be a task space vector defined by [19],

$$x = h(q) \qquad (2)$$

where $h(\cdot) \in R^n \to R^n$ is generally a non-linear transformation describing the relation between joint space and task space. The task-space velocity $\dot{x}$ is related to joint-space velocity $\dot{q}$ as:

$$\dot{x} = J(q)\dot{q} \qquad (3)$$

where $J(q) \in R^{n \times n}$ is the Jacobian matrix from joint space to task space.

Note that if cameras are used to monitor the position of the end-effector, the task coordinates are defined as image coordinates. Let r represents the position of the end-effector in Cartesian coordinates and x represents the vector of image feature parameters [25]. The velocity vector $\dot{x}$ is therefore related to $\dot{q}$ as

$$\dot{x} = J_I(r)\dot{r} = J_I(r)J_e(q)\dot{q}, \qquad (4)$$

where $J_I(r)$ is the image Jacobian matrix [25, 26, 27] and $J_e(q)$ is the manipulator Jacobian matrix of the mapping from joint space to Cartesian space. When a position sensor such as an electromagnetic measurement system or laser tracking systems is used to measure the end effector's position, the task co-ordinates are defined as the Cartesian co-ordinates.

A property of the kinematic equation described by equation (3) is stated as follows:

Property 4 The right hand side of equation (3) is linear in a set of constant kinematic parameters $\theta_k = (\theta_{k1}, \cdots, \theta_{kq})^T$, such as link lengths, link twist angles. Hence, equation (3) can be expressed as,

$$\dot{x} = J(q)\dot{q} = Y_k(q, \dot{q})\theta_k \qquad (5)$$

where $Y_k(q, \dot{q}) \in R^{n \times q}$ is called the kinematic regressor matrix. ◇

It is well known that in joint-space control method, the exact kinematics described by equation (2) is required to solve the inverse kinematics to generate a desired joint position. When the control problem is formulated in Cartesian space, the exact Jacobian matrix or forward kinematics is still required in the control laws. Hence, the robot control system is not able to adapt to uncertainties or unforeseen changes in the kinematics if its exact knowledge is required to control the robot's movements.

3 Approximate Jacobian Setpoint Control of Robots

In this section, approximate Jacobian feedback controllers [32] are presented for setpoint control of robots with uncertain kinematics and dynamics. The controllers do not require the exact knowledge of kinematics and Jacobian matrix that is assumed in the literature of robot control.

3.1 Approximate Jacobian Setpoint Control with Gravitational Force Compensation

First, a simple Approximate Jacobian feedback controller with gravitational force compensation is presented. The control input using an approximate Jacobian matrix $\hat{J}(q) \in R^{m \times n}$ is proposed as [32],

$$\tau = -\hat{J}^T(q)K_p s(e) - B_v \dot{q} + g(q) \qquad (6)$$

where $\hat{J}^T(q)$ is chosen so that

$$\|J^T(q) - \hat{J}^T(q)\| \leq p, \qquad (7)$$

and $e = x - x_d = (e_1, \cdots, e_m)^T$ is a positional deviation from a desired position $x_d \in R^m$, $s(e) = (s_1(e_1), \cdots, s_m(e_m))^T$, p is a positive constant to be defined later, $K_p = k_p I$, $B_v = b_v I$ are positive feedback gains for the task-space position error and joint-space velocity respectively, I is the identity matrix,

$s_i(\cdot)$, $i = 1, \cdots, m$, are saturation functions [19]. The task-space position x is measured by a sensor such as vision system, electromagnetic measurement system or position sensor system.

Let us define a scalar potential function $S_i(\theta)$ where $\theta \in R$ and its derivative $s_i(\theta)$ as shown in Figure 1. The functions $S_i(\theta)$ and $s_i(\theta)$ have the following properties [19]:

(1) $S_i(\theta) > 0$ for $\theta \neq 0$ and $S_i(0) = 0$.
(2) $S_i(\theta)$ is twice continuously differentiable, and the derivative $s_i(\theta) = \frac{dS_i(\theta)}{d\theta}$ is strictly increasing in θ for $|\theta| < \gamma_i$ with some γ_i and saturated for $|\theta| \geq \gamma_i$, i.e. $s_i(\theta) = \pm s_i$ for $\theta \geq +\gamma_i$ and $\theta \leq -\gamma_i$ respectively where s_i is a positive constant.
(3) There exist a constant $\bar{c}_i > 0$, such that,

$$S_i(\theta) \geq \bar{c}_i s_i^2(\theta), \tag{8}$$

for $\theta \neq 0$.

The closed-loop equation of the system is obtained by substituting equation (6) into equation (1),

$$M(q)\ddot{q} + (\frac{1}{2}\dot{M}(q) + S(q,\dot{q}))\dot{q} + \hat{J}^T(q)K_p s(e) + B_v \dot{q} = 0. \tag{9}$$

To carry out the stability analysis for the closed-loop system with the approximate Jacobian controller, a Lyapunov function candidate is defined as

$$V = \tfrac{1}{2}\dot{q}^T M(q)\dot{q} + \alpha \dot{q}^T M(q)\hat{J}^T(q)s(e) + \sum_{i=1}^{m}(k_p + \alpha b_v)S_i(e_i), \tag{10}$$

To show the positive definiteness of the Lyapunov function candidate, we note that

$$\tfrac{1}{4}\dot{q}^T M(q)\dot{q} + \alpha \dot{q}^T M(q)\hat{J}^T(q)s(e) + \sum_{i=1}^{m}(k_p + \alpha b_v)S_i(e_i)$$
$$= \tfrac{1}{4}(\dot{q} + 2\alpha \hat{J}^T(q)s(e))^T M(q)(\dot{q} + 2\alpha \hat{J}^T(q)s(e))$$
$$- \alpha^2 s^T(e)\hat{J}(q)M(q)\hat{J}^T(q)s(e) + \sum_{i=1}^{m}(k_p + \alpha b_v)S_i(e_i)$$
$$\geq \sum_{i=1}^{m}\{k_p \bar{c}_i + \alpha(b_v \bar{c}_i - \alpha \lambda_m)\}s_i^2(e_i)$$

where $\lambda_m \triangleq \lambda_{max}[\hat{J}(q)M(q)\hat{J}^T(q)]$ and $\lambda_{max}[A]$ denotes the maximum eigenvalue of matrix A. Substituting the above equation into equation (10), we have,

$$V \geq \frac{1}{4}\dot{q}^T M(q)\dot{q} + \sum_{i=1}^{m}\{k_p \bar{c}_i + \alpha(b_v \bar{c}_i - \alpha \lambda_m)\}s_i^2(e_i) \geq 0$$

where b_v and α can be chosen so that

$$b_v \bar{c}_i - \alpha \lambda_m > 0. \tag{11}$$

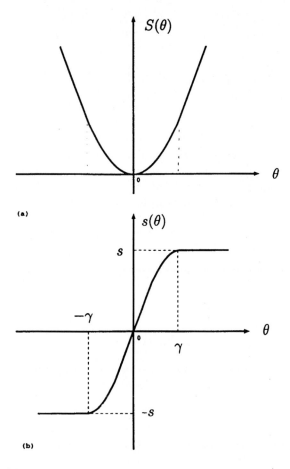

Fig. 1. (a) Quasi-natural potential: $S(\theta)$, (b) derivative of $S(\theta)$: $s(\theta)$

Next, we proceed to show that the time derivative of the Lyapunov function is negative definite in $(\dot{q}, s(e))$. Differentiating V with respect to time and substituting the closed-loop equation (9) into it, we have

$$\frac{d}{dt}V = -W \tag{12}$$

where

$$W = b_v \dot{q}^T \dot{q} + \alpha k_p s^T(e) \hat{J}(q) \hat{J}^T(q) s(e) - (k_p + \alpha b_v) \dot{q}^T (J^T(q) - \hat{J}^T(q)) s(e)$$
$$-\alpha \{s^T(e)\hat{J}(q)(\tfrac{1}{2}\dot{M}(q) - S(q,\dot{q}))\dot{q} + \dot{s}^T(e)\hat{J}(q)M(q)\dot{q} + s^T(e)\dot{\hat{J}}(q)M(q)\dot{q}\}. \tag{13}$$

From the last term on the right-hand side of equation (13), since $s(e)$ is bounded, there exist a constant $c_0 > 0$ such that [19]

$$\alpha|s^T(e)\hat{J}(q)(\frac{1}{2}\dot{M}(q)-S(q,\dot{q}))\dot{q}+\dot{s}^T(e)\hat{J}(q)M(q)\dot{q}+s^T(e)\dot{\hat{J}}(q)M(q)\dot{q}|\le \alpha c_0\|\dot{q}\|^2. \tag{14}$$

Substituting inequality (14) into equation (13) yields

$$\begin{aligned}W \ge{} & \dot{q}^T(b_vI-\alpha c_0I)\dot{q}+\alpha k_p s^T(e)\hat{J}(q)\hat{J}^T(q)s(e)\\ & -(k_p+\alpha b_v)\dot{q}^T(J^T(q)-\hat{J}^T(q))s(e).\end{aligned} \tag{15}$$

Now, letting $\bar{\Delta}\triangleq J^T(q)-\hat{J}^T(q)$, we have

$$W \ge \dot{q}^T(b_vI-\alpha c_0I)\dot{q}+\alpha k_p s^T(e)\hat{J}(q)\hat{J}^T(q)s(e)-(k_p+\alpha b_v)\dot{q}^T\bar{\Delta}s(e). \tag{16}$$

The existence of a $\bar{\Delta}$ so that W is positive definite can be clearly seen from equation (16). In the following development, a sufficient condition to guarantee the positive definiteness of W will be derived. From equation (16), we have

$$W \ge (b_v-\alpha c_0)\|\dot{q}\|^2 - p(k_p+\alpha b_v)\|s(e)\|\cdot\|\dot{q}\|+\alpha k_p\lambda_{\hat{j}}\|s(e)\|^2, \tag{17}$$

where $\lambda_{\hat{j}}=\lambda_{min}[\hat{J}(q)\hat{J}^T(q)]$. Next, we note that

$$-\|s(e)\|\cdot\|\dot{q}\| \ge -\frac{1}{2}(\|s(e)\|^2+\|\dot{q}\|^2). \tag{18}$$

Substituting inequality (18) into equation (17) gives,

$$W \ge b_v l_1\|\dot{q}\|^2+b_v l_2\|s(e)\|^2, \tag{19}$$

where

$$l_1=1-\tfrac{\alpha c_0}{b_v}-\tfrac{p}{2}(\bar{a}+\alpha),\quad l_2=\alpha\bar{a}\lambda_{\hat{j}}-\tfrac{p}{2}(\bar{a}+\alpha), \tag{20}$$

and $\bar{a}=\frac{k_p}{b_v}$. Hence, if

$$\min\{\frac{2(1-\frac{\alpha c_0}{b_v})}{\bar{a}+\alpha},\frac{2\bar{a}\alpha\lambda_{\hat{j}}}{\bar{a}+\alpha}\} > p, \tag{21}$$

and $b_v > \alpha c_0$, then $l_1 > 0$ and $l_2 > 0$ and hence, W is positive definite.

In order to guarantee the stability of the system with approximate Jacobian matrix, the allowable bound p of the Jacobian uncertainty $\bar{\Delta}=\hat{J}^T(q)-J^T(q)$ must satisfy the condition in 21. Figure 2 shows an interesting graphical illustration of the condition.

It can be seen from the condition that both the feedback-gains ratio $\bar{a}=k_p/b_v$ and the absolute value of b_v play an important role in stabilizing the system in the presence of uncertain kinematics. As seen from the condition, increasing b_v will increase $2(1-\frac{\alpha c_0}{b_v})/(\bar{a}+\alpha)$ and hence, result in higher p for the same feedback ratio $\bar{a}$. That is, larger b_v would allow a higher margin of uncertainty of the approximate Jacobian matrix for the same $\bar{a}$. When b_v is so large that $\alpha c_0/b_v$ is negligible, condition (21) reduces to

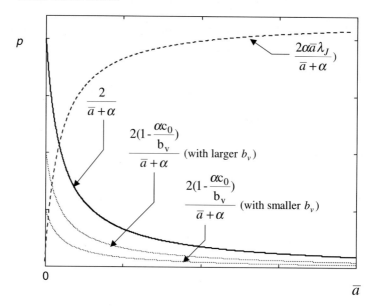

Fig. 2. Variation of p with $\bar{a}$ and b_v (joint-space damping)

$$\min\{\frac{2}{\bar{a}+\alpha}, \frac{2\bar{a}\alpha\lambda_{\hat{j}}}{\bar{a}+\alpha}\} > p. \tag{22}$$

Conversely, if p is small, a smaller controller gain b_v is required. Next, it is interesting to note the effect of $\bar{a}$ on the stability of the system. As seen from the condition, a larger range of $\bar{a}$ can be chosen for a smaller p. In fact, in the extreme case where $p = 0$, $\bar{a}$ could be chosen as any value. However, when p is larger, a narrower range of $\bar{a}$ is allowed. An important and practical conclusion of the result is that, when the system is unstable, redesign of $\hat{J}(q)$ or calibrations may not be necessary as the instability may be due to the reasons that the feedback gain b_v or the feedback gain ratio $\bar{a}$ is not tuned properly. In practice, we should therefore try to stabilize the system or increase the margin of stability first by increasing feedback gains or reducing $\bar{a}$.

We are now in a position to state the following Theorem [32]:

Theorem 1. *The equilibrium state $(x_d, 0)$ of the closed-loop system described by equation (9) is asymptotically stable with uncertain Jacobian matrix $\hat{J}(q)$ if the feedback gains K_p and B_v are chosen to satisfy conditions (11), (21) and $\hat{J}(q)$ is chosen to satisfy condition (7).*

Proof:
Since both V and W are positive definite, from equation (12), we have

$$\frac{d}{dt}V = -W < 0. \tag{23}$$

Hence, V is a Lyapunov function whose time derivative is negative in $(s(e), \dot{q})$. This implies directly the asymptotic stability of the equilibrium state such that, $x_d - x \to 0$, $\dot{q} \to 0$, as $t \to \infty$.

△△△

3.2 Adaptive PD for Uncertain Gravitational Force

Next, approximate Jacobian setpoint controller with uncertain gravitational force compensation is presented. Note that the gravity term can be completely characterized by a set of parameters $\psi = (\psi_1, \cdots, \psi_p)^T$ [19] as

$$g(q) = Z(q)\psi, \qquad (24)$$

where $Z(q) \in R^{n \times p}$ is the gravity regressor. Then, the control input is proposed as [32]

$$\tau = -\hat{J}^T(q)K_p s(e) - B_v \dot{q} + Z(q)\hat{\psi}, \qquad (25)$$

$$\dot{\hat{\psi}} = -LZ^T(q)(\dot{q} + \alpha \hat{J}^T(q)s(e)), \qquad (26)$$

where $Z(q) \in R^{n \times p}$ is the gravity regressor given in equation (24), $K_p = k_p I$, $B_v = b_v I$, α is a positive scalar and $L \in R^{p \times p}$ is a positive definite matrix. Note that the above controller does not require generalized inverse of the Jacobian matrix.

Substituting equations (24) and (25) into equation (1), we have the closed-loop equation,

$$M(q)\ddot{q} + (\tfrac{1}{2}\dot{M}(q) + S(q, \dot{q}))\dot{q} + \hat{J}^T(q)K_p s(e) + B_v \dot{q} + Z(q)\Delta\psi = 0, \quad (27)$$

where $\Delta\psi = \psi - \hat{\psi}$. We are now ready to state the following Theorem:

Theorem 2. *The closed-loop system described by equations (27) and (26) give rise to the asymptotic convergence of $(x(t), \dot{q})$ to $(x_d, 0)$ as $t \to \infty$ if the feedback gains K_p, B_v and $\hat{J}(q)$ are chosen as in Theorem 1.*

Proof:
Let us define a Lyapunov function as:

$$V_1 = V + \frac{1}{2}\Delta\psi^T \bar{L}^{-1} \Delta\psi \qquad (28)$$

where V is defined in equation (10) of the previous subsection. Note that V_1 is positive definite since V is positive definite in $s(e)$ and $\dot{q}$. Differentiate V_1 with respect to time, we have

$$\frac{d}{dt}V_1 = -W \le 0. \qquad (29)$$

Therefore, by LaSalle's invariance Theorem, we have, $x_d - x \to 0$, $\dot{q} \to 0$ as $t \to \infty$.

Remark 1. To overcome this problem of kinematic uncertainty, approximate Jacobian setpoint controllers with task-space damping are first proposed [28]. The proposed controllers [28] require the measurement of a task-space position by a sensor such as vision systems and the task-space velocity is obtained by differentiation of the position. In addition, the controllers require the inverse of Jacobian matrix. In [32], approximate Jacobian feedback controllers with joint-space damping are also proposed. The main advantages of the controllers in [32] are that the task-space velocity and the inverse of the approximate Jacobian matrix are not required.

Remark 2. Another way of compensating for the uncertain gravitational force is the use of an approximate Jacobian PID control law [30, 31]. In this case, the structure of the gravitational force is not required. The result can also be extended to deal with uncertainty in gravity regressor matrix [33].

Remark 3. An adaptive law can be used to update the kinematic parameters of the approximate Jacobian matrix online [34, 35]. By using the adaptive Jacobian setpoint controllers, the stability conditions become simpler as compared to using a fixed approximate Jacobian matrix in [28, 32, 33].

Remark 4. If task-space damping is available as in [29], the stability condition becomes

$$\min\{\frac{2(1-\frac{\alpha c_0}{b_v})}{\bar{a}+2b_{JT}}, 2\alpha\lambda_{\hat{j}}\} > p, \tag{30}$$

where b_{JT} denotes the norm bound for $\hat{J}^{-T}(q)$. A graphical illustration of condition (30) is shown in Figure 3. In order to guarantee the stability of the system with approximate Jacobian matrix and task-space damping, the allowable bound p of the Jacobian uncertainty $\bar{\Delta} = \hat{J}^T(q) - J^T(q)$ must be less than the minimum of the two curves described by $\frac{2(1-\frac{\alpha c_0}{b_v})}{\bar{a}+2b_{JT}}$, and $2\alpha\lambda_{\hat{j}}$. Since there exists an α sufficiently large so that $\alpha > \frac{1}{2b_{JT}\lambda_{\hat{j}}}$, condition (30) can be simplified as

$$p < \frac{2(1-\frac{\alpha c_0}{b_v})}{\bar{a}+2b_{JT}}, \tag{31}$$

which is a simple condition inversely proportional for $\bar{a}$. The condition also implies that when task-space damping is available, the feedback gain ratio $\bar{a}$ can be set smaller.

4 Adaptive Jacobian Tracking Control of Robots

In the previous section, approximate Jacobian setpoint controllers are presented. However, in some applications, it is necessary to specify the motion

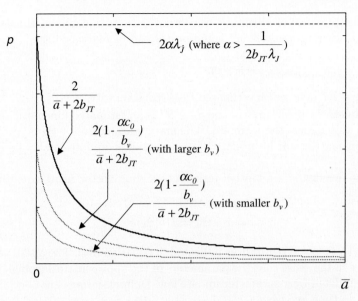

Fig. 3. Variation of p with $\bar{a}$ and b_v (task-space damping)

in much more details than simply stating the desired final position. Thus, a desired trajectory should be specified. In this section, we present an Adaptive Jacobian Tracking Controller for robot with uncertain kinematics and dynamics [36]. The main idea is to introduce an adaptive task-space sliding vector using an estimated task-space velocity. The uncertain kinematic parameters of the estimated task-space velocity and the Jacobian matrix are being updated online by a kinematic parameter update law, using the concept of kinematic regressor that we defined in *Property 4*. The other properties of the robot dynamics described in section 2 are also exploited in designing the adaptive controller.

Let us define a vector $\dot{x}_r \in R^n$ as,

$$\dot{x}_r = \dot{x}_d - \alpha(x - x_d), \tag{32}$$

where x or $x - x_d$ is measured from a position sensor and $x_d \in R^n$ is a desired trajectory specified in task space and $\dot{x}_d = \frac{dx_d}{dt} \in R^n$ is the desired velocity specified in task space. Many commercial sensors are available for measurement of x, such as vision systems, electromagnetic measurement systems, position sensitive detectors or laser tracking systems.

Differentiating equation (32) with respect to time, we have

$$\ddot{x}_r = \ddot{x}_d - \alpha(\dot{x} - \dot{x}_d), \tag{33}$$

where $\dot{x} = \frac{dx}{dt}$ is the task-space velocity and $\ddot{x}_d = \frac{d\dot{x}_d}{dt} \in R^n$ is the desired acceleration in task space.

In the presence of kinematic uncertainty, the parameters of the Jacobian matrix is uncertain and hence equation (5) can be expressed as

$$\hat{\dot{x}} = \hat{J}(q,\hat{\theta}_k)\dot{q} = Y_k(q,\dot{q})\hat{\theta}_k, \tag{34}$$

where $\hat{\dot{x}} \in R^n$ denotes an estimated task-space velocity, $\hat{J}(q,\hat{\theta}_k) \in R^{n \times n}$ is an approximate Jacobian matrix and $\hat{\theta}_k \in R^q$ is an estimated kinematic parameters. In this paper, the estimated kinematic parameters of the approximate Jacobian matrix will be updated by a parameter update law to be defined later.

Next, we define a adaptive task-space sliding vector using equation (34) as,

$$\hat{s}_x = \hat{\dot{x}} - \dot{x}_r = \hat{J}(q,\hat{\theta}_k)\dot{q} - \dot{x}_r, \tag{35}$$

where $\hat{J}(q,\hat{\theta}_k)\dot{q} = Y_k(q,\dot{q})\hat{\theta}_k$ as indicated in equation (34). The above vector is adaptive in the sense that the parameters of the approximate Jacobian matrix will be updated by a parameter update law. Differentiating equation (35) with respect to time, we have,

$$\dot{\hat{s}}_x = \ddot{\hat{x}} - \ddot{x}_r = \hat{J}(q,\hat{\theta}_k)\ddot{q} + \dot{\hat{J}}(q,\hat{\theta}_k)\dot{q} - \ddot{x}_r, \tag{36}$$

where $\ddot{\hat{x}}$ denotes the derivative of $\hat{\dot{x}}$. Next, let

$$\dot{q}_r = \hat{J}^{-1}(q,\hat{\theta}_k)\dot{x}_r, \tag{37}$$

where $\hat{J}^{-1}(q,\hat{\theta}_k)$ is the inverse of the approximate Jacobian matrix $\hat{J}(q,\hat{\theta}_k)$. In this paper, we assume that the robot is operating in a finite task space such that the approximate Jacobian matrix is of full rank. From equation (37), we have

$$\ddot{q}_r = \hat{J}^{-1}(q,\hat{\theta}_k)\ddot{x}_r + \dot{\hat{J}}^{-1}(q,\hat{\theta}_k)\dot{x}_r, \tag{38}$$

where $\dot{\hat{J}}^{-1}(q,\hat{\theta}_k) = -\hat{J}^{-1}(q,\hat{\theta}_k)\dot{\hat{J}}(q,\hat{\theta}_k)\hat{J}^{-1}(q,\hat{\theta}_k)$. Hence, we have an adaptive sliding vector in joint space as,

$$s = \dot{q} - \dot{q}_r = \hat{J}^{-1}(q,\hat{\theta}_k)((\hat{\dot{x}} - \dot{x}_d) + \alpha(x - x_d))$$
$$= \hat{J}^{-1}(q,\hat{\theta}_k)\hat{s}_x \tag{39}$$

and

$$\dot{s} = \ddot{q} - \ddot{q}_r = \hat{J}^{-1}(q,\hat{\theta}_k)\dot{\hat{s}}_x + \dot{\hat{J}}^{-1}(q,\hat{\theta}_k)\hat{s}_x \tag{40}$$

Substituting equations (39) and (40) into equation (1), the equations of motion can be expressed as,

$$M(q)\dot{s} + (\tfrac{1}{2}\dot{M}(q) + S(q,\dot{q}))s$$
$$+ M(q)\ddot{q}_r + (\tfrac{1}{2}\dot{M}(q) + S(q,\dot{q}))\dot{q}_r + g(q) = \tau. \tag{41}$$

From *Property 3*, the last four terms of equation (41) are linear in a set of dynamics parameters θ_d and hence can be expressed as,

$$M(q)\ddot{q}_r + (\frac{1}{2}\dot{M}(q) + S(q,\dot{q}))\dot{q}_r + g(q) = Y_d(q,\dot{q},\dot{q}_r,\ddot{q}_r)\theta_d, \quad (42)$$

where $Y_d(q,\dot{q},\dot{q}_r,\ddot{q}_r)$ is the dynamic regressor matrix.

From equation (42), the dynamic equation (41) can be expressed as,

$$M(q)\dot{s} + (\frac{1}{2}\dot{M}(q) + S(q,\dot{q}))s + Y_d(q,\dot{q},\dot{q}_r,\ddot{q}_r)\theta_d = \tau. \quad (43)$$

In this paper, we propose an adaptive controller based on the approximate Jacobian matrix as,

$$\tau = -\hat{J}^T(q,\hat{\theta}_k)(K_v\Delta\dot{x} + K_p\Delta x) - \hat{J}^T(q,\hat{\theta}_k)K\hat{s}_x$$
$$+ Y_d(q,\dot{q},\dot{q}_r,\ddot{q}_r)\hat{\theta}_d, \quad (44)$$

where

$$\hat{s}_x = Y_k(q,\dot{q})\hat{\theta}_k - \dot{x}_r, \quad (45)$$

and $\Delta\dot{x} = \dot{x} - \dot{x}_d$, $\Delta x = x - x_d$, $K_v \in R^{n \times n}$, $K_p \in R^{n \times n}$ and $K \in R^{n \times n}$ are positive definite matrices. In the proposed controller (44), the first term is an approximate Jacobian transpose feedback law of the task-space velocity and position errors, the next term is an approximate Jacobian transpose feedback law of the adaptive sliding vector (45) and the last term is an estimated control input based on equation (42). The estimated kinematic parameters $\hat{\theta}_k$ of the approximate Jacobian matrix $\hat{J}(q,\hat{\theta}_k)$ are updated by,

$$\dot{\hat{\theta}}_k = L_k Y_k^T(q,\dot{q})(K_v\Delta\dot{x} + K_p\Delta x), \quad (46)$$

and the estimated dynamics parameters $\hat{\theta}_d$ of the dynamic model are updated by,

$$\dot{\hat{\theta}}_d = -L_d Y_d^T(q,\dot{q},\dot{q}_r,\ddot{q}_r)s. \quad (47)$$

where $L_k \in R^{q \times q}$ and $L_d \in R^{p \times p}$ are positive definite matrices.

Although some kinematic parameters appear in $\hat{\theta}_d$, we should adapt on them separately in $\hat{\theta}_k$ to preserve linearity. The linear parameterization of the kinematic parameters is obtained from equation (5). The estimated kinematic parameters $\hat{\theta}_k$ of the approximate Jacobian matrix $\hat{J}(q,\hat{\theta}_k)$ is updated by the parameter update equation (46). The estimated parameters $\hat{\theta}_k$ is then used in the inverse approximate Jacobian matrix $\hat{J}^{-1}(q,\hat{\theta}_k)$ and hence $\dot{q}_r$ and $\ddot{q}_r$ in the dynamic regressor matrix. Note that $\hat{\theta}_k$ (like q and $\dot{q}$) is just part of the states of the adaptive control system and hence can be used in the control variables even if it is nonlinear in the variables (provided that a linear parameterization can be found else where in the system model i.e. equation (5)). Since $\hat{J}(q,\hat{\theta}_k)$ and its inverse $\hat{J}^{-1}(q,\hat{\theta}_k)$, are updated by q and $\hat{\theta}_k$, $\hat{J}(q,\hat{\theta}_k)$

and $\dot{\hat{J}}^{-1}(q,\hat{\theta}_k) = -\hat{J}^{-1}(q,\hat{\theta}_k)\dot{\hat{J}}(q,\hat{\theta}_k)\hat{J}^{-1}(q,\hat{\theta}_k)$ are functions of q, $\dot{q}$, $\hat{\theta}_k$, Δx and $\Delta \dot{x}$ because $\dot{\hat{\theta}}_k$ is described by equation (46).

The closed-loop equation is obtained by substituting equation (44) into equation (43) to give

$$M(q)\dot{s} + (\tfrac{1}{2}\dot{M}(q) + S(q,\dot{q}))s + Y_d(q,\dot{q},\dot{q}_r,\ddot{q}_r)\Delta\theta_d$$
$$+\hat{J}^T(q,\hat{\theta}_k)(K_v\Delta\dot{x} + K_p\Delta x) + \hat{J}^T(q,\hat{\theta}_k)K\hat{s}_x = 0, \qquad (48)$$

where $\Delta\theta_d = \theta_d - \hat{\theta}_d$.

Let us define a Lyapunov-like function candidate as

$$V = \tfrac{1}{2}s^T M(q)s + \tfrac{1}{2}\Delta\theta_d^T L_d^{-1}\Delta\theta_d$$
$$+\tfrac{1}{2}\Delta x^T(K_p + \alpha K_v)\Delta x + \tfrac{1}{2}\Delta\theta_k^T L_k^{-1}\Delta\theta_k, \qquad (49)$$

where $\Delta\theta_k = \theta_k - \hat{\theta}_k$. Differentiating equation (49) with respect to time and using *Property 1*, we have

$$\dot{V} = s^T M(q)\dot{s} + \tfrac{1}{2}s^T\dot{M}(q)s - \Delta\theta_d^T L_d^{-1}\dot{\hat{\theta}}_d,$$
$$+\Delta x^T(K_p + \alpha K_v)\Delta\dot{x} - \Delta\theta_k^T L_k^{-1}\dot{\hat{\theta}}_k \qquad (50)$$

Substituting $M(q)\dot{s}$ from equation (48), $\dot{\hat{\theta}}_k$ from equation (46) and $\dot{\hat{\theta}}_d$ from equation (47) into the above equation, using *Property 2* and equation (39) yields,

$$\dot{V} = -s^T\hat{J}^T(q,\hat{\theta}_k)K\hat{s}_x - s^T\hat{J}^T(q,\hat{\theta}_k)(K_v\Delta\dot{x} + K_p\Delta x)$$
$$+\Delta x^T(K_p + \alpha K_v)\Delta\dot{x} - \Delta\theta_k^T Y_k^T(q,\dot{q})(K_v\Delta\dot{x} + K_p\Delta x)$$
$$= -\hat{s}_x^T K\hat{s}_x - \hat{s}_x^T(K_v\Delta\dot{x} + K_p\Delta x) + \Delta x^T(K_p + \alpha K_v)\Delta\dot{x}$$
$$-\Delta\theta_k^T Y_k^T(q,\dot{q})(K_v\Delta\dot{x} + K_p\Delta x). \qquad (51)$$

From equations (45), (5) and (32), we have

$$\hat{s}_x = \Delta\dot{x} + \alpha\Delta x - Y_k(q,\dot{q})\Delta\theta_k, \qquad (52)$$

where

$$Y_k(q,\dot{q})\Delta\theta_k = J(q)\dot{q} - \hat{J}(q,\hat{\theta}_k)\dot{q} = \dot{x} - \dot{\hat{x}}, \qquad (53)$$

Substituting equation (52) into equation (51) yields,

$$\dot{V} = -(\Delta\dot{x} + \alpha\Delta x - Y_k(q,\dot{q})\Delta\theta_k)^T K(\Delta\dot{x} + \alpha\Delta x - Y_k(q,\dot{q})\Delta\theta_k)$$
$$-\Delta\dot{x}^T K_v\Delta\dot{x} - \alpha\Delta x^T K_p\Delta x \leq 0. \qquad (54)$$

We are now in a position to state the following Theorem:

Theorem 3. *For a finite task space such that the approximate Jacobian matrix is non-singular, the approximate Jacobian adaptive control law (44) and the parameter update laws (46) and (47) for the robot system (1) result in the convergence of position and velocity tracking errors. That is, $x - x_d \to 0$ and $\dot{x} - \dot{x}_d \to 0$, as $t \to \infty$. In addition, the estimated task-space velocity converges to the actual task-space velocity, i.e. $\hat{\dot{x}} \to \dot{x}$, $t \to \infty$.*

Proof:
Since $M(q)$ is uniformally positive definite, V in equation (49) is positive definite in s, Δx, $\Delta \theta_k$ and $\Delta \theta_d$. Since $\dot{V} \leq 0$, V is also bounded, and therefore s, Δx, $\Delta \theta_k$ and $\Delta \theta_d$ are bounded vectors. This implies that $\hat{\theta}_k$, $\hat{\theta}_d$ are bounded, x is bounded if x_d is bounded, and $\hat{s}_x = \hat{J}(q, \hat{\theta}_k)s$ is bounded as seen from equation (39). Since Δx is bounded, $\dot{x}_r$ in equation (32) is also bounded if $\dot{x}_d$ is bounded. Therefore, $\dot{q}_r$ in equation (37) is also bounded if the approximate Jacobian matrix is non singular. From equations (39), $\dot{q}$ is bounded and the boundedness of $\dot{q}$ means that $\dot{x}$ is bounded since the Jacobian matrix is bounded. Hence, $\Delta \dot{x}$ is bounded and $\ddot{x}_r$ in equation (33) is also bounded if $\ddot{x}_d$ is bounded. From equation (46), $\dot{\hat{\theta}}_k$ is therefore bounded since Δx, $\Delta \dot{x}$, $\dot{q}$ are bounded and $Y_k(\cdot)$ is a trigonometric function of q. Therefore, $\ddot{q}_r$ in equation (38) is bounded. From the closed-loop equation (48), we can conclude that $\dot{s}$ is bounded. The boundedness of $\dot{s}$ imply the boundedness of $\ddot{q}$ as seen from equation (40). From equation (36), $\dot{\hat{s}}_x$ is therefore bounded. Finally, differentiating equation (52) with respect to time and re-arranging yields,

$$\Delta \ddot{x} + \alpha \Delta \dot{x} = \dot{\hat{s}}_x + \dot{Y}_k(q, \dot{q}, \ddot{q})\Delta \theta_k - Y_k(q, \dot{q})\dot{\hat{\theta}}_k, \tag{55}$$

which means that $\Delta \ddot{x} = \ddot{x} - \ddot{x}_d$ is also bounded.

To apply Barbalat's lemma, let us check the uniform continuity of $\dot{V}$. Differentiating equation (54) with respect to time gives,

$$\ddot{V} = -2\hat{s}_x^T K \dot{\hat{s}}_x - 2\Delta \dot{x}^T K_v \Delta \ddot{x} - 2\alpha \Delta x^T K_p \Delta \dot{x}. \tag{56}$$

This shows that $\ddot{V}$ is bounded since $\hat{s}_x$, $\dot{\hat{s}}_x$ and Δx, $\Delta \dot{x}$, $\Delta \ddot{x}$ are all bounded. Hence, $\dot{V}$ is uniformly continuous. Using Barbalat's lemma, we have $\Delta x = x - x_d \to 0$, $\Delta \dot{x} = \dot{x} - \dot{x}_d \to 0$, and $\hat{s}_x \to 0$ as $t \to \infty$. Finally, from equation (53), the convergence of Δx, $\Delta \dot{x}$ and $\hat{s}_x$ imply that $\dot{x} - \hat{\dot{x}} \to 0$, as $t \to \infty$.

△△△

Remark 5. It is interesting to note that the stability of the closed-loop system can still be ensured if K_v (or K_p) in the controller (44) and the parameter update law (46) is set as zero. Substituting $K_v = 0$ into V in equation (49) and $\dot{V}$ in equation (54), we have

$$\begin{aligned}V = &\tfrac{1}{2}s^T M(q)s + \tfrac{1}{2}\Delta \theta_d^T L_d^{-1} \Delta \theta_d \\ &+ \tfrac{1}{2}\Delta x^T K_p \Delta x + \tfrac{1}{2}\Delta \theta_k^T L_k^{-1} \Delta \theta_k,\end{aligned} \tag{57}$$

and
$$\dot{V} = -\hat{s}_x^T K \hat{s}_x - \alpha \Delta x^T K_p \Delta x \leq 0. \tag{58}$$

Remark 6. If the controller gain K in equation (44) is set as zero, then
$$\tau = -\hat{J}^T(q,\hat{\theta}_k)(K_v \Delta \dot{x} + K_p \Delta x) + Y_d(q,\dot{q},y_r,\dot{y}_r)\hat{\theta}_d, \tag{59}$$
then τ simply consists of an approximate impedance controller (or approximate Jacobian PD controller [28]) and an adaptive dynamic compensation term. From equation (54), we have,
$$\dot{V} = -\Delta \dot{x}^T K_v \Delta \dot{x} - \alpha \Delta x^T K_p \Delta x \leq 0. \tag{60}$$
Hence, Δx and $\Delta \dot{x}$ still converge to zero as $t \to \infty$ but the convergence of $\hat{s}_x$ and $\dot{x} - \hat{\dot{x}}$ cannot be ensured. However, the control gain K in equation (44) is useful in system identification to obtain the kinematics parameters. As seen from equation (53), when $\dot{x} - \hat{\dot{x}} \to 0$, $Y_k(q,\dot{q})\Delta\theta_k \to 0$ and hence the kinematics parameters converge if the "persistent excitation" (P.E.) condition is satisfied. Alternatively, the kinematic parameter update equation (46) can be modified as,
$$\dot{\hat{\theta}}_k = -L_k P_k Y_k^T(q,\dot{q}) P(\hat{\dot{x}} - \dot{x}) + L_k Y_k^T(q,\dot{q})(K_v \Delta \dot{x} + K_p \Delta x), \tag{61}$$
where P is a symmetric positive definite matrix. This add to $\dot{V}$ minus the P square norm of $Y_k(q,\dot{q})\Delta\theta_k$ and hence the convergence of $\hat{\theta}_k$ can be achieved even if K is set as zero.

Remark 7. From equation (52), the adaptive sliding vector can be expressed as:
$$\hat{s}_x = \Delta \dot{x} + \alpha \Delta x + Y_k(q,\dot{q})\hat{\theta}_k - Y_k(q,\dot{q})\theta_k. \tag{62}$$
Hence, the sign of the parameter update laws in equations (46) and (47) are different because the last term in equation (44) is positive while the last term in equation (62) is negative. Therefore, the sign of equation (46) is positive.

Remark 8. In the presence of kinematic uncertainty, inverse kinematics cannot be used to derive the desired trajectory in joint space. In addition, when the dynamics equation is expressed in task space by using equation (3) and its derivative, we have
$$M(q)J^{-1}(q)\ddot{x} + (-M(q)J^{-1}(q)\dot{J}(q) + B + \frac{1}{2}\dot{M}(q) + S(q,\dot{q}))J^{-1}(q)\dot{x} + g(q) = \tau$$

The above equation cannot be expressed in a form as in Property 3 because $J^{-1}(q)$ is not linear in the unknown kinematic parameters. In addition, the mapping between force and torque using Jacobian transpose is also uncertain due to the unknown kinematic parameters. Therefore, the standard adaptive controller by Slotine and Li [3] cannot be applied directly to overcome the uncertainty in both kinematics and dynamics. Hence, in the presence of kinematic uncertainty, the adaptive method [3] results in tracking error or even unstable response in the end-effector's motion. The nonlinearity and uncertainty of the robot kinematics pose a difficult and challenging adaptive tracking control problem which remains unsolved for almost two decades.

Remark 9. The first tracking convergent adaptive controller for robots with uncertainties in kinematics and dynamics was considered in[36]. The main idea was to introduce an adaptive sliding vector based on estimated task-space velocity, so that kinematic and dynamic adaptation can be performed concurrently. A novel dynamics regressor using the estimated kinematics parameters was also proposed. It was shown that the end-effector's position converges to the desired position even when the kinematics and Jacobian matrix are uncertain. The proposed controller requires the measurement of a task-space velocity or differentiation of the task-space position which is very noisy. In [37], an adaptive Jacobian tracking control law is proposed to eliminate the need for task-space velocity while creating task-space damping. The approach is based on filtered differentiation of the measured visual task-space position. To avoid singularities associated the Euler angles representation, adaptive Jacobian tracking controller based on unit quaternion is proposed [38].

5 Conclusion

This paper presents several approximate Jacobian controllers for robots with uncertain kinematics and dynamics. Both setpoint and tracking control problems are formulated based on Lyapunov analysis. It has been shown that the end effector is able to follow a desired motion even when the kinematics and Jacobian matrix are uncertain. This gives the robot a high degree of flexibility in dealing with unforseen changes and uncertainties in its kinematics and dynamics.

References

1. M. Takegaki and S. Arimoto, "A new feedback method for dynamic control of manipulators," *ASME J. of Dynamic Systems, Measurement and Control*, vol. 102, pp. 119–125, 1981.
2. S. Arimoto and F. Miyazaki, "Asymptotic stability of feedback control for robot manipulators," in *Proc. of IFAC Symposium on Robot Control*, (Barcelona), pp. 447 – 452, 1985.
3. J. J. E. Slotine and W. Li, "On the adaptive control of robot manipulators," *Int. J. Robotics Research*, no. 6, pp. 49–59, 1987.

4. J. J. E. Slotine and W. Li, "Adaptive manipulator control: A case study," *IEEE Transactions on Automatic Control*, vol. AC-33, no. 11, pp. 995 – 1003, 1988.
5. J. J. Craig, P. Hsu, and S. S. Sastry, "Adaptive control of mechanical manipulators," *Int. J. Robotics Research*, vol. 6, no. 2, pp. 16–28, 1987.
6. R. H. Middleton and G. C. Goodwin, "Adaptive computed torque control for rigid link manipulators," *Systems and Control Letter*, vol. 10, pp. 9–16, 1988.
7. D. E. Koditschek, "Adaptive techniques for mechanical systems," in *Fifth Yale Workshop on Applications of Adaptive Systems Theory*, (New Haven, CT), pp. 259–265, 1987.
8. J. T. Wen and D. Bayard, "New class of control laws for robotic manipulators -part 2. adaptive case," *International Journal of Control*, vol. 47, no. 5, pp. 1387–1406, 1988.
9. B. Paden and R. Panja, "A globally asymptotically stable 'PD+' controller for robot manipulator," *International Journal of Control*, vol. 47, no. 6, pp. 1697–1712, 1988.
10. R. Kelly and R. Carelli, "Unified approach to adaptive control of robotic manipulators," in *Proc. 27th IEEE Conf. on Decision and Control*, 1988.
11. R. Ortega and M. W. Spong, "Adaptive motion control of rigid robots: a tutorial," *Automatica*, vol. 25, no. 6, pp. 877–888, 1989.
12. N. Sadegh and R. Horowitz, "Stability and robustness analysis of a class of adaptive controllers for robotic manipulators," *International Journal of Robotics Research*, vol. 9, no. 3, pp. 74–92, 1990.
13. G. Niemeyer and J.J.E. Slotine, "Performance in Adaptive Manipulator Control," *Int. J. Robotics Research*, vol. 10, no. 2, 1991.
14. P. Tomei, "Adaptive PD controller for robot manipulators," *IEEE Transactions on Robotics and Automation*, vol. 7, pp. 565 – 570, 1991.
15. R. Kelly, "Comments on adaptive PD controller for robot manipulators," *IEEE Transactions on Robotics and Automation*, vol. 9, pp. 117 – 119, 1993.
16. H. Berghuis, R. Ortega and H. Nijmeijer, "A robust adaptive robot controller", *IEEE Transactions on Robotics and Automation*, vol. 9, no. 6, pp 825 - 830, 1993.
17. L. L. Whitcomb, A. Rizzi, and D.E. Koditschek, "Comparative Experiments with a New Adaptive Controller for Robot Arms", *IEEE Transactions on Robotics and Automation*, vol. 9, no. 1, pp 59-70, 1993.
18. K.W. Lee and H. Khalil, "Adaptive output feedback control of robot manipulators using high gain observer", *Int. J. Control*, vol. 67, no. 6, 1997
19. S. Arimoto, *Control Theory of Nonlinear Mechanical Systems - A Passivity-Based and Circuit-Theoretic Approach*. Oxford: Clarendon Press, 1996.
20. S. Arimoto, "Robotics research toward explication of everyday physics," *International Journal of Robotic Research*, vol. 18, no. 11, pp. 1056–1063, 1999.
21. T. Yoshikawa, "Force control of robot manipulators," in *Proc. IEEE Conf. on Robotics and Automation*, (San Francisco), pp. 220–226, 2000. Invited session on robot control.
22. A. Bicchi, "Hands for dexterous manipulation and robust grasping: a difficult road toward simplicity," *IEEE Trans. on Robotics and Automation*, vol. 16, no. 6, pp. 652 – 662, 2000.
23. F. L. Lewis, C. T. Abdallah, and D. M. Dawson, *Control of Robot Manipulators*. New York: Macmillan Publishing Company, 1993.
24. L. Sciavicco and B. Siciliano, *Modelling and control of robot manipulators*. New York: Springer-Verlag, 2000.

25. S. Hutchinson, G. Hager and P. Corke, "A tutorial on visual servo control," *IEEE Trans. on Robotics and Automation*, vol. 12, no. 5, pp. 651 – 670, 1996.
26. L. E. Weiss, A. C. Sanderson and C. P. Neuman, "Dynamic sensor-based control of robots with visual feedback," *IEEE Trans. on Robotics and Automation*, vol. RA-3, no. 5, pp. 404 – 417, 1987.
27. B. Espiau, F. Chaumette and P. Rives, "A new approach to visual servoing in robotics", *IEEE Trans. on Robotics and Automation*, Vol. 8 , No. 3, pp 313 - 326, 1992.
28. C. C. Cheah, S. Kawamura, and S. Arimoto, "Feedback control for robotic manipulators with an uncertain jacobian matrix," *Journal of Robotic System*, vol. 12, no. 2, pp. 119–134, 1999.
29. C. C. Cheah, M. Hirano, S. Kawamura and S. Arimoto, "Approximate Jacobian Control with Task-space Damping for Robot Manipulators", *IEEE Transactions on Automatic Control*, Vol. 49, No. 5, pp 752- 757, May 2004.
30. C. C. Cheah, S. Kawamura, S. Arimoto and K. Lee, "PID Control for Robotic Manipulator with Uncertain Jacobian Matrix", *Proc. of IEEE Int. Conference on Robotics and Automation* , (Detroit, Michigan), pp 494-499, 1999.
31. C.Q. Huang, X.G. Wang and Z.Q. Wang, "A class of transpose Jacobian-based NPID regulators for robot manipulators with uncertain kinematics", *Journal of Robotic Systems*, vol. 19, no. 11, 527-539, 2002.
32. C. C. Cheah, M. Hirano, S. Kawamura, and S. Arimoto, "Approximate jacobian control for robots with uncertain kinematics and dynamics," *IEEE Trans. on Robotics and Automation*, vol. 19, no. 4, pp. 692–702, 2003.
33. H. Yazarel and C. C. Cheah, "Task-space adaptive control of robotic manipulators with uncertainties in gravity regressor matrix and kinematics," *IEEE Trans. on Automatic Control*, vol. 47, no. 9, pp. 1580 – 1585, 2002.
34. C. C. Cheah, "Approximate Jacobian Robot Control with Adaptive Jacobian Matrix", *Proc. of IEEE Int. Conference on Decision and Control*, (Hawaii, USA), pp 5859-5864, 2003.
35. W. E. Dixon, "Adaptive regulation of amplitude limited robot manipulators with uncertain kinematics and dynamics," in *Proc. of American Control Conference*, (Boston, USA), pp. 3939–3844, 2004.
36. C. C. Cheah, C. Liu and J.J.E. Slotine, "Approximate Jacobian Adaptive Control for Robot Manipulators", *Proc. of IEEE Int. Conference on Robotics and Automation*, (New Orleans, USA), pp 3075-3080, 2004.
37. C. C. Cheah, C. Liu and J.J.E. Slotine, "Adaptive Tracking Control for Robots with Unknown Kinematic and Dynamic Properties", *International Journal of Robotic Research*, Vol. 25, No. 3, pp 283-296, 2006.
38. D. Braganza, W. E. Dixon, D. M. Dawson, and B. Xian, Tracking Control for Robot Manipulators with Kinematic and Dynamic Uncertainty, *Proc. of IEEE Conference on Decision and Control*, (Seville, Spain), pp 5293-5297, 2005,

Adaptive Visual Servoing of Robot Manipulators *

Yun-Hui Liu and Hesheng Wang

Department of Automation and Computer Aided Engineering,
The Chinese University of Hong Kong, Shatin, NT, Hong Kong
yhliu@acae.cuhk.edu.hk

Summary. A novel adaptive controller is presented in this Chapter for image-based dynamic control of a robot manipulator when the intrinsic and extrinsic parameters of the camera and the position coordinates of the feature points are unknown. Both the fixed camera and eye-in-hand camera configurations are considered. The key idea lies in the use of a depth-independent image Jacobian matrix to map the visual signals onto the joint space of the robot manipulator. By virtue of the depth-independent image Jacobian matrix, it is possible to linearly parameterize the closed loop dynamics of the system by the uncalibrated camera parameters and the unknown feature coordinates. A new adaptive algorithm, different from the Slotine and Li algorithm, has been proposed to estimate the unknown parameters and coordinates on-line. The asymptotic stability of the system under the control of the proposed method is rigorously proved by the Lyapunov theory with the nonlinear robot dynamics fully taken into account.

1 Introduction

Human heavily relies on visual feedback in motion control. To mimic human intelligence, visual servoing has been proposed for motion control of robot manipulators. Visual servo control leads to greater flexibility to robot operations because vision is the most convenient, effective and natural sensor to acquire information of the external environment.

1.1 Image-based and Position-based Visual Servoing

Visual servoing has been extensively studied in robotics since the early 1990s and various methods have been developed. Existing methods can be classified

---

* This work was supported in part by the Hong Kong Research Grant Council under the grant CUHK4217/03E and by the National Science Foundation of China under projects 60334010 and 60475029. Y. H. Liu is also with National University of Defense Technology as a visiting professor.

into two basic schemes, namely position-based control [9][11][19][33][39][41] and image-based control[1]-[8]. A position-based approach first uses an algorithm to estimate the 3-D position and orientation of the robot manipulator or the feature points from the images and then feeds the estimated position/orientation back to the robot controller. Since 3-D position/orientation estimation from images is subject to big noises, position-based methods are weak to disturbances and measurement noises. An image-based approach selects a set of feature points on the robot or the target object and directly employs their projection signals on the image plane of the camera in robot control. Since the feedback signals are projection errors on the image plane, image-based controllers are considered more robust to disturbances than position-based methods.

1.2 Fixed Camera and Eye-in-Hand Configurations

There are two possible configurations, namely eye-in-hand configuration and fixed camera configuration, to set up a vision system for visual servo control of robot manipulators. In a fixed camera setup (Fig. 1) [2][3][18][23][37][39], the camera is fixed at a position near the manipulator. In this case, the camera does not move with the robot manipulator and its objective is to monitor motion of the robot. In an eye-in-hand setup (Fig. 2)[1][7][14][30][43], the camera is mounted at the end-effector of the robot manipulator so that it moves with the manipulator. The camera is to measure information of objects in the surrounding environment. While each configuration has its own advantages and drawbacks, they both can be found in real applications. Eye-in-hand systems can be widely found in research laboratories and are being used in tele-operated inspection systems in hazardous environments such as nuclear factories. Fixed camera setups are used in vision-based pick-and-place at manufacturing lines, robots assisted by networked cameras, etc.

1.3 Kinematics-based and Dynamic Visual Servoing

Existing methods can be classified into kinematics-based [17][19][31][32][35] and dynamic methods [3]-[6][9][18][22][23][39][44]. Kinematics-based methods do not consider the nonlinear dynamics of the robot manipulator and design the controller based on the kinematics only. An underlying assumption here is that the robot can accurately generate desired control inputs to visual servo loop, which are usually velocity commands of the joints or the end-effector. It is well known that the nonlinear robot forces have significant impact on robot motion, especially when the robot manipulator moves at high speed. Neglecting them not only decays the control accuracy but also results in instability. In a rigorous sense, the stability is not guaranteed for all kinematics-based controllers. Kinematics-based controllers are suitable for slow robot motion only.

Dynamic visual servoing takes into account the nonlinear robot dynamics in controller design and stability analysis. The nonlinear centrifugal and Coriolis forces are compensated for in the control loop. Dynamic controllers guarantee the stability and are suitable for both slow and fast robot motions. Compared to kinematics-based methods, work on dynamic visual servoing is relatively limited, though the importance has been recognized for a long time by many researchers. This is mainly due to the difficulties in incorporating the nonlinear forces in controller design and the existence of the nonlinear scaling factors corresponding to the reciprocal of the depths of the feature points in the perspective projection. Since it is difficult to measure or estimate on-line the depths of the feature points, most dynamic controllers are subject to planar motions of robot manipulators only.

1.4 Calibrated and Uncalibrated Visual Servoing

In image-based visual serovoing, the image Jacobian or interaction matrix is widely employed to map visual signals onto the joint space of the robot manipulator. The image Jacobian depends nonlinearly on the intrinsic parameters of the camera, including the focal length, coordinates of the principal point, and the scaling factors, and the extrinsic parameters, e.g. the homogenous transform matrix between the vision system and the robot manipulator. The accuracy of the camera parameters is crucial to the control performance. The works mentioned in the previous section assumed that the camera parameters are accurately calibrated. Although tremendous efforts have been made to camera calibration in computer vision and many algorithms have been proposed, obtaining accurate camera parameters is still a costly process.

To avoid camera calibration, people attempted to directly employ an uncalibrated vision system. Major efforts were directed to on-line estimation of the image Jacobian. Hosoda and Asada [17] and Jagersand et al. [20] proposed an algorithm for one-line estimation of the image Jacobian using the Broyden updating formula. Piepmeier et al. [37] proposed a dynamic quasi-Newton method for the same purpose. Ruf et al. [38] investigated on-line calibration of the unknown parameters. Recently, Malis[31] proposed a novel method which is invariant to changes of the camera intrinsic parameters. However, this method cannot cope with unknown extrinsic parameters. The approach proposed by Lu et al. [29] is for on-line calculation of the exterior orientation only. It should be pointed out that works mentioned above are kinematics-based approaches.

Works on dynamic visual servoing with unknown camera parameters are limited to planar robots or plane motion. Xiao et al. [42] studied a vision-based hybrid control of a robot manipulator whose endpoint moves on a plane using an uncalibrated fixed camera. The adaptive controllers proposed by Kelly [21][22], Hsu et al. [18], Astolfi et al. [2], and Bishop and Spong [4] are suitable for planar manipulators only. The controller developed in our early work [39] is for position-based visual servoing when the extrinsic parameters of the camera are unknown but the intrinsic parameters are calibrated. Therefore, there is no visual servo controller that is able to cope with uncalibrated cameras as well as the nonlinear dynamics of robot manipulators in general 3-D motion.

1.5 Outline of This Chapter

This chapter summarizes our recent work [24]-[28] on dynamic visual servoing with guaranteed dynamic stability to cope with uncalibrated cameras and the 3-D motion of robot manipulators. A new adaptive controller will be presented for controlling the projections of a number of feature points with unknown 3-D positions to desired positions on the image plane of an uncalibrated camera. When designing the controller, we fully consider the nonlinear robot dynamics. Both the fixed camera and the eye-in-hand configurations are addressed in this chapter.

This new dynamic visual servo controller is developed on the basis of two major ideas: the proposal of a depth-independent image Jacobian or interaction matrix [25][26] and the development of a new adaptive algorithm for on-line estimation of the camera parameters and the unknown coordinates of the feature points. The reason why existing methods cannot cope with 3-D robot motion is because the image Jacobian or interaction matrix cannot be linearly parameterized by the unknown parameters. This bottleneck problem can be solved by using the depth-independent image Jacobian matrix. The use of the depth-independent image Jacobian matrix enables the unknown camera parameters to appear linearly in the closed-loop dynamics. Therefore, an adaptive algorithm, similar to that proposed by Slotine and Li [40], could be used to estimate the unknown camera parameters and position coordinates on-line. However, since the problem addressed is an output adaptive control problem, the asymptotic convergence of the image errors cannot be achieved by the Slotine-Li algorithm only. Our adaptive algorithm combines the Slotine-Li method with on-line gradient descending minimization process of the estimated projection errors of the feature points. We have employed the Lyapunov theory to rigorously prove asymptotic convergence of the image errors when the nonlinear dynamics of the robot manipulator are fully taken into account.

This chapter is organized as follows. Section 2 will review the kinematics and the depth-independent image Jacobian matrix. Section 3 will present image-based control of a single feature point using an uncalibrated camera. In Section 4, we will discuss the extension of the controller to the problem of controlling multiple feature points. Section 5 presents the experimental results and Section 6 concludes the major results in this chapter.

2 Kinematics and Dynamics

In this section, we review perspective projection, intrinsic and extrinsic parameters of a camera, kinematics and dynamics of robots under visual servoing.

Before the review, we give the notations adopted in this chapter. A matrix and a vector are represented by a bold capital letter and a bold lower case letter, respectively. Let an italic letter represent a scalar quantity. A matrix, or vector, or scalar accompanied with a bracket(t) represents that its value varies with time. Furthermore, $\mathbf{I}_{k \times k}$ and $\mathbf{0}_{k \times l}$ denote the $k \times k$ identity matrix and the $k \times l$ zero matrix, respectively.

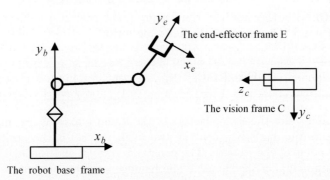

Fig. 1. A fixed camera setup for visual servoing

2.1 Perspective Projection and Camera Parameters

First, consider the fixed camera setup (Fig. 1). Three coordinate frames, namely the robot base frame, the end-effector frame, and the camera frame, have been established to represent the kinematic relation between the vision system and the manipulator. Denote the joint angle of the manipulator by a $n \times 1$ vector $\mathbf{q}(t)$, where n is the number of degrees of freedom. For simplicity, we first consider a single feature point whose homogeneous coordinates with respect to the end-effector frame are denoted by a 4×1 vector $\mathbf{x}$. Note that vector $\mathbf{x}$ is a constant vector. Denote the coordinates of the feature point with respect to the camera frame by $^c\mathbf{x}(t)$. Let $\mathbf{T}$ be the 4 by 4 homogeneous transform matrix from the robot base frame to the camera frame. Note that $\mathbf{T}$ is a constant and unknown matrix:

$$\mathbf{T} = \begin{pmatrix} \mathbf{R} & \mathbf{p} \\ \mathbf{0}_{1 \times 3} & 1 \end{pmatrix}, \qquad (1)$$

where $\mathbf{R}$ is the 3 by 3 rotation matrix and $\mathbf{p}$ denotes the 3 by 1 translational vector.

Denote the homogenous transform matrix of the end-effector frame with respect to the robot base frame by $\mathbf{T}_e(\mathbf{q})$. The coordinates $^c\mathbf{x}(t)$ and $\mathbf{x}$ are related by

$$^c\mathbf{x}(t) = \mathbf{T}\mathbf{T}_e(\mathbf{q}(t))\mathbf{x}. \qquad (2)$$

Denote by $\mathbf{y}(t) = \begin{pmatrix} u(t) \\ v(t) \\ 1 \end{pmatrix}$ the homogenous coordinates of the projection of the feature point on the image plane. Under the perspective projection model,

$$\mathbf{y}(t) = \frac{1}{{}^c z(\mathbf{q}(t))} \mathbf{\Omega T T}_e(\mathbf{q}(t))\mathbf{x}, \tag{3}$$

where ${}^c z(\mathbf{q}(t))$ is the depth of the feature point with respect to the camera frame. $\mathbf{\Omega}$ is a 3×4 matrix in the following form:

$$\mathbf{\Omega} = \begin{pmatrix} \alpha & -\alpha \cot \varphi & u_0 & 0 \\ 0 & \frac{\gamma}{\sin \varphi} & v_0 & 0 \\ 0 & 0 & 1 & 0 \end{pmatrix}, \tag{4}$$

where α and γ are the scalar factors of the u and v axes of the image plane. φ represents the angle between the two axes. (u_0, v_0) is the position of the principal point of the camera. The constants α, γ, φ, and (u_0, v_0) are the intrinsic parameters of the camera. The homogenous transformation matrix $\mathbf{T}$ is called the extrinsic parameter of the camera. Denote by $\mathbf{M}$ the product of the matrices $\mathbf{\Omega}$ and $\mathbf{T}$, i.e.,

$$\mathbf{M} = \mathbf{\Omega T} = \begin{pmatrix} \alpha \mathbf{r}_1^T - \alpha \cot \varphi \mathbf{r}_2^T + u_0 \mathbf{r}_3^T & \alpha p_x - \alpha \cot \varphi p_y + u_0 p_z \\ \frac{\gamma}{\sin \varphi} \mathbf{r}_2^T + v_0 \mathbf{r}_3^T & \frac{\gamma}{\sin \varphi} p_y + v_0 p_z \\ \mathbf{r}_3^T & p_z \end{pmatrix}, \tag{5}$$

where $\mathbf{r}_i^T$ denotes the i-th row vector of the rotation matrix, and (p_x, p_y, p_z) are the coordinates of the translation vector $\mathbf{p}$. The matrix $\mathbf{M}$ is called the *perspective projection matrix* and its dimension is 3×4. Note that this matrix depends on the intrinsic and extrinsic parameters, independent of the position of the feature point. Eq. (3) can be rewritten as

$$\mathbf{y}(t) = \frac{1}{{}^c z(\mathbf{q}(t))} \mathbf{M T}_e(\mathbf{q}(t))\mathbf{x}. \tag{6}$$

The depth of the feature point is given by

$$^c z(\mathbf{q}(t)) = \mathbf{m}_3^T \mathbf{T}_e(\mathbf{q}(t))\mathbf{x}, \tag{7}$$

where $\mathbf{m}_3^T$ denotes the third row vector of the perspective projection matrix $\mathbf{M}$.

Consider the eye-in-hand setup in Fig. 2. In this case, the camera moves with the robot manipulator. Similarly, we set up three coordinate frames, namely the robot base frame, the end-effector frame, and the camera frame. Denote the homogenous coordinates of the feature point w.r.t. the robot base and the camera frames by $\mathbf{x}$ and ${}^c \mathbf{x}$, respectively. Note that the $\mathbf{x}$ is an unknown constant vector. Denote the homogenous transform matrix of the end-effector frame with respect to the base frame by $\mathbf{T}_e(\mathbf{q})$, which can be

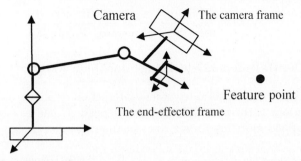

Fig. 2. The eye-in-hand setup for visual servoing

calculated from the kinematics of the manipulator. Denote the homogeneous transformation matrix of the camera frame with respect to the end-effector frame by $\mathbf{T_c(q)}$. From coordinate transformation, we have

$$^c\mathbf{x}(t) = \mathbf{T_c^{-1}T_e^{-1}(q}(t))\mathbf{x}. \tag{8}$$

Then the projection of the feature point on the image plane is given by

$$\begin{aligned}\mathbf{y}(t) &= \frac{1}{^cz(\mathbf{q}(t))}\Omega {}^c\mathbf{x}(t) \\ &= \frac{1}{^cz(\mathbf{q}(t))}\Omega\mathbf{T_c^{-1}T_e^{-1}(q}(t))\mathbf{x}. \\ &= \frac{1}{^cz(\mathbf{q}(t))}\mathbf{MT_e^{-1}(q}(t))\mathbf{x}\end{aligned} \tag{9}$$

The depth of the feature point is given by

$$^cz(\mathbf{q}(t)) = \mathbf{m}_3^T\mathbf{T_e^{-1}(q}(t))\mathbf{x}. \tag{10}$$

Similar to eq. (5), the perspective projection matrix is given by

$$\mathbf{M} = \Omega\mathbf{T}_c^{-1}. \tag{11}$$

It is important to note that the rank of the perspective projection matrix $\mathbf{M}$ is 3 [12]. When the camera parameters are not calibrated, the perspective projection matrix $\mathbf{M}$ should be estimated from coordinates of the feature points and their projections on the image plane. Note that there are 12 components in the matrix. Since two equations correspond to one feature point, six feature points are necessary to estimate the perspective projection matrix. However, the following property should be noted [25]:

Property 1: *Given the world coordinates of a sufficient number of feature points and their projections on the image plane, the perspective projection matrix $\mathbf{M}$ can be determined only up to a scale.*

This can be easily explained by the fact that the matrix $\alpha\mathbf{M}$ for any nonzero α results in the same image as matrix $\mathbf{M}$ does because

$$\mathbf{y}(t) = \frac{\mathbf{MT}_e(\mathbf{q}(t))\mathbf{x}}{\mathbf{m}_3^T \mathbf{T}_e(\mathbf{q}(t))\mathbf{x}} = \frac{\alpha \mathbf{MT}_e(\mathbf{q}(t))\mathbf{x}}{\alpha \mathbf{m}_3^T \mathbf{T}_e(\mathbf{q}(t))\mathbf{x}}. \tag{12}$$

2.2 The Depth-Independent Image Jacobian Matrix

Now, we present the depth-independent interaction or image Jacobian matrix proposed in our early work. The depth-independent image Jacobian matrix plays one of the most important roles in the controller design and stability analysis.

First, consider the fixed camera setup. By differentiating eq. (6), we obtain the following velocity relationship:

$$\begin{aligned}\dot{\mathbf{y}}(t) &= \frac{1}{{}^c z(\mathbf{q}(t))} (\mathbf{M}\frac{d}{dt}\{\mathbf{T}_e(\mathbf{q}(t))\mathbf{x}\} - \mathbf{y}(t){}^c\dot{z}(\mathbf{q}(t))) \\ &= \frac{1}{{}^c z(\mathbf{q}(t))} \underbrace{\mathbf{A}(\mathbf{y}(t))\frac{\partial}{\partial \mathbf{q}}\{\mathbf{T}_e(\mathbf{q}(t))\mathbf{x}\}}_{\mathbf{D}(\mathbf{q}(t),\mathbf{y}(t))} \dot{\mathbf{q}}(t) \end{aligned}, \tag{13}$$

where the matrix $\mathbf{A}(\mathbf{y}(t))$ is a 3×4 matrix in the following form:

$$\mathbf{A}(\mathbf{y}(t)) = \mathbf{M} - \mathbf{y}(t)\mathbf{m}_3^T = \begin{pmatrix} \mathbf{m}_1^T - u(t)\mathbf{m}_3^T \\ \mathbf{m}_2^T - v(t)\mathbf{m}_3^T \\ \mathbf{0}_{1\times 3} \end{pmatrix}. \tag{14}$$

The term $\frac{\partial}{\partial \mathbf{q}}\{\mathbf{T}_e(\mathbf{q}(t))\mathbf{x}\}$ is the Jacobian matrix of the robot manipulator, which is denoted by $\mathbf{J}(\mathbf{q}(t),\mathbf{x})$. It should be noted that $\frac{1}{{}^c z(\mathbf{q}(t))}\mathbf{A}(\mathbf{y}(t))\mathbf{J}(\mathbf{q}(t),\mathbf{x})$ is the image Jacobian matrix. The matrix $\mathbf{A}(\mathbf{y}(t))$ is independent on the depth of the feature point, we call it *depth-independent interaction matrix*[25][26]. The matrix $\mathbf{A}(\mathbf{y}(t))\mathbf{J}(\mathbf{q}(t),\mathbf{x})$ is called the *depth-independent image Jacobian matrix* denoted by $\mathbf{D}(\mathbf{q}(t),\mathbf{y}(t))$. It should be noted that the components of matrix $\mathbf{A}(\mathbf{y}(t))$ are linear to the components of the perspective projection matrix $\mathbf{M}$.

Proposition 1: *The depth-independent interaction matrix $\mathbf{A}(\mathbf{y}(t))$ has a rank of 2* [25].

Proof: This can be simply proved. Assume that the rank of the matrix $\mathbf{A}(\mathbf{y}(t))$ is smaller than 2. In this case, there are nonzero scalars λ_1 and λ_2 such that

$$\lambda_1(\mathbf{m}_1 - u(t)\mathbf{m}_3) + \lambda_2(\mathbf{m}_2 - v(t)\mathbf{m}_3) = 0. \tag{15}$$

Equation (15) can be written as follows:

$$\lambda_1 \mathbf{m}_1 + \lambda_2 \mathbf{m}_2 - (\lambda_1 u(t) + \lambda_2 v(t))\mathbf{m}_3 = 0. \tag{16}$$

If the coefficient $(\lambda_1 u(t)+\lambda_2 v(t))$ is not equal to zero, equation (16) means that the row vectors of the perspective projection matrix $\mathbf{M}$ are linearly dependent.

If the coefficient is equal to zero, vectors $\mathbf{m}_1$ and $\mathbf{m}_2$ are linearly dependent. Therefore, the rank of the matrix $\mathbf{M}$ is smaller than 3. This is against to the fact that the matrix $\mathbf{M}$ has a rank of 3. Consequently, the rank of the matrix $\mathbf{A}(\mathbf{y(t)})$ is 2.

By differentiating the depth in eq. (7)

$$^c\dot{z}(\mathbf{q}(t)) = \underbrace{\mathbf{m}_3^T \frac{\partial}{\partial \mathbf{q}}\left\{\mathbf{T}_e^{(\mathbf{q}}(t))\mathbf{x}\right\}}_{\mathbf{h}(\mathbf{q}(t))} \dot{\mathbf{q}}(t). \tag{17}$$

For the eye-in-hand case, we can define the depth-independent interaction matrix similarly by differentiating eq. (9):

$$\begin{aligned}\dot{\mathbf{y}}(t) &= \frac{1}{^cz(\mathbf{q}(t))}(\mathbf{M}\dot{\mathbf{T}}_e^{-1}(\mathbf{q}(t))) - \mathbf{y}(t)^c\dot{z}(\mathbf{q}(t))\\ &= \frac{1}{^cz(\mathbf{q}(t))}\mathbf{A}(\mathbf{y}(t))\frac{\partial}{\partial \mathbf{q}}\left\{\mathbf{T}_e^{-1}(\mathbf{q}(t))\mathbf{x}\right\}\dot{\mathbf{q}}(t),\end{aligned} \tag{18}$$

where the depth-independent interaction matrix $\mathbf{A}(\mathbf{y}(t))$ has the same form as that in eq. (14).

Remark 1: *It should be noted that the matrix $\frac{\partial}{\partial \mathbf{q}}\left\{\mathbf{T}_e^{-1}(\mathbf{q}(t))\mathbf{x}\right\}$ is related to the Jacobian matrix of the manipulator. For non zero vector $\mathbf{x}$, this matrix is always of full rank. Furthermore, it is not necessary to take the inverse of the manipulator Jacobian to calculate this matrix.*

2.3 Robot Dynamics

Assume that the robot manipulator addressed is a serial arm with revolute joints. The dynamic equation of the robot manipulator has the following standard form

$$\mathbf{H}(\mathbf{q}(t))\ddot{\mathbf{q}}(t) + (\frac{1}{2}\dot{\mathbf{H}}(\mathbf{q}(t)) + \mathbf{C}(\mathbf{q}(t),\dot{\mathbf{q}}(t)))\dot{\mathbf{q}}(t) + \mathbf{g}(\mathbf{q}(t)) = \boldsymbol{\tau}, \tag{19}$$

where $\mathbf{H}(\mathbf{q(t)})$ is the positive-definite symmetric inertia matrix. $\mathbf{C}(\mathbf{q(t)},\dot{\mathbf{q}}(\mathbf{t}))$ is a skew-symmetric matrix such that for any vector $\mathbf{s}$

$$\mathbf{s}^T\mathbf{C}(\mathbf{q}(t),\dot{\mathbf{q}}(t))\,\mathbf{s} = 0. \tag{20}$$

The term $\mathbf{g}(\mathbf{q(t)})$ represents the gravitational force, and $\boldsymbol{\tau}$ is the joint torque of the robot manipulator. The first and second terms on the left side of eq. (19) represent the inertial force, and the Coriolis and centrifugal forces, respectively. It should be noted that the inertial, Coriolis and centrifugal forces are highly nonlinear. Kinematics-based controllers are designed with an assumption that the robot moves slowly so that the nonlinear forces can be neglected. In a strict sense, the stability of the system is not guaranteed. To guarantee the stability rigorously, we need to consider the nonlinear dynamic forces in the controller design.

3 Adaptive Dynamic Visual Servoing

3.1 Problem Statement

This chapter addresses image-based visual servoing of robot manipulators with uncalibrated cameras. In the fixed camera setup, the problem is to control motion of the robot manipulator properly so that a number of feature points marked on the end-effector can be moved to desired positions on the image plane. In the eye-in-hand setup, the problem is to control motion of the robot manipulator so that the projections of a number of feature points in the environment onto the image plane can be moved to desired positions. For simplicity, we first address the problem of controlling a single feature point and extend the discussion to problem of multiple feature points.

Denote the desired position of the feature point on the image plane by $\mathbf{y_d}$, which is a constant vector. According to whether the position of the feature point with respect to the robot manipulators is given, we define the following three control problems:

Problem 1: Image-based Visual Servoing with Unknown Feature Points: *Assume that the intrinsic and extrinsic parameters have been accurately calibrated, but the coordinates $\mathbf{x}$ of the feature point with respect to the robot manipulator is unknown. Design a proper joint input $\boldsymbol{\tau}(t)$ of the robot manipulator such that the image $\mathbf{y(t)}$ of the feature point is convergent to the desired value $\mathbf{y_d}$ as time approaches to the infinity.*

Problem 2: Image-based Visual Servoing with Uncalibrated Camera: *Assume that the intrinsic and extrinsic parameters of the camera are not calibrated, but the coordinates $\mathbf{x}$ of the feature point with respect to the robot manipulator is known. Design a proper joint input $\boldsymbol{\tau}(t)$ of the robot manipulator such that the image $\mathbf{y(t)}$ of the feature point is convergent to the desired value $\mathbf{y_d}$ as time approaches to the infinity.*

Problem 3: Image-based Visual Servoing with Uncalibrated Camera and Unknown Feature Points: *Assume that the intrinsic and extrinsic parameters of the camera and the coordinates $\mathbf{x}$ of the feature point with respect to the robot manipulator are all unknown. Design a proper joint input $\boldsymbol{\tau}(t)$ of the robot manipulator such that the image $\mathbf{y(t)}$ of the feature point is convergent to the desired value $\mathbf{y_d}$ as time approaches to the infinity.*

Problem 3 is more general than the other two problems because it is subject to more unknown parameters. By fixing some of the parameters, Problem 3 can be changed to Problem 1 or Problem 2, respectively. Therefore, in the following our efforts will be focused to the controller design for Problem 3.

It should be pointed out that we will not limit motion of the robot manipulator to planar motion. The robot manipulator moves in a 3D space. Furthermore, the nonlinear dynamics of the robot manipulator must be taken into account in the controller design. Since the kinematics equation of the fixed camera setup is inherently the same as that of the eye-in-hand setup, for simplicity, our discussion will be mainly made on the dynamic visual servoing of robot manipulators using an uncalibrated fixed camera.

3.2 The Parameters Vector

It should be noted that the camera parameters and the position vector $\mathbf{x}$ are unknown. List all possible products of the components of matrix $\mathbf{M}$ with components of vector $\mathbf{x}$ by a vector $\boldsymbol{\theta}$, which depends only on the unknown parameters including the camera intrinsic and extrinsic parameters and the unknown coordinates $\mathbf{x}$ of the feature point. We call vector $\boldsymbol{\theta}$ the unknown parameters vector. Since one component of the matrix $\mathbf{M}$ can be fixed to one, the vector $\boldsymbol{\theta}$ consists of 38 unknown parameters.

Property 2: *The product* $\mathbf{A}(\mathbf{y}(t))\mathbf{J}(\mathbf{q}(t),\mathbf{x})\boldsymbol{\rho}$ *of the depth-independent image Jacobian matrix with any vector $\boldsymbol{\rho}$, can be linearly parameterized by the unknown parameters ,i.e.*

$$\mathbf{A}(\mathbf{y}(t))\mathbf{J}(\mathbf{q}(t),\mathbf{x})\boldsymbol{\rho} = \mathbf{P}(\mathbf{q}(t),\mathbf{y}(t),\boldsymbol{\rho})\,\boldsymbol{\theta}, \tag{21}$$

where $\mathbf{P}(\mathbf{q}(t),\mathbf{y}(t),\boldsymbol{\rho})$ is the regressor matrix without *depending on the unknown parameters*.

This property is the foundation for us to design the adaptive algorithm for estimating the unknown parameters on-line.

3.3 Controller Design

The controller is designed on the basis of two ideas. First, we use the depth-independent image Jacobian matrix to map the image errors onto the joint space of the robot manipulator. Second, the controller employs an adaptive algorithm to estimate the unknown parameters $\boldsymbol{\theta}$ on-line. The image error of the feature point is obtained by measuring the difference between its current position and the desired one on the image plane:

$$\Delta \mathbf{y}(t) = \mathbf{y}(t) - \mathbf{y}_d, \tag{22}$$

where $\Delta \mathbf{y}(t)$ is the 3×1 image error vector whose third component is always zero. Denote a time-varying estimation of the unknown parameters $\boldsymbol{\theta}$ by $\hat{\boldsymbol{\theta}}(t)$. Using the estimated parameters, we propose the following controller:

$$\boldsymbol{\tau}(t) = \mathbf{g}(\mathbf{q}(t)) - \mathbf{K}_1 \dot{\mathbf{q}}(t) - \left\{ \hat{\mathbf{D}}^T(\mathbf{q}(t),\mathbf{y}(t)) + \frac{1}{2}\hat{\mathbf{h}}^T(\mathbf{q}(t))\Delta\mathbf{y}^T(t) \right\} \mathbf{B}\Delta\mathbf{y}(t). \tag{23}$$

The first term is the gravity compensator. The second term represents a velocity damping in the joint space. The third term is the image error feedback. The matrix $\hat{\mathbf{D}}(\mathbf{q}(t),\mathbf{y}(t))$ represents an estimation of the depth-independent image Jacobian matrix calculated using the estimated parameters. $\hat{\mathbf{h}}(\mathbf{q}(t))$ is an estimation of the vector $\mathbf{h}(\mathbf{q}(t))$. $\mathbf{K}_1$ is a positive-definite gain matrix and $\mathbf{B}$ is a positive definite position gain matrix.

It should be pointed out that no inverse of the manipulator Jacobian or the depth-independent image Jacobian matrix is used in the controller (23).

It is important to note that the depth factor $1/^c z(\mathbf{q}(t))$ does not appear in controller. The quadratic form of $\Delta \mathbf{y}(t)$ in eq. (23) is to compensate for the effect caused by the removal of the depth factor. Using the depth-independent image Jacobian matrix and including the quadratic term differentiates our controller from other existing ones. By substituting the control law (23) into the robot dynamics (19), we obtain the following closed loop dynamics:

$$\begin{aligned}
&\mathbf{H}(\mathbf{q}(t))\ddot{\mathbf{q}}(t) + (\tfrac{1}{2}\dot{\mathbf{H}}(\mathbf{q}(t)) + \mathbf{C}(\mathbf{q}(t),\dot{\mathbf{q}}(t)))\dot{\mathbf{q}}(t) = -\mathbf{K}_1 \dot{\mathbf{q}}(t) \\
&-[\mathbf{D}^T(\mathbf{q}(t),\mathbf{y}(t)) + \tfrac{1}{2}\mathbf{h}(\mathbf{q}(t))\Delta \mathbf{y}^T(t)]\mathbf{B}\Delta \mathbf{y}(t) \\
&-[(\hat{\mathbf{D}}^T(\mathbf{q}(t),\mathbf{y}(t)) - \mathbf{D}^T(\mathbf{q}(t),\mathbf{y}(t))) \\
&+ \tfrac{1}{2}(\hat{\mathbf{h}}(\mathbf{q}(t)) - \mathbf{h}(\mathbf{q}(t)))\Delta \mathbf{y}^T(t)]\mathbf{B}\Delta \mathbf{y}(t).
\end{aligned} \quad (24)$$

From the Property 2, the last term in eq. (24) can be represented as a linear form of the estimation errors of the parameters as follows:

$$\begin{aligned}
&[(\hat{\mathbf{D}}^T(\mathbf{q}(t),\mathbf{y}(t)) - \mathbf{D}^T(\mathbf{q}(t),\mathbf{y}(t))) \\
&+ \tfrac{1}{2}(\hat{\mathbf{h}}(\mathbf{q}(t)) - \mathbf{h}(\mathbf{q}(t)))\Delta \mathbf{y}^T(t)]\mathbf{B}\Delta \mathbf{y}(t) = -\mathbf{Y}(\mathbf{q}(t),\mathbf{y}(t))\Delta \boldsymbol{\theta}(t),
\end{aligned} \quad (25)$$

where $\Delta \boldsymbol{\theta}(t) = \hat{\boldsymbol{\theta}}(t) - \boldsymbol{\theta}$, representing the estimation error and $\mathbf{Y}(\mathbf{q}(t),\mathbf{y}(t))$ is a regressor matrix without depending on the unknown parameters.

3.4 Estimation of the Camera Parameters

This subsection presents a new adaptive algorithm to estimate the unknown parameters on-line. This algorithm combines the Slotine-Li algorithm with an on-line minimization process based on an idea similar to structure from motion in computer vision.

Structure from motion is a problem of constructing the 3-D coordinates (structure) of fixed feature points and recovering motion of the camera from multiple images captured by a moving camera. In our case, the camera is fixed, but the feature point is moving. Furthermore, the motion of the feature point can be measured by the joint angles of the robot manipulator. The problem here is to construct the three dimensional coordinates of the feature point with respect to the end-effector frame and to estimate the perspective projection matrix of the camera from the images of the feature point at a sequence of positions when the robot is moving (Fig. 3). We can formalize this as a problem of finding a solution of the estimated matrix $\hat{\mathbf{M}}(t)$ and coordinates $\hat{\mathbf{x}}(t)$ that minimizes the Frobenius norm,

$$\mathbf{E} = \sum_{j=1}^{l} \left\| {}^c\hat{z}(\mathbf{q}(t_j))\mathbf{y}(t_j) - \hat{\mathbf{M}}(t)\mathbf{T_e}(\mathbf{q}(t_j))\hat{\mathbf{x}}(t) \right\|^2, \quad (26)$$

from images of the feature point at a sequence of positions when the robot moves. The symbol l denotes the number of images captured by the camera at different configurations of the manipulator. Here t_j represents the time instant when the j-th image was captured. Note that the l images can be selected from the trajectory of the robot manipulator. For simplicity, define

$$\mathbf{e}(t,t_j) = {}^c\hat{z}(t_j)\mathbf{y}(t_j) - \hat{\mathbf{M}}(t)\mathbf{T}_e(\mathbf{q}(t_j))\hat{\mathbf{x}}(t). \tag{27}$$

The vector $\mathbf{e}(t,t_j)$ is called the *estimated projection error* of the feature point, which can be written as follows:

$$\begin{aligned}\mathbf{e}(t,t_j) &= \mathbf{y}(t_j)({}^c\hat{z}(t_j) - {}^cz(t_j)) - (\hat{\mathbf{M}}(t)\mathbf{T}_e(\mathbf{q}(t_j))\hat{\mathbf{x}}(t) - \mathbf{M}\mathbf{T}_e(\mathbf{q}(t_j))\mathbf{x})\\ &= \mathbf{y}(t_j)(\hat{\mathbf{m}}_3(t)\mathbf{T}_e(\mathbf{q}(t_j))\hat{\mathbf{x}}(t) - \mathbf{m}_3\mathbf{T}_e(\mathbf{q}(t_j))\mathbf{x})\\ &\quad - (\hat{\mathbf{M}}(t)\mathbf{T}_e(\mathbf{q}(t_j))\hat{\mathbf{x}}(t) - \mathbf{M}\mathbf{T}_e(\mathbf{q}(t_j))\mathbf{x}).\end{aligned} \tag{28}$$

From Property 2, the equation (28) can be written as follows:

$$\mathbf{e}(t,t_j) = \mathbf{W}(\mathbf{q}(t_j),\mathbf{y}(t_j))\Delta\boldsymbol{\theta}(t). \tag{29}$$

From the results in computer vision [12], we have

Proposition 3: *If a sufficient number of images of the feature point on its trajectory due to motion of the robot manipulator have been captured, the equation*

$$\mathbf{W}(\mathbf{q}(t_j),\mathbf{y}(t_j))\Delta\boldsymbol{\theta}(t) = 0, \quad \forall j = 1,2,...,l, \tag{30}$$

implies one of the following two cases:

(1) the estimated parameters differ from real values by a scale, i.e. there exists a scalar λ such that

$$\hat{\boldsymbol{\theta}}(t) = \lambda\boldsymbol{\theta}. \tag{31}$$

(2) The estimated parameters are zero.

As shown in [27], the trivial solution $\hat{\boldsymbol{\theta}}(t) = \mathbf{0}$ can be simply avoided by fixing one of the components of the perspective projection matrix to a particular value. The following adaptive rule is proposed to update the estimation of the parameters:

$$\frac{d}{dt}\hat{\boldsymbol{\theta}}(t) = -\boldsymbol{\Gamma}^{-1}\{\mathbf{Y}^T(\mathbf{q}(t),\mathbf{y}(t))\dot{\mathbf{q}}(t) + \sum_{j=1}^{l}\mathbf{W}^T(\mathbf{q}(t_j),\mathbf{y}(t_j))\mathbf{K}_3\mathbf{W}(\mathbf{q}(t_j),\mathbf{y}(t_j))\Delta\boldsymbol{\theta}(t)\}, \tag{32}$$

where $\boldsymbol{\Gamma}$ and $\mathbf{K}_3$ are positive-definite and diagonal gain matrices.

Remark 2: *Although the estimation error $\Delta\boldsymbol{\theta}(t)$ explicitly appears in the eq. (32), it is not necessary to use the unknown estimation error to calculate $\mathbf{W}(\mathbf{q}(t_j),\mathbf{y}(t_j))\Delta\boldsymbol{\theta}(t)$, which can be calculated by eq. (27).*

Remark 3: *In the adaptive algorithm (32), in addition to the regressor term (the first term on the right hand side), there is an term corresponding to on-line minimization of the Frobenius norm of the estimated projection errors to guarantee the convergence of the estimated parameters to the real values up to a scale.*

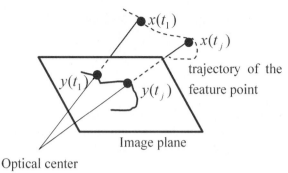

Fig. 3. Images of the feature point at a sequence of positions

3.5 Stability Analysis

This section analyzes the stability of the proposed adaptive controller on the basis of the nonlinear robot dynamics. For simplicity, we assume that the feature point is visible during the motion so that its depth with respect to the camera frame is always positive.

Theorem 1: *If a sufficient number of images captured by the camera during motion of the robot manipulator are employed to estimate the unknown parameters using the adaptive algorithm (32), the proposed controller (23) gives rise to the asymptotic convergence of the image error, i.e.*

$$\lim_{t\to\infty} \Delta\mathbf{y}(t) = \mathbf{0}. \quad (33)$$

Furthermore, the estimated parameters $\hat{\boldsymbol{\theta}}(t)$ are convergent to the real values up to a scale.

Proof: To prove the convergence of the image error, we introduce the following non-negative function:

$$v(t) = \frac{1}{2}\{\dot{\mathbf{q}}^T(t)\mathbf{H}(\mathbf{q}(t))\dot{\mathbf{q}}(t) + {}^c z(\mathbf{q}(t))\Delta\mathbf{y}^T(t)\mathbf{B}\Delta\mathbf{y}(t) + \Delta\boldsymbol{\theta}^T(t)\boldsymbol{\Gamma}\Delta\boldsymbol{\theta}(t)\}. \quad (34)$$

Note that the depth ${}^c z(\mathbf{q}(t))$ is always positive. Multiplying the $\dot{\mathbf{q}}^T(t)$ from the left to the closed loop dynamics (24) results in

$$\begin{aligned}
&\dot{\mathbf{q}}^T(t)\mathbf{H}(\mathbf{q}(t))\ddot{\mathbf{q}}(t) + \tfrac{1}{2}\dot{\mathbf{q}}^T(t)\dot{\mathbf{H}}(\mathbf{q}(t))\dot{\mathbf{q}}(t) = -\dot{\mathbf{q}}^T(t)\mathbf{K}_1\dot{\mathbf{q}}(t)\\
&\quad -\dot{\mathbf{q}}^T(t)\mathbf{D}^T(\mathbf{q}(t),\mathbf{y}(t))\mathbf{B}\Delta\mathbf{y}(t)\\
&\quad -\tfrac{1}{2}\dot{\mathbf{q}}^T(t)\mathbf{h}^T(\mathbf{q}(t))\Delta\mathbf{y}^T(t)\mathbf{B}\Delta\mathbf{y}(t)\\
&\quad +\dot{\mathbf{q}}^T(t)\mathbf{Y}(\mathbf{q}(t),\mathbf{y}(t))\Delta\boldsymbol{\theta}(t).
\end{aligned} \quad (35)$$

From equation (13), we have

$$\dot{\mathbf{q}}^T(t)\mathbf{D}^T(\mathbf{q}(t),\mathbf{y}(t)) = {}^c z(\mathbf{q}(t))\dot{\mathbf{y}}^T(t) = {}^c z(\mathbf{q}(t))\Delta\dot{\mathbf{y}}^T(t). \quad (36)$$

By multiplying the $\Delta\boldsymbol{\theta}^T(t)$ from the left to the adaptive rule (32), we obtain

$$\Delta\boldsymbol{\theta}^T(t)\boldsymbol{\Gamma}\Delta\dot{\boldsymbol{\theta}}(t) = -\Delta\boldsymbol{\theta}(t)\mathbf{Y}^T(\mathbf{q}(t),\mathbf{y}(t))\dot{\mathbf{q}}(t) \\ - \sum_{j=1}^{l} \Delta\boldsymbol{\theta}^T(t)\mathbf{W}^T(t_j)\mathbf{K}_3\mathbf{W}(t_j)\Delta\boldsymbol{\theta}(t). \tag{37}$$

Differentiating the function $v(t)$ in (34) results in

$$\begin{aligned}\dot{v}(t) = \;& \dot{\mathbf{q}}^T(t)(\mathbf{H}(\mathbf{q}(t))\ddot{\mathbf{q}}(t) + \tfrac{1}{2}\dot{\mathbf{H}}(\mathbf{q}(t))\dot{\mathbf{q}}(t)) \\ & + \Delta\boldsymbol{\theta}^T(t)\boldsymbol{\Gamma}\Delta\dot{\boldsymbol{\theta}}(t) + {}^c z(\mathbf{q}(t))\Delta\mathbf{y}^T(t)\mathbf{B}\Delta\dot{\mathbf{y}}(t) \\ & + \tfrac{1}{2}{}^c \dot{z}(\mathbf{q}(t))\Delta\mathbf{y}^T(t)\mathbf{B}\Delta\mathbf{y}(t).\end{aligned} \tag{38}$$

From equation (17), we have ${}^c\dot{z}(\mathbf{q}(t)) = \mathbf{h}(\mathbf{q}(t))\dot{\mathbf{q}}(t)$. By combining the equations (35-38), we have

$$\dot{v}(t) = -\dot{\mathbf{q}}^T(t)\mathbf{K}_1\dot{\mathbf{q}}(t) - \sum_{j=1}^{m}\Delta\boldsymbol{\theta}^T(t)\mathbf{W}^T(t_j)\mathbf{K}_3\mathbf{W}(t_j)\Delta\boldsymbol{\theta}(t). \tag{39}$$

Therefore, $v(t)$ is a non-increasing function so that it is upper bounded. From equation (34), bounded $v(t)$ directly implies that the joint velocity, the image errors, and the estimation errors are all bounded. Then, the joint acceleration $\ddot{\mathbf{q}}(t)$ is bounded from the closed-loop dynamics (24) and so is $\dot{\hat{\boldsymbol{\theta}}}(t)$ from the adaptive algorithm (32). Consequently, the joint velocity $\dot{\mathbf{q}}(t)$ and the estimated parameters $\hat{\boldsymbol{\theta}}(t)$ are uniformly continuous. From the Barbalat Lemma, we conclude that

$$\lim_{t\to\infty} \dot{\mathbf{q}}(t) = \mathbf{0}, \tag{40}$$

$$\lim_{t\to\infty} \mathbf{W}(\mathbf{x}(t_j),\mathbf{y}(t_j))\Delta\boldsymbol{\theta} = \mathbf{0}. \tag{41}$$

When $\mathbf{W}(t_j)\Delta\boldsymbol{\theta}(t) = \mathbf{0}$ and the number l of images is sufficiently large (>19), the estimated parameters are convergent to the real values up to a scale. This means that the estimated projection matrix $\hat{\mathbf{M}}(t)$ has a rank of 3.

To prove the convergence of the image error, we further consider the equilibrium points of the system. From the closed-loop dynamics (24) of the robot manipulator, at the equilibrium point we have:

$$\mathbf{J}^T(\mathbf{q}(t),\hat{\mathbf{x}}(t))\underbrace{(\hat{\mathbf{A}}^T(\mathbf{y}(t)) + \frac{1}{2}\hat{\mathbf{m}}_3(t)\Delta\mathbf{y}^T(t))}_{\mathbf{Q}(\hat{\boldsymbol{\theta}}(t),\mathbf{y}(t))}\mathbf{B}\Delta\mathbf{y}(t) = \mathbf{0}. \tag{42}$$

Note that $\Delta\mathbf{y}(t)$ is a 3 dimensional vector whose third component is zero. $\mathbf{B}$ is a 3 by 3 matrix. $\mathbf{Q}(\hat{\boldsymbol{\theta}}(t),\mathbf{y}(t))$ is a 4 by 3 matrix whose fourth row is due to the homogenous coordinates. For non-zero estimated coordinates $\hat{\mathbf{x}}(t)$, it is possible to prove that the rank of the matrix $\mathbf{J}(\mathbf{q}(t),\hat{\mathbf{x}}(t))$ is 3 when the rank of the manipulator Jacobian matrix is larger than or equal to 3. Therefore, at the equilibrium points we have:

$$\mathbf{Q}(\hat{\boldsymbol{\theta}}(t),\mathbf{y}(t))\mathbf{B}\Delta\mathbf{y}(t) = 0. \tag{43}$$

Note that

$$\mathbf{Q}(\hat{\boldsymbol{\theta}}(t), \mathbf{y}(t)) = \begin{pmatrix} \hat{\mathbf{m}}_1^T(t) + (0.5\Delta u(t) - u(t))\hat{\mathbf{m}}_3^T(t) \\ \hat{\mathbf{m}}_2^T(t) + (0.5\Delta v(t) - v(t))\hat{\mathbf{m}}_3^T(t) \\ \mathbf{0}_{1\times 4} \end{pmatrix}^T. \qquad (44)$$

The rank of the estimated perspective projection matrix $\hat{\mathbf{M}}(t)$ is 3 when the estimated parameters are convergent to the real values up to a scale. By using a similar proof to that for Proposition 1, it is possible to demonstrate that the matrix $\mathbf{Q}(\hat{\boldsymbol{\theta}}(t), \mathbf{y}(t))$ has a rank of 2. From eq. (43), it is obvious that $\Delta \mathbf{y}(t) = \mathbf{0}$ at the equilibrium point. Consequently, the image error is convergent to zero as the time approaches to the infinity.

It should be pointed out that the estimated parameters are not convergent to the true values. If a sufficient number of positions of the feature point are selected for the parameter estimation, the parameters are convergent to the true values up to a scale. The convergence of the image errors does not mean the convergence of the end-effector to the desired position. To regulate the end-effector in 3-D space, the feature points must be properly selected. The selection of the feature points is beyond the scope of this chapter.

3.6 Discussions on Stability Conditions

Theorem 1 states that the condition for the asymptotic convergence of the image error is that the number l of images must be sufficiently large. We look into the conditions for each of the problem stated in the subsection 3.1.

We first consider Problem 1. Since the camera is calibrated, the perspective project matrix is known and hence the rank of the matrix $\mathbf{Q}(\hat{\boldsymbol{\theta}}(t), \mathbf{y}(t))$ is always 2. Therefore, if the Jacobian matrix of the robot manipulator is of a rank greater than 2, the asymptotic convergence is always guaranteed. Furthermore, the estimated parameters corresponding to $\hat{\mathbf{x}}(t)$ are also convergent to the real values if the number l of images is larger than 1. This can be easily explained by the fact that $\mathbf{e}(t, t_j) = 0$ at two points means the estimated position of the feature point satisfies the triangle constraint. In other words, once the projections of the feature point on the images captured by a calibrated camera at two points are given, the position of the feature point can be uniquely determined.

The condition for asymptotic stability in Problem 2 is more complicated than that for Problem 1. In this problem, the parameters to be estimated are 11 parameters corresponding to the normalized components of the perspective projection matrix. Six images of the image are necessary to solve the unknown parameters when $\mathbf{e}(t, t_j) = 0$ for all the images. Our early work [25] proved that the asymptotic convergence of the image error can be guaranteed if 5 images are so selected that it is not possible to find three projections of the feature point which are collinear on the image plane. It should be noted that in this problem the estimated parameters are not convergent to the true values.

In Problem 3, there are total 38 unknown parameters. To solve the parameters by assuming that the estimated projection errors $\mathbf{e}(t, t_j)$ is zero, the number l of images must not be smaller than 19. Since our objective is to force the feature point to the desired position instead of calculating the unknown parameters, the number l could be made smaller. The detailed condition needs to be further investigated in future.

4 Uncalibrated Visual Servoing for Multiple Feature Points

We extend the proposed adaptive controller to position control of a number of feature points. Denote the number of the feature points by S. Denote the homogenous coordinates of feature point i on the image plane by $\mathbf{y}_i(t)$ and its coordinates with respect to the end-effector frame by constant vector $\mathbf{x}_i$. The two coordinates are related by the perspective projection:

$$\mathbf{y}_i(t) = \frac{1}{{}^c z_i(\mathbf{q}(t))} \mathbf{M} \mathbf{T}_e(\mathbf{q}(t)) \mathbf{x}_i, \qquad (45)$$

where ${}^c z_i(\mathbf{q}(t))$ denotes the depth of the feature point i with respect to the camera frame. The velocity relationship is given by

$$\begin{aligned}\dot{\mathbf{y}}_i(t) &= \frac{1}{{}^c z_i(\mathbf{q}(t))} (\mathbf{M} \frac{\partial}{\partial \mathbf{q}} \{\mathbf{T}_e(\mathbf{q}(t)) \mathbf{x}_i\} \dot{\mathbf{q}}(t) - \mathbf{y}_i(t) {}^c \dot{z}_i(\mathbf{q}(t))) \\ &= \frac{1}{{}^c z_i(\mathbf{q}(t))} \mathbf{A}_i(\mathbf{y}_i(t)) \mathbf{J}_i(\mathbf{q}(t), \mathbf{x}_i) \dot{\mathbf{q}}(t),\end{aligned} \qquad (46)$$

where the depth-independent interaction matrix $\mathbf{A}_i(\mathbf{y}_i(t))$ is similar to that in eq. (14). It should be noted that the Jacobian matrix $\mathbf{J}_i(\mathbf{q}(t), \mathbf{x}_i)$ of the robot manipulator is different for different feature points. We design a controller similar to that in eq. (23) as follows:

$$\tau = \mathbf{g}(\mathbf{q}(t)) - \mathbf{K}_1 \dot{\mathbf{q}}(t) -$$

$$(\mathbf{J}_1^T(\mathbf{q}(t), \hat{\mathbf{x}}_1), \mathbf{J}_2^T(\mathbf{q}(t), \hat{\mathbf{x}}_2), ..., \mathbf{J}_S^T(\mathbf{q}(t), \hat{\mathbf{x}}_S)) \begin{pmatrix} \mathbf{Q}_1(\hat{\boldsymbol{\theta}}(t), \mathbf{y}_1(t)) \mathbf{B}_1 \Delta \mathbf{y}_1(t) \\ \mathbf{Q}_2(\hat{\boldsymbol{\theta}}(t), \mathbf{y}_2(t)) \mathbf{B}_2 \Delta \mathbf{y}_2(t) \\ ... \\ \mathbf{Q}_S(\hat{\boldsymbol{\theta}}(t), \mathbf{y}_S(t)) \mathbf{B}_S \Delta \mathbf{y}_S(t) \end{pmatrix}, \qquad (47)$$

where

$$\mathbf{Q}_i(\hat{\boldsymbol{\theta}}(t), \mathbf{y}_i(t)) = \hat{\mathbf{A}}_i^T(\mathbf{y}(t)) + \frac{1}{2} \hat{\mathbf{m}}_3(t) \Delta \mathbf{y}_i^T(t). \qquad (48)$$

By substituting the control law (46) into the robot dynamics (19) and noting Property 2, we obtain the following closed loop dynamics:

$$\mathbf{H}(\mathbf{q}(t))\ddot{\mathbf{q}}(t) + \{\tfrac{1}{2}\dot{\mathbf{H}}(\mathbf{q}(t)) + \mathbf{C}(\mathbf{q}(t),\dot{\mathbf{q}}(t))\}\dot{\mathbf{q}}(t) = \mathbf{K}_1\dot{\mathbf{q}}(t) -$$
$$\left(\mathbf{J}_1^T(\mathbf{q}(t),\hat{\mathbf{x}}_1), \mathbf{J}_2^T(\mathbf{q}(t),\mathbf{x}_2), ..., \mathbf{J}_S^T(\mathbf{q}(t),\mathbf{x}_S)\right) \begin{pmatrix} \mathbf{Q}_1(\boldsymbol{\theta}(t),\mathbf{y}_1(t))\mathbf{B}_1\Delta\mathbf{y}_1(t) \\ \mathbf{Q}_2(\boldsymbol{\theta}(t),\mathbf{y}_2(t))\mathbf{B}_2\Delta\mathbf{y}_2(t) \\ ... \\ \mathbf{Q}_S(\boldsymbol{\theta}(t),\mathbf{y}_S(t))\mathbf{B}_S\Delta\mathbf{y}_S(t) \end{pmatrix}$$
$$+ \mathbf{Y}(\mathbf{q}(t),\mathbf{y}_1(t),...,\mathbf{y}_S(t))\Delta\boldsymbol{\theta}(t). \tag{49}$$

The following adaptive algorithm, similar to that in eq. (32), is proposed to estimate the unknown parameters:

$$\frac{d}{dt}\hat{\boldsymbol{\theta}}(t) = -\boldsymbol{\Gamma}^{-1}\{\mathbf{Y}^T(\mathbf{q}(t),\mathbf{y}_1(t),...,\mathbf{y}_S(t))\dot{\mathbf{q}}(t) \\ + \sum_{i=1}^{S}\sum_{j=1}^{l}\mathbf{W}^T(\mathbf{q}(t_j),\mathbf{y}_i(t_j))\mathbf{K}_3\mathbf{e}_i(t,t_j)\}, \tag{50}$$

where $\mathbf{y}_i(t_j)$ denotes the projection of the feature point i on the j-th image selected for the parameter adaptation. The number l must be so selected that $l > 19$. The estimation error $\mathbf{e}_i(t,t_j)$ is similar to that in eq. (27).

Theorem 2: *The adaptive controller (47) and (50) guarantees that the position errors of the feature points are convergent to positions satisfying:*

$$\left(\mathbf{J}_1^T(\mathbf{q}(t),\hat{\mathbf{x}}_1), ..., \mathbf{J}_S^T(\mathbf{q}(t),\hat{\mathbf{x}}_S)\right) \begin{pmatrix} \mathbf{Q}_1(\hat{\boldsymbol{\theta}}(t),\mathbf{y}_1(t))\mathbf{B}_1\Delta\mathbf{y}_1(t) \\ ... \\ \mathbf{Q}_S(\hat{\boldsymbol{\theta}}(t),\mathbf{y}_S(t))\mathbf{B}_S\Delta\mathbf{y}_S(t) \end{pmatrix} = \mathbf{0}. \tag{51}$$

Proof: The proof is similar to that for Theorem 1. Introduce the following non-negative function:

$$v(t) = \tfrac{1}{2}\{\dot{\mathbf{q}}^T(t)\mathbf{H}(\mathbf{q}(t))\dot{\mathbf{q}}(t) \\ + \sum_{i=1}^{S}{}^cz_i(\mathbf{q}(t))\Delta\mathbf{y}_i^T(t)\mathbf{B}_i\Delta\mathbf{y}_i(t) + \Delta\boldsymbol{\theta}^T(t)\boldsymbol{\Gamma}\Delta\boldsymbol{\theta}(t)\}. \tag{52}$$

By multiplying the $\dot{\mathbf{q}}^T(t)$ and $\Delta\boldsymbol{\theta}^T(t)$ from the left to the closed loop dynamics (49), and the adaptive rule (50), respectively, we obtain

$$\dot{v}(t) = -\dot{\mathbf{q}}^T(t)\mathbf{K}_1\dot{\mathbf{q}}(t) \\ - \sum_{i=1}^{S}\sum_{j=1}^{m}\Delta\boldsymbol{\theta}^T(t)\mathbf{W}^T(\mathbf{q}(t_j),\mathbf{y}_i(t_j))\cdot\mathbf{K}_3\mathbf{W}(\mathbf{q}(t_j),\mathbf{y}_i(t_j))\Delta\boldsymbol{\theta}(t). \tag{53}$$

By using a similar logic to that in the proof of Theorem 1, it is possible to prove that $\ddot{\mathbf{q}}(t)$ and $\dot{\hat{\boldsymbol{\theta}}}(t)$ are bounded, and thus that $\dot{\mathbf{q}}(t)$ and $\hat{\boldsymbol{\theta}}(t)$ are uniformly continuous. On the other hand, from the closed-loop dynamics (49) the equilibrium points must satisfy eq. (51). The Barbalat's Lemma states the convergence of the image errors to the equilibrium points satisfying (51).

It is not possible to conclude the convergence of the image errors to zero from eq. (51) only. The image errors are convergent to zero only when the rank of the coefficient matrix

$$\left(\mathbf{J}_1^T(\mathbf{q}(t),\hat{\mathbf{x}}_1), \mathbf{J}_2^T(\mathbf{q}(t),\mathbf{x}_2), ..., \mathbf{J}_S^T(\mathbf{q}(t),\hat{\mathbf{x}}_S)\right)$$

is equal to $2S$. This implies that the number n of degrees of freedom of the manipulator must be larger than or equal to $2S$. When all the feature points are on a rigid end-effector, it is possible to conclude the convergence of the image errors of the feature points to zero. The detailed discussion is referred to our early work [25].

Finally, note that although the discussion has been made for the fixed camera configuration, the controller for the eye-in-hand configuration can be designed similarly because the kinematics and dynamics are inherently the same.

5 Experiments

5.1 Experimental Setup

To verify the performance of the proposed controller, we have implemented it in a 3 DOF robot manipulator at the Networked Sensors and Robotics Laboratory of The Chinese University of Hong Kong (Fig. 4). The manipulator is driven by Maxon brushed DC motors at the three joints. The powers of the motors at the first, second and third joints are 20 watt, 10 watt, and 10watt, respectively, and the corresponding gear ratios are 480:49, 12:1, and 384:49, respectively. The gear ratios and the motor powers are all relatively small, so the nonlinear forces of the robot manipulator have relatively strong effect on its motion. The moment inertia about its vertical axis of the first link of the manipulator is 0.005kgm^2, the masses of the second and third links are 0.167 kg, and 0.1 kg, respectively. The second and third link lengths are, 0.145m and 0.1285m, respectively. Three incremental optical encoders with a resolution of 2000 pulses/turn are employed to measure the joint angles of the manipulator. The joint velocities are obtained by differentiating the measured joint angles. We use a Ptgrey camera at the rate of 120 fps to capture the image. An Intel Pentium IV PC is employed to process the image captured by the frame grabber and to extract the image features. The sampling period of the control loop is 13ms.

5.2 Visual Servoing using a Fixed Camera

In this first experiment, we addressed Problem 2 in a fixed camera setup (Fig. 1). In other words, we assumed that both the intrinsic and extrinsic parameters of the fixed camera are unknown, but the positions of the feature points with respect to the end-effector frame are given. We considered three feature points whose coordinates with respect to the end-effector frame are $\mathbf{x}_1 = (-0.02, 0, 0.021)^T$m, $\mathbf{x}_2 = (-0.05, 0.0105, 0.018)^T$m, and $\mathbf{x}_3 = (-0.05 - 0.0105, 0.018)^T$m, respectively. Fig. 5 shows the initial and

Fig. 4. The experimental setup

desired positions of the feature points on the image plane. Figure 6(a) illustrates the trajectories of the feature points on the image plane and Fig. 6(b) plots the profiles of the image errors. The profiles of the estimated parameters are demonstrated in Fig. 7. The experimental results confirm the asymptotic convergence of the image errors of the feature points to zero. The residual image errors are within one pixel. In this experiment, we employed three current positions of the feature points in minimizing the estimated projection errors in the adaptive algorithm. The control gains used the experiments are $\mathbf{B} = 0.00035\,\mathbf{I}_{2\times 2}$ and $\mathbf{K}_1 = \text{diag}\{5,\ 200,\ 60\}$. The adaptive gains were $\mathbf{K}_3 = 0.00001\,\mathbf{I}_{2\times 2}$ and

$$\boldsymbol{\Gamma} = \text{diag}\{0.0005,\ 0.0005,\ 0.0005\ 0.0005,\ 0.0005,\ 0.0005 \\ 0.0005,\ 0.0005,\ 50,\ 50,\ 50\}.$$

The true values and initial estimations of the camera intrinsic parameters are shown in Table I. The true camera extrinsic parameters were not available. The initial extrinsic parameters are

$$\hat{\mathbf{T}}(0) = \begin{bmatrix} 0.97 & -0.26 & 0 & 0.1 \\ 0 & 0 & -1 & 0.1 \\ 0.26 & 0.97 & 0 & 3 \\ 0 & 0 & 0 & 1 \end{bmatrix}.$$

5.3 Visual Servoing Using an Eye-in-Hand Camera

The second experiment is to verify the performance of the proposed method in uncalibrated visual servoing using an eye-in-hand camera (Fig. 8). The control

Table 1. TABLE I The Camera's Intrinsic Parameters

	Real parameters	Initial estimations
α(pixel)	1806	2000
γ(pixel)	1812	2000
u_0(pixe)	282	300
v_0(pixel)	249	300

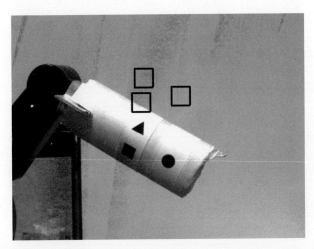

Fig. 5. The initial and desired (squared) positions of the feature points

problem addressed here is Problem 3, i.e. both the camera parameters and coordinates of the feature points are unknown. We first set initial positions of the feature point and then move the robot manipulator to another position and record its desired position of the feature point on the image plane. Figure 9 shows the initial and desired position of the feature point on the image plane. Fig. 10 plots the profiles of the image errors under control of the proposed adaptive controller. The experimental results clearly demonstrate the asymptotic convergence of the image error to zero. The control gains used in the experiments are $\mathbf{K_1} = 15$, $\mathbf{B} = 0.00003$, $\mathbf{K_3} = 0.01$, $\mathbf{\Gamma} = 5000\,\mathbf{I}$. The initial estimated transformation matrix of the end-effector frame respect to the vision frame is

$$\hat{\mathbf{T}}_c(0) = \begin{bmatrix} 0 & 0 & 1 & 0 \\ 0 & -1 & 0 & 0 \\ 1 & 0 & 0 & 1 \\ 0 & 0 & 0 & 1 \end{bmatrix}.$$

The initial estimated position of the feature point is $\hat{\mathbf{x}}(0) = \begin{bmatrix} 1 & -0.1 & 0.2 \end{bmatrix}^T$. The real and initial estimated intrinsic parameters are $a_u = 871$, $a_v = 882$, $u_0 = 381$, $v_0 = 278$, and $\hat{a}_u(0) = 1000$, $\hat{a}_v(0) = 1000$, $\hat{u}_0(0) = 300$, $\hat{v}_0(0) = 300$, respectively.

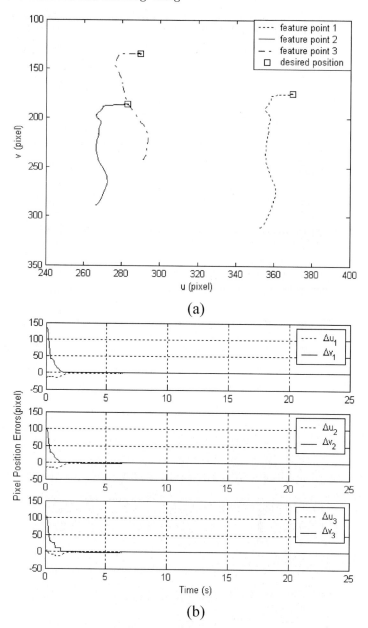

Fig. 6. Experimental results for three feature points: (a) trajectories of the feature points on the image plane, and (b) the image errors

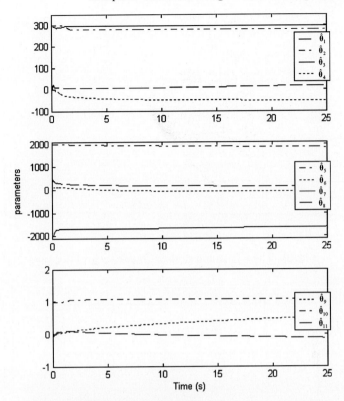

Fig. 7. The estimated parameters

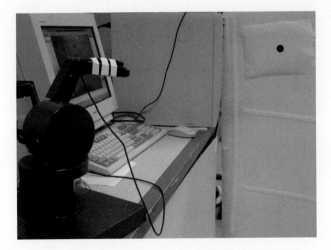

Fig. 8. The experiment setup system

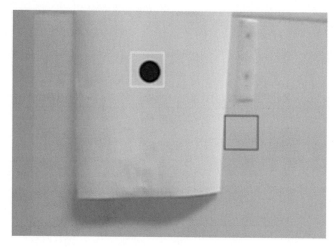

Fig. 9. The initial and desired (black square) positions

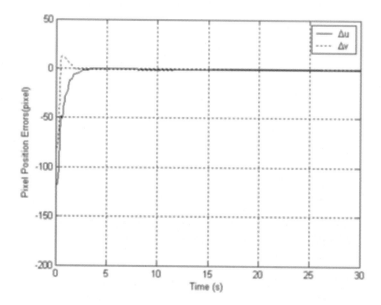

Fig. 10. The image errors of the feature point

The experimental results confirm that the proposed adaptive controller yield asymptotic stability of the vision-based control system for both the fixed camera and the eye-in-hand setups when the camera parameters and position of the feature points are unknown.

6 Conclusions

This chapter presents a novel adaptive controller for dynamic visual seroving of robot manipulators in general 3-D motion when the intrinsic and extrinsic parameters of the camera and the positions of the feature points are not known. Both the fixed camera and the eye-in-hand configurations were addressed. The basic idea of our controller lies in the use of the depth-independent image Jacobian matrix to map the image errors onto the joint space so that the dynamic system can be linearly parameterized by the unknown parameters. The adaptive algorithm combines the Slotine-Li algorithm with an on-line minimization process derived from the structure from motion problem in computer vision in order to estimate the unknown parameters on-line. The asymptotic stability of the system under the control of the adaptive controller is proved by the Lyapunov theory based on the fully nonlinear dynamics of the robot manipulator. The superior performance of the proposed method has been verified by experiments on a 3 DOF robot manipulator. In the future, we will address the robustness of the controller to measurement errors of the visual position and velocity. Since the sampling rate of a vision system is usually slow, the visual velocity measurement is subject to big noises. Designing a controller without using the visual velocity is one of our future topics.

References

1. P. K. Allen, A. Timcenko, B. Yoshimi, and P. Michelman, "Automated tracking and grasping of a moving object with a robotic hand-eye system," *IEEE Trans. on Robotics and Automation*, vol. 9, no. 2, pp. 152-165, 1993.
2. A. Astolfi, L. Hsu, M. Netto, and R. Ortega, "Two solutions to the adaptive visual servoing problem," *IEEE Trans. on Robotics and Automation*, vol. 18, no.3, pp. 387-392, 2002.
3. B. Bishop, S. Hutchinson, and M. Spong, "Camera modeling for visual servo control applications," *Mathematical and Computer Modeling*, vol. 24, no. 5/6, pp.79-102, 1996.
4. B. Bishop and M.W. Spong, "Adaptive calibration and control of 2D monocular visual servo system," *Proc. of IFAC Symp. Robot Control*, pp. 525-530, 1997.
5. R. Carelli, O. Nasisi, and B. Kuchen, "Adaptive robot control with visual feedback," *Proc. of the American Control Conf.*, pp. 1757-1760, 1994.
6. C C. Cheah, M. Hirano, S. Kawamura, and S. Arimoto, "Approximate Jacobian control for robots with uncertain kinematics and dynamics," *IEEE Trans. on Robotics and Automation*, vol. 19, no. 4, pp. 692 – 702, 2003.

7. F. Chaumette, "Potential problems of stability and convergence in image based and position based visual servoing." *In Conflux of vision and control LNCIS*, vol. 237, pp. 66–78, 1998.
8. J. Chen, D. M. Dawson, W. E. Dixon, and A. Behal, "Adaptive homography-based visual servo tracking," *Proc. of the IEEE Int. Conf. on Intelligent Robots and Systems*, pp. 230-235, 2003.
9. L. Deng;, F. Janabi-Sharifi, and W. J. Wilson, "Stability and robustness of visual servoing methods," *Prof. of IEEE Int. Conf. on Robotics and Automation*, pp. 1604 – 1609, 2002.
10. B. Espiau, F. Chaumette, and P. Rives, "A new approach to visual servoing in Robotics," *IEEE Trans. on Robotics and Automation*, vol. 8, no. 3, pp. 313-326, 1992.
11. H. H. Fakhry and W. J. Wilson, "Modified resolved acceleration controller for position-Based visual servoing," *Mathematical and Computer Modeling*, vol. 24, no. 5/6, pp. 1-9, 1996.
12. D. A Forsyth and J. Ponce, "Computer vision: a modern approach," Prentice Hall, 2003.
13. E. Grosso, G. Metta, A. Oddera, and G. Sandini, "Robust visual servoing in 3-D reaching tasks," *IEEE Trans. on Robotics and Automation*, vol. 12, no. 5, pp. 732-742, 1996.
14. K. Hashimoto, T. Kimoto, T. Ebine, and H. Kimura, "Manipulator control with image-based visual servo," *Proc. of IEEE Int. Conf. on Robotic and Automation*, pp. 2267-2272, 1991.
15. K. Hashimoto, T. Ebine and H. Kimura, "Dynamic visual feedback control for a hand-eye manipulator", *Proc. of IEEE Int. Conf. on Robotics and Automation*, pp. 1863-1869, 1992.
16. J. Hespanha, Z. Dodds, G. D. Hager, and A. S. Morse, "What can be done with an uncalibrated stereo system," *Proc. of IEEE Int. Conf. on Robotics and Automation*, pp. 1366-1372, 1998.
17. K. Hosoda and M. Asada, "Versatile visual servoing without knowledge of true Jacobain," *Proc. of IEEE/RSJ Int. Conf. on Intelligent Robots and Systems*, pp. 186-191, 1994.
18. L. Hsu and P. L. S. Aquino, "Adaptive visual tracking with uncertain manipulator dynamics and uncalibrated camera," *Proc. of IEEE Int. Conf. on Decision and Control*, pp. 1248-1253, 1999.
19. S. Hutchinson, G. D. Hager, and P. I. Corke, "A tutorial on visual servo control," *IEEE Trans. on Robotics and Automation*, vol. 12, no. 5, pp. 651-670, 1996.
20. M. Jagersand, O. Fuentes, and R. Nelson, "Experimental evaluation of uncalibrated visual servoing for precision manipulation," *Proc. of Int. Conf. on Robotics and Automation*, pp. 2874-2880, 1997.
21. R. Kelly, "Robust asymptotically stable visual servoing of planar manipulator," *IEEE Trans. on Robotics and Automation*, vol. 12, no. 5, pp.759-766, 1996.
22. R. Kelly, F. Reyes, J. Moreno, and S. Hutchinson, "A two-loops direct visual control of direct-drive planar robots with moving target," *Proc. of IEEE Int. Conf. on Robotics and Automation*, pp. 599-604, 1999.
23. D. Kim, A. A. Rizzi, G. D. Hager and D. E. Koditschek, "A robust' convergent visual servoing system," *Proc. of IEEE/RSJ Int. Conf. on Intelligent Robots and Systems*, pp. 348-353, 1995.

24. Y. H. Liu, K. Kitagaki, T. Ogasawara and S. Arimoto, "Model-based adaptive hybrid control for manipulators under multiple geometric constraints," *IEEE Trans. on Control Systems Technology*, vol. 7, no. 1, pp. 97-109, 1999.
25. Y. H. Liu, H. S. Wang, C. Wang and K. K. Lam, Uncalibrated visual servoing of robots using a depth-independent interaction matrix, *IEEE Trans. on Robotics*, 2006.
26. Y. H. Liu, H. S. Wang and K. Lam, "Dynamic visual servoing of robots in uncalibrated environments", *Proc of IEEE Int. Conf. on Robotics and Automation* Barcelona pp.3142-3147, 2005.
27. H. Wang and Y. H. Liu, "Adaptive image-based trajectory tracking of robots", *Proc of IEEE Int. Conf. on Robotics and Automation*, pp.2564-2569, 2005.
28. Y. H. Liu, H. Wang and D. Zhou, "Dynamic tracking of manipulators using visual feedback from an uncalibrated fixed camera," *Proc. of IEEE Int. Conference on Robotics and Automation*, 2006.
29. C.-P. Lu, E. Mjolsness, and G. D. Hager, "Online computation of exterior orientation with application to hand-eye calibration," *Mathematical and Computer Modeling*, vol. 24, no. 5/6, pp. 121-143, 1996.
30. E. Malis, F. Chaumette, and S. Boudet, "Positioning a coarse-calibrated camera with respect to an unknown object by 2D 1/2 visual servoing," *Proc. of IEEE Int. Conf. on Robotics and Automation*, pp. 1352-1359, 1998.
31. E. Malis, "Visual servoing invariant to changes in camera-intrinsic parameters," *IEEE Trans. on Robotics and Automation*, vol. 20, no.1, pp. 72-81, 2004.
32. E. Malis and F. Chaumette, "Theoretical improvements instability analysis of a new class of model-free visual servoing methods," *IEEE Trans. on Robotics and Automation*, vol. 18, no. 2, pp. 176-186, 2002
33. P. Martinet and J. Gallice, "Position based visual servoing Using a non-linear approach," *Proc. of IEEE/RSJ Int. Conf. on Intelligent Robots and Systems*, pp. 531-536, 1999.
34. Maruyama, A. and Fujita, M, "Robust visual servo control for planar manipulators with the eye-in-hand configurations," *Proc. of IEEE Conf. on Decision and Control*, pp. 2551 – 2552, 1997.
35. N. P. Papanikolopoulos and P. K. Khosla, "Adaptive robotic visual tracking: theory and experiments," *IEEE Trans. on Automatic Control*, vol. 38, no. 3, pp. 429-445, 1993.
36. N. P. Papanikolopoulos, B. J. Nelson, and P. K. Khosla, "Six degree-of-freedom hand/eye visual tracking with uncertain parameters," *IEEE Trans. on Robotics and Automation*, vol. 11, no. 5, pp. 725-732, 1995.
37. J. A. Piepmeier, G. V. McMurray and H. Lipkin, "Uncalibrated dynamic visual servoing," *IEEE Trans. on Robotics and Automation*, vol. 20, no.1, pp. 143-147, 2004.
38. A. Ruf, M. Tonko, R. Horaud, and H.-H. Nagel, "Visual tracking of an end-effector by adaptive kinematic prediction," *Proc. of IEEE/RSJ Int. Conf. on Intelligent Robots and Systems*, pp. 893-898, 1997.
39. Y. Shen, D. Sun, Y. H. Liu and K. Li, "Asymptotic trajectory tracking of manipulators using uncalibrated visual feedback," *IEEE/ASME Trans. on Mechatronics*, vo. 8, no. 1, pp. 87-98, 2003.
40. J. J. Slotine and W. Li, "On the adaptive control of robot manipulators," *Int. J. Robotics Research*, vol. 6, pp. 49-59, 1987.

41. W. J. Wilson, C. C. W. Hulls, and G. S. Bell, "Relative end-effector control using Cartesian position based visual servoing," *IEEE Trans. on Robotics and Automation,* vol. 12, no. 5, pp. 684-696, 1996.
42. D. Xiao, B. Ghosh, N. Xi, and T. J. Tarn, "Intelligent robotic manipulation with hybrid position/force control in an uncalibrated workspace," *Proc. of 1998 IEEE Int. Conf. on Robotics and Automation,* pp. 1671-1676, 1998.
43. B. H. Yoshimi and P. K. Allen, Active, "Uncalibrated visual servoing," *Proc. of IEEE Int. Conf. on Robotics and Automation,* pp. 156-161, 1994.
44. E. Zergeroglu, D. M. Dawson, M. S. de Queiroz, and S. Nagarkatti, "Robust visual-servo control of robot manipulators in the presence of uncertainty," *Proc. of the 38th IEEE Int. Conf. on Decision and Control,* pp. 4137-4142, 1999.

Orthogonalization Principle for Dynamic Visual Servoing of Constrained Robot Manipulators

Vicente Parra-Vega and Emmanuel Dean-Leon

Mechatronics Division - Research Center for Advanced Studies (CINVESTAV)
IPN Av. 2508, San Pedro Zactenco, Mexico City, 07300, Mexico.
Email:(vparra,edean)@cinvestav.mx

Summary. A monocular visual servoing scheme for constrained robots is considered in this chapter. Inspired by the Orthogonalization Principle (OP) introduced by Suguru Arimoto in the context of robot force control, a Visual Orthogonalization Principle (VOP) is proposed and a novel control scheme for adaptive image-based visual servoing is presented. The scheme guarantees a global *exponential convergence* for the image-based position-velocity and contact forces even when the robot parameters are considered unknown. The stability of the new control scheme is tested under experiments. The experimental results comply to the theoretical considerations.

1 Introduction

Since the publication of the work [1] in 1977 the problem of robot force control represented a tremendous challenge in the robotics and control community for over 15 years [2], [3], [4]. Many control schemes were proposed to deal with the problem on how to exert force by a robot manipulator over a rigid surface while simultaneously moving its end-effector along this surface. The main difficulty in this problem stemmed from the fact that when the robot is in contact to a rigid surface, modeled by an implicit equation, it is geometrically constrained. This can be modeled by the algebraic differential equations (DAE). Recognizing this fact, a new controller based on the DAE formulation was introduced in [5]. There, it was proposed to project the n degrees of freedom robot dynamics into two orthogonal subspaces complements, one related to the force signals and the other related to the position signals, to derive a global, asymptotically stable simultaneous force-position controller. However, this approach results in a computationally expensive control scheme.

It was not until 1992 that a physically-based principle was introduced [6] to formally solve this problem in the powerful settings of the passivity-based control theory. The principle is based the fundamental interpretation, made by Suguru Arimoto, of what physically happens when the end-effector of a rigid body system with n degrees of freedom comes in contact to, and establish motion over, a rigid surface. This principle was coined as the "Orthogonalization Principle" (OP) and applied to the control of robot manipulators constrained by a rigid surface modeled by an implicit equation. In [6], the OP was introduced to produce a simple nominal reference based on two orthogonal complements without projecting and decomposing the robot dynamics. The result was a local asymptotically stable force-position controller. Later on, this basic principle allowed the synthesis of motion control for such complex systems as cooperating robot arms [7] and multi-fingered robotic hands [8]. In this chapter, an extension of the OP is introduced to tackle the problem of the passivity-based monocular adaptive visual servoing for constrained robot manipulators. It is shown that the extended OP produces second order sliding modes, guaranteeing a global exponential tracking.

2 The Orthogonalization Principle: Robot Force Control

2.1 Constrained Robot Dynamics

Consider a robot whose gripper maintains a stable contact to an infinitely rigid surface. According to [6], this system is modeled by the following nonlinear differential algebraic equations

$$H(q)\ddot{q} + C(q,\dot{q})\dot{q} + g(q) = \tau + J_{\varphi+}^T(q)\lambda \tag{1}$$

$$\varphi(q) = 0 \tag{2}$$

where the generalized joint position $q \in \Re^n$ and the joint velocity $\dot{q} \in \Re^n$. In (1), matrix $H(q) \in \Re^{n \times n}$ stands for the robot inertial matrix; $C(q,\dot{q})\dot{q} \in \Re^n$ stands for the vector of centripetal and Coriolis torques; $g(q) \in \Re^n$ is the vector of gravitational torques; $J_{\varphi+}(q) = \frac{J_\varphi(q)}{J_\varphi(q)J_\varphi^T(q)} \in \Re^{n \times m}$ is the constrained normalized Jacobian of the kinematic constraint $\varphi(q) = 0$ (rigid surface with continuous gradient); $\lambda \in \Re^m$ is the constrained Lagrangian multiplier for m contact points (magnitude of the contact force); $\varphi(q) = 0 \in \Re^m$ models the surface (for m independent contact points), and finally $\tau \in \Re^n$ stands for the vector of the joint torques.

2.2 The Orthogonalization Principle

According to the forward kinematic mapping $X = f(q) \in \Re^n$. Since $\varphi(q) = g(X) \equiv 0$, we have

$$\varphi(q) = 0 \to \frac{d}{dt}\varphi(q) = \frac{\partial g(X)}{\partial X}\frac{\partial X(f(q))}{\partial q}\frac{dq}{dt} \equiv 0$$

$$\Longrightarrow \frac{d}{dt}\varphi(q) = J_g(x) J_x(q)\dot{q}$$

$$= J_\varphi(q)\dot{q} \equiv 0 \qquad (3)$$

This means that $J_\varphi(q)$ is orthogonal to $\dot{q}$ in the joint space. Thus, $\dot{q}$ belongs to the kernel of $J_\varphi(q)$. However, it is well known from the classical mechanics that the vector of the generalized velocities lies in the tangent space at the contact point. Therefore, $\dot{q} = Q\dot{q}$, where $Q \in \Re^{n\times n}$ stands for the generator of the null space of $J_\varphi(q)$, with $J_\varphi(q)$ being orthogonal to Q. In words, the OP states that $\dot{q}$ can be decomposed of the direct summation of two components, one in the velocity subspace Q and the other in the force subspace $J_\varphi(q)$. The nominal reference for $\dot{q}$ can be constructed similarly. Therefore, a unique orthogonalized velocity joint error signal can be introduced to build a unique open loop error dynamics depending on both the velocity and force error signals. This is the key idea of the seminal paper [6]. Using the OP in the physical interpretation of Arimoto [6], one can build a unique nominal reference in terms of the two orthogonal errors and avoid decomposing the full nonlinear robot dynamics as proposed in [5].

2.3 Passivity of Constrained Robot Dynamics

The integral of the dot product of $\dot{q}$ and τ yields

$$\int_{t_0}^{t_f} \dot{q}^T \tau = E(t_f) - E(t_0) - \underbrace{\dot{q}^T J_{\varphi+}^T(q)\lambda}_{\text{zero}} \leq -E(t_0)$$

where $E(t)$ is the total energy of the robot. Note that the antisymmetry of $[\dot{H}(q) - (C(q,\dot{q}) + C(q,\dot{q})^T)]$ is used in the derivation of this equation. The passivity of the robot dynamics is established then from the joint velocity input $\dot{q}$ to the torque output τ [1]. Since in the force control the objective is the convergence of the position/velocity tracking errors simultaneously with the force tracking error, a unique error signal at the velocity level must be established to conform to the passivity inequality in the closed-loop control. This unique error signal is introduced [6] via the so-called nominal reference $\dot{q}_r$ based on the OP:

$$\dot{q}_r = \dot{q}_v + \dot{q}_f \equiv Q\dot{q}_{ev} + J_\varphi^T \dot{q}_{ef} \qquad (4)$$

Equation (4) depends on the orthogonal nominal references for the velocity $\dot{q}_v$ and force $\dot{q}_f$. Since $\dot{q} = Q\dot{q}$, the extended error surface can be defined follows

---

[1] Notice that if the robot dynamics equation includes a viscous friction term, then the dissipativity is established.

$$S_q = \dot{q} - \dot{q}_r \equiv Q(\dot{q} - \dot{q}_{ev}) - J_\varphi^T \dot{q}_{ef} \tag{5}$$

To design a passivity-based controller, the linearity of the left hand side of the robot dynamics in a set of its physical parameters can be used to define:

$$H(q)\ddot{q}_r + C(q,\dot{q})\dot{q}_r + g(q) = Y_r \Theta \tag{6}$$

Here, $Y_r = Y_r(q, \dot{q}, \dot{q}_r, \ddot{q}_r) \in \Re^{n \times p}$ is the dynamic regressor matrix, and $\Theta \in \Re^p$ stands for the unknown (but constant) vector of the robot parameters. Adding and subtracting (6) to (1) produces the open loop error equation

$$H(q)\dot{S}_q = -C(q,\dot{q})S_q + \tau + J_{\varphi+}^T(q)\lambda - Y_r \Theta$$

Now, if

$$\tau = -K_d S_q + Y_r \Theta - J_{\varphi+}^T(\lambda_d - \eta \Delta F) \tag{7}$$

where $\Delta F = \int_{t_0}^{t_f} (\lambda(\sigma) - \lambda_d(\sigma)) d\sigma$, $\lambda_d(t)$ is the desired magnitude of the contact force, K_d is a symmetric positive definite 2×2 feedback gain, and $\eta > 0$, the simultaneous local asymptotic convergence of $\lambda \to \lambda_d$ and $\dot{q} \to \dot{q}_{ev}$ is assured. If $\dot{q}_{ev} = \dot{q}_d - \Omega(q - q_d)$ for $\Omega \in \Re_+^{2 \times 2}$, then $q \to q_d$, where q_d stands for the desired motion of the end-effector on the surface [6].

To extend the previous result to the visually driven force control one needs to address two problems. The first one is how to redesign (4) in terms of image coordinates, and the second one is how to produce a visual-based control law (7) to guarantee the simultaneous convergence of the contact force error and the visual coordinate errors in the presence of the robot parametric uncertainties. To this end, we introduce the Visual Orthogonalization Principle (VOP).

3 The Visual Force Control Problem

Similar to the standard force control problem, the OP naturally arises in visual servoing of constrained robot manipulators; the joint space orthogonalization is preserved since the video camera does not introduces any additional dynamics. The problem now is how to synthesize the joint torque input in terms of the desired visual trajectories and guarantee that the contact force error and the visual position error converge simultaneously. This is a very important problem in modern applications, wherein non-invasive sensors, like CCD cameras, are used to guide the system under human surveillance and supervision. In this case, the robot moves along the surface and the camera captures its motion by the optical flow. To solve the control problem at hand, the OP has to be reformulated in the context of the image-based visual coordinates to incorporate visual errors to the nominal reference. This gives rise to the VOP. Note that this type of robot control tasks involves significant difficulties. It stands as a robot control paradigm that surpasses traditional schemes in robot control and sensor fusion, thus requiring new theoretical frameworks. In our problem, the VOP fuses generalized sensors (measurement of encoder, tachometer, moment and force), and non-generalized sensors (CCD camera) through the nominal reference $\dot{q}_r$. To continue the explanation, it is necessary to briefly review the dynamic visual servoing.

3.1 Visual Servoing

Visual servoing is an ill-posed control scheme because measurements from the camera do not deliver directly the state of the system, and thus cannot be modified directly by the control input to the robot. Besides, the optical flow (velocity of the visual landmarks) is not orthogonal to the joint torque input. To make clear the choice between the position-based and image-based visual servoings, the following features of these techniques [9] should be noted:

1. **Position-based servoing**: The image coordinates are transformed into generalized coordinates to compute the control laws. This approach is prone to errors due to this transformation and is computationally difficult.
2. **Image-based servoing**: The target to be tracked is captured and the computed error in the image plane is obtained. Then, the joint control input is synthesized to ensure asymptotic behavior of the visual error. This approach is robust to the camera calibration since the tracking error remains in the visual coordinates.

The image-based visual servoing is more practical because, in addition to the arguments pointed in item 2, the user can input the desired position directly in the image task space, i.e., directly from the image she/he sees. The research on visual servoing started with the pioneering work [10] and so far several authors extended the scope [11]~[18]. In 1993, the authors of [11] proposed a model and an adaptive control scheme for an eye-in-hand system where the depth of each feature was estimated at each sampling time. In [12], the authors introduced a new technique called a visual compliance that was achieved by a hybrid vision/position control structure. Some authors included the nonlinear robot dynamics in the control design [14]~[17]. Some of them modeled the vision system by a simple rotation matrix [14], others proposed a variety of techniques for the off-line camera calibration [13], and only a few approaches were aimed at the more important problem of the on-line calibration under the closed loop control. Specifically, for a *fixed camera* configuration, the authors of [15] considered a more representative model of the camera-robot system to design a control law that compensates for unknown intrinsic camera parameters but requires the exact knowledge of the camera orientation.

Later, the authors of [16] presented a redesigned control law that also takes into account uncertainties in the camera orientation. The control law features the local asymptotic stability but requires the perfect knowledge of the robot gravitational terms, and the error of the estimation of camera orientation is restricted to $(-90^o, 90^o)$. Further developments were presented in [13], wherein a position tracking control scheme with an on-line adaptive calibration of the camera that guaranteed the global asymptotic position tracking was presented. Nevertheless, this approach requires the knowledge of the robot dynamics and the desired trajectories need to be persistently excited. In [17], the authors designed an adaptive camera calibration control law that compensates for the uncertain camera parameters and the entire robot-camera system, achieving the global *asymptotic* position tracking. Recently, a robust and continuous joint PID-like controller was introduced in [18]. This scheme guarantees the exponential convergence of the image-based tracking errors, in spite of the lack of the knowledge of the camera and robot parameters. In comparison to the approaches considered above, it does not presents limitations on the camera orientation.

Despite of the availability of various approaches considered in this section, none of them fuses the force information for the image-based tracking of the constrained robot systems. The robot control problem is still elusive, though [19] lights the path for the passivity-based dynamic tracking in visual servoing schemes. However, it fails when the camera angle is close to π [19].

3.2 Fusing Visual and Joint Signals

When only the sensors associated with the generalized coordinates are involved in robot force control, the OP unobtrusively provides a harmonious unique error signal, combining the position and contact force errors. However, when the robot tasks involve also non-generalized sensors[2], the control law must deal with the multisensor fusion of the force and joint encoders signals along with the visual information. Therefore, in order to implement a sensor fusion-based controller, a careful and judicious analysis of the robot nonlinear dynamics, sensors behavior, and the contact tasks is required. To continue, let us review briefly some visual-based force servoing schemes.

3.3 Visual Force Servoing

The reference [21] focuses on the sensor fusion of the force and visual landmarks. The authors of [22] study a visual contour tracking in a structured environment. In [23], the authors present an adaptive robot controller to realize contact tasks in an unknown environment. In [23] it is assumed that the movement of the camera-manipulator system is slow and the mapping from the joint space to the image space is constant, which severely limits the system performance. Along similar developments, the paper [20] presents a computed torque scheme for an uncalibrated environment, which requires the

---

[2] For example a CCD camera. The non-generalized sensors do not directly measure the state variables of the robot dynamic equations.

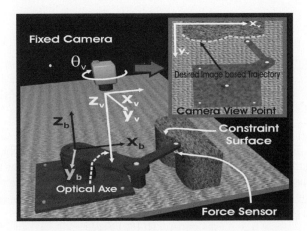

Fig. 1. Visual servoing of the contact task.

exact knowledge of the robot dynamics and relies upon a very complex control law. The control laws in [23] and [20] require complex computations and do not fully solve the control problem posed above.

4 Dynamics of the Visually Driven Constrained Robot

Here, a monocular visual servoing scheme is presented[3]. In order to design a proper *nominal reference* for the joint velocities $\dot{q}_r$, the direct and inverse robot kinematics, based on the static pin hole, with thin lens without aberration camera model [9], is used. Let the direct kinematics of the robot be

$$x_b = f(q) \tag{8}$$

where $x_b \in \Re^2$ represents the 2D position of robot end-effector in the robot work space and $f(\cdot) : \Re^2 \to \Re^2$. The differential kinematics of the robot manipulator is defined as follows

$$\dot{x}_b = J(q)\dot{q} \tag{9}$$

It relates the Cartesian velocities $\dot{x}_b \in \Re^2$ to the joint space velocities $\dot{q} \in \Re^2$. The visual position $x_v = [u, v]^T \in \Re^2$ of the robot end-effector in the image (screen) space[4] is given as follows [9]

---

[3] Explicitly, this means that the 2D case is considered and the image plane is parallel to the 2D work plane of the nonlinear DAE robot. Notice, however, that the extension to the 3D case, though straightforward, requires a stereo camera model with additional considerations for the (pseudo)inverse of the differential kinematic mapping.

[4] The subscript v of x_v denotes *visual* from the *visual space* notation.

$$x_v = \alpha h(z) R(\theta_v) x_b + \beta \qquad (10)$$

where $\alpha = \text{diag}[\alpha_u, \alpha_v] \in \Re_+^{2\times 2}$ is the scale factor; $h(z) = \frac{\xi}{\xi-z} < 0, z > \xi$. Here, ξ stands for the focal distance, and $R(\theta_v) \in SO(2)$ is the upper left 2×2 matrix of $R_3(\theta_v) \in SO(3)$; $\beta \in \Re_+^2$, depends on the intrinsic and extrinsic parameters[5] of the camera. The differential kinematic model of the camera is defined as follows

$$\dot{x}_v = \alpha h(z) R(\theta_v) \dot{x}_b \qquad (11)$$

where $\dot{x}_v \in \Re^2$ determines the visual robot end-effector velocity, i.e. the *visual flow*. Notice that the the transformation $\alpha h(z) R(\theta)$ is constant and it maps statically the robot *Cartesian velocities* $\dot{x}_b$ into the *visual flow* $\dot{x}_v$. By using (9), equation (11) becomes

$$\dot{x}_v = \alpha h(z) R(\theta_v) J(q) \dot{q} \qquad (12)$$

It relates the *visual flow* $\dot{x}_v$ to the *joint velocity vector* $\dot{q}$. Thus, in terms of the visual velocities[6] the inverse differential kinematics (12) becomes

$$\dot{q} = J_{Rinv} \dot{x}_v \qquad (13)$$

where $J_{Rinv} = J(q)^{-1} R(\theta)^{-1} h(z)^{-1} \alpha^{-1}$. This relation is useful in designing the nominal reference for the joint velocities $\dot{q}_r$.

4.1 Visual Orthogonalization Principle (VOP)

Since the robot end-effector is in contact to the constrain surface, $\varphi(q) = 0 \, \forall t$ and the OP explains the implications of $\frac{d}{dt} \varphi(q) = 0$. Therefore, using (13) we obtain

$$J_\varphi(q) \dot{q} = \dot{q} \equiv J_\varphi J_{Rinv} \dot{x}_v \doteq 0$$

This means that $J_\varphi(q)$ is orthogonal to the optical flow $\dot{x}_v$ mapped into the joint space. Clearly, there exists an orthogonal projection Q of $J_\varphi(q)$ which spans the tangent space at the contact point between the end-effector and the surface $\varphi(q) = 0$. In other words, (13) and $\dot{q} = Q\dot{q}$ leads to

$$\dot{q} = Q J_{Rinv} \dot{x}_v \qquad (14)$$

From (4) and (14), the nominal reference for VOP becomes

$$\dot{q}_r = Q J_{Rinv} \dot{x}_r + \dot{q}_f \qquad (15)$$

Finally

$$\dot{q}_r = Q J_{Rinv} \dot{x}_r + \varGamma_{F_2} J_\varphi^T(q) \dot{q}_{rf} \qquad (16)$$

---

[5] The focal distance, the depth of the field, the translation of the camera center to the image center, and the distance between the optical axis and the robot base.
[6] The entries of $J_{Rinv} \in \Re^{2\times 2}$ are functions of the robot and camera parameters.

where $\Gamma_{F_2} > 0$ is a 2×2 feedback gain. Let the *nominal visual reference* for the velocities be

$$\dot{x}_r = \dot{x}_{vd} - \Psi \Delta x_v + S_{vd} - \Gamma_{v_1} \int_{t_0}^{t} S_{v\delta}(\zeta) d\zeta - \Gamma_{v_2} \int_{t_0}^{t} \text{sign}\left[S_{v\delta}(\zeta)\right] d\zeta \quad (17)$$

where $\dot{x}_{vd}$ stands for the desired visual velocity trajectory, and $\Delta x_v = x_v - x_{vd}$ is the visual position tracking error, $\Psi > 0$ a positive definite $n \times n$ feedback gain, and $\Gamma_{v_i} = \Gamma_{v_i}^T \in \Re_+^{n \times n}, i = 1, 2$. Let the *visual error surface* be

$$S_{v\delta} = \underbrace{(\Delta \dot{x}_v + \Psi \Delta x_v)}_{S_v} - \underbrace{S_v(t_0) e^{-\kappa_v t}}_{S_{vd}} \quad (18)$$

where $\Delta \dot{x}_v = \dot{x}_v - \dot{x}_{vd}$ defines the visual velocity tracking error, $\kappa_v > 0$. Now, consider the following *nominal force reference*

$$\dot{q}_{rf} = \Delta F - S_{Fd} + \Gamma_{F_1} \int_{t_0}^{t} S_{F\delta}(\zeta) d\zeta + \Gamma_{F_2} \int_{t_0}^{t} \text{sign}\left[S_{F\delta}(\zeta)\right] d\zeta \quad (19)$$

for the *force error surface*

$$S_{F\delta} = \underbrace{\Delta F}_{S_F} - \underbrace{S_F(t_0) e^{-\kappa_F t}}_{S_{Fd}} \quad (20)$$

where

$$\Delta F = \int_{t_0}^{t} \Delta \lambda(\zeta) d\zeta, \quad \Delta \lambda = \lambda - \lambda_d,$$

$\Delta \lambda$ is the force tracking error and λ_d is the desired contact force; $\kappa_F > 0$, and $\Gamma_{F_i} = \Gamma_{F_i}^T \in \Re_+^{m \times m}, i = 1, 2$. Using equations (16), (17), (19) and (14), we obtain the following representation for the *joint error surface* $S_q = \dot{q} - \dot{q}_r$

$$S_q = QJ_{Rinv}(\dot{x}_v - \dot{x}_r) - \Gamma_{F_2} J_\varphi^T(q) \dot{q}_{rf}$$
$$= QJ_{Rinv} S_{vv} - \Gamma_{F_2} J_\varphi^T(q) S_{vF} \quad (21)$$

where

$$S_{vv} = S_{v\delta} + \Gamma_{v_1} \int_{t_0}^{t} S_{v\delta}(\zeta) d\zeta + \Gamma_{v_2} \int_{t_0}^{t} \text{sign}\left[S_{v\delta}(\zeta)\right] d\zeta \quad (22)$$

$$S_{vF} = S_{F\delta} + \Gamma_{F_1} \int_{t_0}^{t} S_{F\delta}(\zeta) d\zeta + \Gamma_{F_2} \int_{t_0}^{t} \text{sign}\left[S_{F\delta}(\zeta)\right] d\zeta \quad (23)$$

S_{vv} stands for the *extended visual manifold*, and S_{vF} stands for the *extended force manifold*.

Remark 1. Notice that S_q is composed of two orthogonal complements. The first, $QJ_{Rinv} S_{vv}$, depends on the image coordinate errors, and the second, $\Gamma_{F_2} J_\varphi^T(q) S_{vF}$, depends on the integral of the contact force errors. Thus, the tracking errors $(\Delta x_v, \Delta \dot{x}_v)$ and ΔF can be controlled independently within a single control loop.

4.2 Global Decomposition of Joint Space

Consider the following partition of the joint space q [5], [24]

$$q = [q_1, q_2]^T \tag{24}$$

where $q_1 \in \Re^m$, and $q_2 \in \Re^{n-m}$. Since the constraint $\varphi(q) \in \Re^m$, there are m dependent states which are defined in (24) as q_1. This partition is not arbitrary. Thus, to identify q_1 the Jacobian of the restriction $J_\varphi(q)$ and the Gauss decomposition are used in order to define a non-singular matrix $m \times m$. The generalized coordinates defined by the choice of this matrix are indeed q_1 [25]. According to the implicit function theorem, there exist, locally, an open group $O \in \Re^{n-m}$ and a function $\Omega : \Re^{n-m} \to \Re^m$ such that

$$q_1 = \Omega(q_2) \tag{25}$$

Then, $\varphi(q) = \varphi(\Omega(q_2), q_2) = 0 \ \forall q_2 \in O$. Using the time derivative of (2) and its partitioning (24), we obtain

$$J_\varphi(q)\dot{q} = [J_{\varphi_1}(q)\dot{q}_1 + J_{\varphi_2}(q)\dot{q}_2] \equiv 0 \tag{26}$$

where $J_{\varphi_1}(q)\dot{q}_1 \in \Re^m$ and $J_{\varphi_2}(q)\dot{q}_2 \in \Re^m$. Solving (26) for $\dot{q}_1$ defines

$$\dot{q}_1 = \Omega_{q_2}\dot{q}_2, \quad \text{where} \quad \Omega_{q_2} = -[J_{\varphi_1}(q)]^{-1} J_{\varphi_2}(q)$$

for $\Omega_{q_2} : \Re^{n-m} \to \Re^m$. Then, joint velocities are built upon the independent coordinates

$$\dot{q} = Q\dot{q}_2, \quad \text{where} \quad Q = [\Omega_{q_2}, \ I_{n-m}]^T \tag{27}$$

and $Q \in \Re^{n \times (n-m)}$ is a full column matrix of rank $(n-m)$. Then, Q is well posed, rank $(\varphi(q)) = m$ and $(J_{\varphi_1}(q))^{-1}$ exists in the finite workspace of (1). Notice again that Q spans the tangent plane at the contact point, and, therefore, $J_\varphi(q)$ and Q are the orthogonal complements. i.e., $QJ_\varphi^T(q) = 0$. Therefore, $J_{\varphi_1}(q) \in \ker(Q)$, and the space $\Re^n$ is decomposed into the two orthogonal subspaces, $\Re^n = \Re(J_\varphi) \oplus \Re(Q)$.

Remark 2. Using a generalization of the implicit function theorem, we can state that Q is well posed $\forall q \in \Omega_q$, where $\Omega_q = \{q | \text{rank}(J(q)) = n, \forall q \in \Re^n\}$ stands for the robot workspace free of singularities. This defines a global decomposition for the 2D case. However, for the 3D case this approach will require an efficient algorithm to compute the independent coordinates on line because the solution of the implicit equation may be not unique. Note, however, that the numerical stability of the decomposition is preserved [25].

4.3 Open Loop Error Equation

Due to the fact that the linear parametrization $Y_r \Theta$ depends on $\ddot{q}_r = f(\ddot{x}_r, \ddot{q}_{rf})$, the computation of $\ddot{q}_{rf}$ and $\ddot{x}_r$ gives

$$\ddot{x}_r = \ddot{x}_{vd} - \Psi \Delta \dot{x}_v + \dot{S}_{vd} - \Gamma_{v_1} S_{v\delta} - \Gamma_{v_2} sign(S_{v\delta}) \qquad (28)$$

$$\ddot{q}_{rf} = \Delta \dot{F} - \dot{S}_{dF} + \Gamma_{F_1} S_{F\delta} + \Gamma_{F_2} sign(S_{F\delta}) \qquad (29)$$

which introduces discontinuous terms. To avoid introducing high frequency discontinuous signals, we need to get rid of discontinuous signals in $Y_r \Theta$. To this end, add and subtract $\tanh(\mu_v S_{v\delta})$ and $\tanh(\mu_F S_{F\delta})$ to $\ddot{q}_r$, assuming that $\mu_F > 0, \mu_v > 0$. Then, $\ddot{q}_r$ becomes

$$\ddot{q}_r = \ddot{q}_{rc} + Q\Gamma_{v_2} z_v - J_\varphi^T(q) \Gamma_{F_2} z_F \qquad (30)$$

with $z_v = \tanh(\mu_v S_{v\delta}) - sign(S_{v\delta})$ and $z_F = \tanh(\mu_F S_{F\delta}) - sign(S_{F\delta})$, and

$$\ddot{q}_{rc} = Q J_{Rinv} \ddot{x}_{rc} + \dot{Q} J_{Rinv} \dot{x}_{rc} + Q \dot{J}_{Rinv} \dot{x}_{rc} + \Gamma_{F_2} J_\varphi^T(q) \ddot{q}_{rfc} + \Gamma_{F_2} \dot{J}_\varphi^T(q) \dot{q}_{rfc}$$

for

$$\ddot{x}_{rc} = \ddot{x}_{vd} - \Psi \Delta \dot{x}_v + \dot{S}_{vd} - \Gamma_{v_1} S_{v\delta} - \Gamma_{v_2} \tanh(\mu_v S_{v\delta}) \qquad (31)$$

$$\ddot{q}_{rfc} = \Delta \dot{F} - \dot{S}_{dF} + \Gamma_{F_1} S_{F\delta} + \Gamma_{F_2} \tanh(\mu_F S_{F\delta}) \qquad (32)$$

Therefore, the linear parametrization (6) becomes

$$H(q)\ddot{q}_r + C(q,\dot{q})\dot{q}_r + g(q) = Y_c \Theta_c + H(Q\Gamma_{v_2} z_v - J_\varphi^T(q) \Gamma_{F_2} z_F) \qquad (33)$$

In this formulation, $Y_c = Y_r(q, \dot{q}, \dot{q}_r, \ddot{q}_{rc})$ is continuous since $(\dot{q}_r, \ddot{q}_{rc}) \in \mathcal{C}^1$, where $Y_c \Theta_c = H(q)\ddot{q}_{rc} + C(q,\dot{q})\dot{q}_{rc} + g(q)$. Adding and subtracting (33) to (1), we obtain the following open loop error equation

$$H(q)\dot{S}_q = \tau - C(q,\dot{q})S_q + J_{\varphi+}^T(q)\lambda - Y_c \Theta_c + H(Q\Gamma_{v_2} z_v - J_\varphi^T(q) \Gamma_{F_2} z_F) \qquad (34)$$

We are ready to present the main result.

5 Control Design

Assume that $(x_v, \dot{x}_v)$ can be measured by the camera, $(q, \dot{q})$ can be measured by, respectively, encoders and tachometers, and (λ, F) can be measured by a force sensor. Assume also that the desired image trajectory is free of singularities, i.e., $(x_{vd}, \dot{x}_{vd}) \in \Omega_x$, for $\Omega_x = \{x_v | rank(J_{Rinv}) = 2, \forall x_v \in \Re^2\}$, and $\lambda_d \in \mathcal{C}^1$. Then, we have the following theorem.

Theorem 1. *Assume that the initial conditions and the desired trajectories belong to $\Omega_T = [\Omega_q, \Omega_x]$, for $\Omega_q = \{q | rank(J(q)) = 2, \forall q \in \Re^2\}$, and consider the robot dynamics (1)-(2) with the following visual adaptive force-position control law*

$$\tau = -K_d S_q + Y_c \hat{\theta}_b + J_{\varphi+}^T(q) [-\lambda_d + \eta \Delta F]$$

$$+ \Gamma_{F_2} J_{\varphi+}^T(q) \left[\tanh(\mu_F S_{F\delta}) + \eta \int_{t_o}^{t} sgn[S_{F\delta}(\zeta)] d\zeta \right] \qquad (35)$$

$$\dot{\hat{\theta}}_b = -\Gamma Y_c^T S_q \qquad (36)$$

where $\hat{\theta}_b$ is the online estimate of the robot parameters, $\Gamma = \Gamma^T \in \Re_+^{p \times p}$, $K_d = K_d^T \in \Re_+^{n \times n}$, and $\eta > 0$. If K_d is large enough and the errors of initial conditions are small enough, and if

$$\Gamma_{v_2} \geq \left\| \frac{d}{dt} \left[Q_\theta^{\#} S_q \right] \right\|, \quad \Gamma_{F_2} \geq \left\| \frac{d}{dt} \left[J_\varphi^{\#} S_q \right] \right\|$$

where $Q_\theta^{\#} = R_\alpha(\theta_v) J(q) \left(Q^T Q \right)^{-1} Q^T$, and $J_\varphi^{\#}(q) = \left(-\Gamma_{F_1} J_\varphi J_\varphi^T(q) \right)^{-1} J_\varphi$, the global exponential convergence of the visual and force tracking errors is guaranteed for any value of the rotational camera angle Θ_v.

Proof. The proof is based on the Lyapunov stability theory along with the variable structure control theory for second order sliding modes. A brief outline of the proof can be stated as follows:

- Part I: Boundedness of the closed loop trajectories. In this part, the passivity from the joint velocity error input to the torque output is established. If the viscous friction is considered, then the dissipativity is established. This implies that the boundedness of the closed loop signals is proved.
- Part II: Second order sliding modes. Once the boundedness of the input signals is proved, the sliding mode regime for the visual and force subspaces needs to be induced. The proper gains are selected in this part.
- Part III: Exponential convergence of the tracking errors. A proper selection of the gains guarantees the sliding mode for each subspace for all time. Then, we prove that each sliding mode induces the exponential convergence of the visual tracking errors and the force tracking errors for all time.

The details of the proof are given in the Appendix.∎

6 Discussions

6.1 Robustness issues

The closed loop system gives rise to two sliding modes. It is well known that the sliding modes are extraordinary robust to parametric uncertainties for certain classes of bounded unmodeled dynamics.

6.2 Well-Posed Inverse Jacobian

Apparently there can be a problem with $J(q(t))^{-1}$. The visual position exponentially converges to the desired visual position without overshoot, i.e., $x_v(t) \to x_{vd}(t)$, $x_{vd}(t) \in \Omega_x \Longrightarrow x_v(t) \in \Omega_x$. However, it does not guarantee that $J(q(t))^{-1}$ is always well posed only because the joint position q converges to the desired joint position q_d with an exponential envelope. The joint position q may experience a short transient and as a consequence $J(q)$

may loose its rank. However, since $q(t)$ converges to $q_d(t)$ locally, it means that $J(q(t)) \to J(q_d(t))$ within Ω_q. Consequently $J(q(t))^{-1}$ is locally well posed, i.e., $\forall t$ rank $\left[J(q(t))^{-1}\right] = 2$. In addition, in visual servoing tasks it is customary to design the desired trajectories to be within Ω_x, and therefore within Ω_q, away from singular joint configurations.

6.3 Smooth Controller

The continuous function $tanh(*)$ is used instead of $sign(*)$ in the control law without jeopardizing the second order sliding mode. Moreover, notice that the $sign(*)$ is not required to induce a second order sliding mode. This is in contrast to the first order sliding modes theory.

6.4 The Control Structure

The control law features a low computational cost and is easy to implement. The structure of the control law is, basically, similar to that presented in [6], [24] except for the camera information processing part.

7 Experimental System

A planar robot with two degrees of freedom (see Fig. 2) is used in our experiments. The robot and camera parameters are listed in, respectively, Tables 1 and 2. The control feedback gains are listed in Table 3.

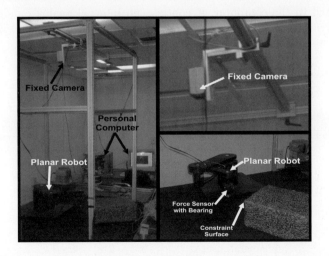

Fig. 2. Experimental setup

7.1 Hardware

Direct-drive Yaskawa AC servomotors SGM-08A314 and SGM-04U3B4L with 2048 pulse encoders are directly coupled to the links of the robot. Digital drives (Yaskawa servopacks SGD-08AS and SGDA-04AS) are integrated into the robot control system. A six-axes force-moment sensor 67M25A-I40-200N12 by JR3 Inc., provided with a DSP Based Interface System for PCI bus, is mounted on the end-effector of the robot. A rigid aluminum probe, with a bearing SKF R8-2Z in its tip, is attached to the end-effector as shown in Fig. 2. The robot task is to move its tool-tip along a specified trajectory over the steel surface while exerting a specified profile of the force normal to the surface. A fixed SONY DFW-VL500 CCD camera is used to capture the position of the robot end effector in the image space (measured in *pixels*). The robot is initialized with a high gain PD control. The inertial frame of the whole system is at the base of the robot, and the contact surface is an XZ plane located at $y = 122\ pixels$.

7.2 Software

A 2.2 GHz personal computer, running on Debian GNU/Linux 3.1 (kernel 2.4-27) with RTAI patch operating system (rtai 3.1.0-4) is used in the experiments. This PC implements two real-time concurrent processes. The first one communicates with the camera via IEEE1394 protocol and controls the acquisition of the robot end-effector position in the image space at a sampling rate of 30 Hz. The second process computes the torque output for the servopacks and runs at a sampling rate of 1 KHz. The communication between the processes is done by allocating a shared memory. A data acquisition board is connected to the computer. It contains an internal analog quadrature encoder interface, 14 bit analog resolution outputs, and digital I/O ports. The

Table 1. Robot Parameters.

Parameter	Mass	Length	Center of Mass	Inertia
Link 1	$7.1956\ Kg$	$0.4\ m$	$0.1775\ m$	$0.2779\ Kgm^2$
Link 2	$1.8941\ Kg$	$0.3\ m$	$0.0979\ m$	$0.02339\ Kgm^2$

Table 2. Camera Parameters.

Parameter	Value
Rotation angle θ_v	90
Scale factor α	$99500\ pixel/m$
Depth field of view z	$1.6\ m$
Camera offset β	$[-335, -218]^T\ pixel$
Focal length ξ	$0.08\ m$

velocity is computed using a dirty Euler numerical differentiation formula filtered with a low pass second order Butterworth filter with a cutoff frequency of 20Hz.

7.3 Control Task

The initial configuration of the robot is shown in the camera image space in Fig. 3.A. Also depicted there are the path of the robot's free motion and the direction of the constrained movement. The control task consists of the following three steps.

1. The end-effector is requested to move until it makes contact with the surface as shown in Fig. 3.B. The free motion time interval is $[0,3]\,s$. Next, within $t \in [3,5]\,s$ the end-effector establishes a stable contact with the constraint surface.
2. The tool-tip exerts a desired profile of the force normal to the surface (from 0 to 7.5 N) while moving forward along the X axis from 403 $pixels$ to 254 $pixels$ (see Fig. 3.C). This is done in the time span $t \in [5,10]\,s$
3. In the time interval $t \in [10,15]\,s$ the exerted force is incremented from 7.5 to 15 N, while moving the tool-tip (see Fig. 3.C) backward along the X axis from 254 $pixels$ to 403 $pixels$.

The desired position and force are both designed with

$$\Phi(t) = P(t)\left[\mathbf{X}_f - \mathbf{X}_i\right] + \mathbf{X}_i, \tag{37}$$

where $P(t)$ is a fifth order polynomial that satisfy $P(t_i) = 0, P(t_f) = 1$ and $\dot{P}(t_i) = \dot{P}(t_f) = 0$. The subscript i and f denote the initial and the final moments, respectively. At the first stage of the control task (free motion), the control law (35)-(36) is used with $J_\varphi^T(q) = 0$ and $Q = I$. The stability of this free motion control scheme is proved in [26].

7.4 Experimental Results

The performance of the simultaneous force and position tracking is illustrated in Fig. 4, 5, 6, and 7. Figure 10 gives an image of the visual tracking of the robot's end effector. The motion of the robot's end-effector in the image space

Table 3. Feedback Gains.

Gain	Value	Gain	Value	Gain	Value	Gain	Value
K_d	$\begin{bmatrix} 14 & 0 \\ 0 & 1.8 \end{bmatrix}$	κ_v	20	$\Gamma_{v_{(1,2)}}$	$\begin{bmatrix} 8 & 0 \\ 0 & 8 \end{bmatrix}$	Γ	1
Ψ	$\begin{bmatrix} 5 & 0 \\ 0 & 5 \end{bmatrix}$	κ_F	20	$\Gamma_{F_{(1,2)}}$	3	η	2.8

is shown in Fig. 5 and 6. Note that the image coordinated system is rotated by θ_v degrees (in this case 90°). The maximum tracking error is 1 *pixel* (near to $0.20mm$). The tracking performance can be improved by using a sub-pixel resolution.

Figure 8 shows the joint torques. As can be seen from Fig. 8, the control output is not saturated. The torque noise in the free motion segment is due to the fact that the control gains are tuned for the position-force control task. These gains are high during the free motion time, and this causes the high response observed in Fig. 8. Fig. 9 depicts the exponential envelope of the Cartesian tracking errors. Fig.4 shows the exerted force profile. As can be seen, from $t = 0s$ to $t = 3s$ the robot's end effector is in free motion (the contact force is near to $0N$) until it makes contact with the surface (an overshoot in the contact force is presented due to contact transition). The end-effector remains in that state 2 more seconds. The applied force is smoothly increased from $0N$ to $7.5N$ while the end-effector moves forward along the X axis in the time interval $[5\ 10]s$. Then, in the time interval $[10\ 15]s$ the applied force is increased from $7.5N$ to $15N$ while the end-effector moves backward. The seemingly high frequency in the force response can be explained by the precision of the sensors. The movement task requires a very precise control,

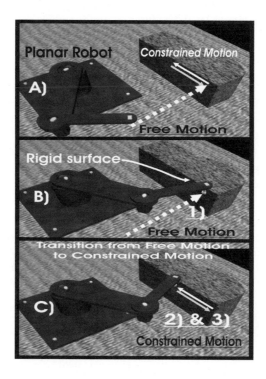

Fig. 3. Experimental phases

but the sensor resolution is limited to 1 *pixel* and the JR3 force sensor noise is $\pm 2N$[7].

8 Conclusions

A novel scheme for the adaptive image-based visual servoing of constrained robots was proposed in this chapter. The new scheme is based on the Visual Orthogonalization Principle (VOP). The main feature of the control scheme is the ability to fuse the *image coordinates* and the *integral of contact forces*. The scheme guarantees a global *exponential convergence* for the image-based position-velocity and the contact forces even when the robot parameters are considered unknown. The experimental results confirm the stability of the control scheme. The novel control scheme can improved to deal with the uncertainties in the description of the constraint surface, the robot Jacobian, and the friction forces. The scheme can be used in a number of control tasks employing the dynamic visual servoing. These task include the cooperative control of multiple robot arms and multi-fingered robotic hands. It can be also used in the control of biped walking machines[8].

Appendix: Proof of Theorem 1

The closed loop dynamics (35)~(36) and (34) yields

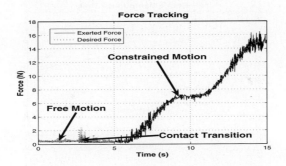

Fig. 4. Force Tracking

---

[7] Better plots can be obtained by simply reducing the desired visual velocity or by increasing $\lambda_d(t)$.
[8] The examples can be found in www.manyrob.cinvestav.mx.

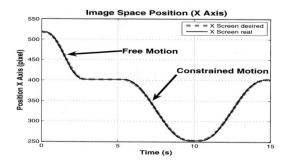

Fig. 5. End effector x position in image space (pixels).

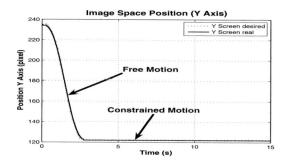

Fig. 6. End effector position in image space (pixels).

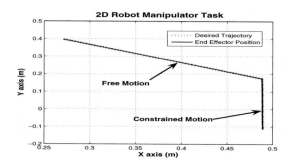

Fig. 7. Cartesian robot task in robot space (m).

$$H(q)\dot{S}_q = -\{K_d + C(q,\dot{q})\}S_q - Yc\Delta\theta_b + J^T_{\varphi+}(q)\left[\Delta\lambda + \Gamma_{F_2}\tanh(\mu_F S_{F\delta})\right]$$
$$+ \eta J^T_{\varphi+}(q)\left[\Delta F + \Gamma_{F_2}\int_{t_0}^{t} sgn\left(S_{F\delta}\left(\zeta\right)\right)d\zeta\right] \tag{38}$$
$$\Delta\dot{\theta}_b = \Gamma Y^T_c S_q \tag{39}$$

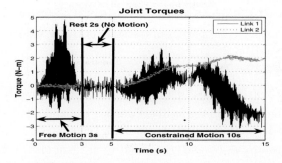

Fig. 8. Input torques.

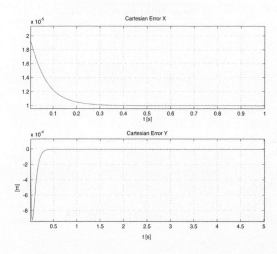

Fig. 9. Cartesian tracking error with exponential envelope.

where $\Delta\theta_b = \theta_b - \hat{\theta}_b$. The proof is organized in three parts.

Part I. Boundedness of the closed loop trajectories. Consider the time derivative of the following *Lyapunov* candidate function

$$V = \frac{1}{2}\left[S_q^T H(q) S_q + \Gamma_{F_2} S_{vF}^T S_{vF} + \Delta\theta_b^T \Gamma^{-1} \Delta\theta_b\right] \quad (40)$$

Along the solutions of (38)-(39) we have

$$\dot{V} \leq -K_d \left\|\hat{S}_q\right\|_2^2 - \Gamma_{F_2} \|S_{vF}\| + \|\hat{S}_q\|\psi \quad (41)$$

where $S_q^T \left(\frac{1}{2}\dot{H}(q) - C(q,\dot{q})\right) S_q = 0$ and ψ is a functional depending on the state and error manifolds [27]. If K_d and Γ_{F_2} are large enough and the errors in initial conditions are small enough, we conclude the semi-negative definiteness of (41) outside of the hyperball $\varepsilon_0 = \{S_q | \dot{V} \leq 0\}$ centered at the origin. Therefore, the following properties of the state of the closed loop system take place

$$(S_q, S_{vF}) \in \mathcal{L}_\infty \to (\|S_{vv}\|, \|S_{vF}\|) \in \mathcal{L}_\infty \tag{42}$$

Then, $\left(S_{v\delta}, \int_{t_0}^{t} \text{sign}\left(S_{v\delta}\left(\zeta\right)\right) d\zeta\right) \in \mathcal{L}_\infty$ and since the desired trajectories are differentiable functions and the feedback gains are bounded, we have $(\dot{q}_r, \ddot{q}_r) \in \mathcal{L}_\infty$. The right hand side of (38) shows that there exists $\varepsilon_1 > 0$ such that $\left\|\dot{S}_q\right\| \leq \varepsilon_1$. Since $S_q \in \mathcal{L}_2$ and J_{Rinv} and Q are bounded, then $QJ_{Rinv}S_{vv}$ is bounded. Since $\varphi(q)$ is smooth and lies in the reachable robot space and $S_{vF} \to 0$, then $J_\varphi^T(q) \Gamma_{F_2} S_{vF} \to 0$. Now, taking into account that $\dot{S}_q$ is bounded, then $\frac{d}{dt}(J_{Rinv}QS_{vv})$ and $\frac{d}{dt}\left(J_\varphi^T(q) \Gamma_{F_2} S_{vF}\right)$ are bounded (this is possible because $\dot{J}_\varphi^T(q)$ is bounded and so is $\dot{Q}$). All these conclusions prove that there exist constants $\varepsilon_2 > 0$ and $\varepsilon_3 > 0$ such that

$$\left|\dot{S}_{vv}\right| < \varepsilon_2, \left|\dot{S}_{vF}\right| < \varepsilon_3$$

So far, the analysis shows only the stability of all the closed-loop signals. Now we prove the appearance of the sliding modes. To this end, we have to prove that for the properly selected feedback gains Γ_{v_2} and Γ_{F_2} the trajectories of the visual position and force converge to zero. This can be done if we can prove that the sliding modes are established in the visual subspace Q and in the force subspace $J_\varphi^T(q)$.

Part II: Second order sliding modes.

Part II.a: Sliding modes for the velocity subspace. From (21) we obtain

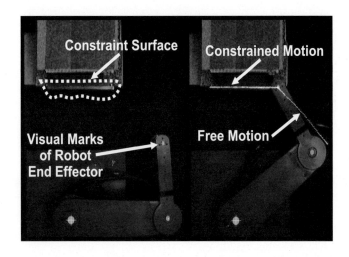

Fig. 10. Camera view point.

$$(Q^T Q)^{-1} Q^T S_q \equiv J_{Rinv} S_{vv} \tag{43}$$

By multiplying (43) by $\alpha h(z) R(\theta) J(q)$ and substituting it into (22), we have

$$Q_\theta^\# S_q = S_{v\delta} + \Gamma_{v_1} \int_{t_0}^t S_{v\delta}(\zeta)\, d\zeta + \Gamma_{v_2} \int_{t_0}^t \text{sign}(S_{v\delta}(\zeta))d\zeta \tag{44}$$

Taking the time derivative of (44), and multiplying it by $S_{v\delta}^T$, we have

$$S_{v\delta}^T \dot{S}_{v\delta} = -\gamma_{v_2} S_{v\delta}^T \text{sign}(S_{v\delta}) - \Gamma_{v_1} S_{v\delta}^T S_{v\delta} + S_{v\delta}^T \frac{d}{dt}\left[Q_\theta^\# S_q\right]$$

$$\leq -\mu_v |S_{v\delta}| - \Gamma_{v_1} \|S_{v\delta}\|^2 \tag{45}$$

where $\mu_v = \Gamma_{v_2} - \varepsilon_4$, and $\varepsilon_4 \geq \left|\frac{d}{dt}\left[Q_\theta^\# S_q\right]\right|$. Thus, we obtain the sliding condition $\Gamma_{v_2} > \varepsilon_4$. Therefore, $\mu_v > 0$ in (45) guarantees a sliding mode at $S_{v\delta} = 0$ when $t_v = \frac{|S_{v\delta}(t_0)|}{\mu_v}$. However, notice that for any initial condition $S_{v\delta}(t_0) = 0$ we have $t_v = 0$, which implies that the sliding mode at $S_{v\delta}(t) = 0$ is guaranteed for all time.

Part II.b: Sliding modes for the force subspace. In much the same way as has been done in Part II.a, we process equation (21) to obtain

$$J_\varphi^\#(q) S_q = S_{F\delta} + \Gamma_{F_1} \int_{t_0}^t S_{F\delta}(\zeta))d\zeta + \Gamma_{F_2} \int_{t_0}^t \text{sign}(S_{F\delta}(\zeta))\, d\zeta \tag{46}$$

Taking the time derivative of (46) and multiplying it by $S_{F\delta}^T$, (46), we have

$$S_{F\delta}^T \dot{S}_{F\delta} = -\Gamma_{F_2} |S_{F\delta}| - \Gamma_{F_1} S_{F\delta}^T S_{F\delta} + S_{F\delta}^T \frac{d}{dt}\left(J_\varphi^\#(q) S_q\right) \tag{47}$$

$$\leq -\Gamma_{F_2} |S_{F\delta}| - \Gamma_{F_1} \|S_{F\delta}\|^2 + |S_{F\delta}| \frac{d}{dt}\left(J_\varphi^\#(q) S_q\right) \tag{48}$$

$$\leq -\mu_F |S_{F\delta}| - \Gamma_{F_1} \|S_{F\delta}\|^2 \tag{49}$$

where $\mu_F = \Gamma_{F_2} - \varepsilon_5$, and $\varepsilon_5 \geq \frac{d}{dt}\left[J_\varphi^\#(q) S_q\right]$. If $\Gamma_{F_2} > \varepsilon_5$, then a sliding mode at $S_{F\delta}(t) = 0$ is induced at $t_f \leq \frac{|S_{F\delta}(t_0)|}{\mu_F}$, but $S_{F\delta}(t_0) = 0$ and thus $S_{F\delta}(t_0) = 0$ is guaranteed for all time.

Part III: Exponential convergence of tracking errors.

Part III.a: Visual tracking errors. Since a sliding mode exists for all time at $S_{v\delta}(t) = 0$, then we have

$$S_v = S_{vd} \quad \forall t \to \Delta \dot{x}_v = -\Psi \Delta x_v + S_v(t_0) e^{-\kappa_v t}$$

This implies that the visual tracking errors globally and exponentially converge to zero and $x_v \to x_{vd}, \dot{x}_v \to \dot{x}_{vd}$. Therefore, in the image space the end-effector reaches the desired position x_{vd} with the desired velocity $\dot{x}_{vd}$.

Part III.b: Force tracking errors. Since a sliding mode at $S_{F\delta}(t) = 0$ is induced for all time, we have $\Delta F = \Delta F(t_0) e^{-\kappa_F t}$. From this we obtain $\Delta \dot{F} \equiv \Delta \lambda = -\kappa_F \Delta F(t_0) e^{-\kappa_F t}$, showing the global convergence of the force tracking errors. Thus λ reaches λ_d exponentially fast. **QED.**

References

1. Whitney D (1997) Force feedback control of manipulator fine motions. ASME Journal of Dynamic Systems, Measurement and Control 99:91–97
2. Whitney D (1985) Historical perspective and state of the art in robot force control. In: Proc. IEEE Int. Conf. on Robotics and Automation 262–268
3. De Schutter J, Bruyninckx H, Zhu WH, Spong M (1997) Force control: A bird's eye view. In: Proc. IEEE CSS/RAS Int. Workshop on Control Problems in Robotics and Automation: Future Directions
4. Arimoto S (1996) Control Theory of Non-linear Mechanical Systems: A Passivity-based and Circuit-theoretic Approach. Oxford Engineering Science Series, Oxford University Press
5. McClamroch H, Wang D (1988) Feedback stabilizationand tracking of constrained robots. IEEE Transactions on Automatic Control 33(5):419–426
6. Arimoto S, Liu YH, Parra-Vega V (1992) Design of model-based adaptive controllers for robot manipulators under geometric constraints. In: Proc. The Japan-USA Symposium on Flexible Automation, USA 615–621
7. Liu YH, Arimoto S, Parra-Vega V, Kitagaki K (1997) Decentralized adaptive control of multiple manipulators in co-operation. International Journal of Control 67(5):649–673
8. Parra-Vega V, Rodriguez-Angeles A, Arimoto S, Hirzinger G (2001) High precision constrained grasping with cooperative adaptive handcontrol. Journal of Intelligent and Robotic Systems 32(3):235–254
9. Hutchinson S, Hager G, Corke P (1996) A tutorial on visual servo control. IEEE Transactions on Robotics and Automation 12(5):651–670
10. Weiss L, Sanderson A, Neuman C (1987) Dynamic sensor based control of robots with visual feedback. IEEE Journal of Robotics and Automation 3(5):404–417
11. Papanikolopoulos N, Khosla P, Kanade T (1993) Visual tracking of a moving target by a camera mounted on a robot: A combination of control and vision. IEEE Journal of Robotics and Automation 9:14–35
12. Castano A, Hutchinson S (1994) Visual compliance: Task directed visual servo control. IEEE Transactions on Robotics and Automation 10(3):334-342
13. Bishop B, Spong M (1997) Adaptive calibration and control of 2D monocular visual servo system. In: Proc. IFAC Symposium on Robot Control, Nantes, France 525–530.
14. Miyazaki F, Masutani Y (1990), Robustness on sensory feedback control based on imperfect jacobian, In: Robotics Research: The Fifth Int Symp, Miura H, Arimoto S (eds.), MIT, Cambridge, MA, 201–208.
15. Kelly R, Marquez A (1995) Fixed-eye direct visual feedback control of planar robots. Journal of System Engineering 4(5):239-248
16. Kelly R (1996) Robust asymptotically stable visual servoing of planar robots. IEEE Transactions on Robotics and Automation 12(5):759-766
17. Zergeroglu E, Dawson D, de Queiroz M, Setlur P (2003) Robust visual-servo control of robot manipulators in the presence of uncertainty. Journal of Robotic Systems 20(2):93-106
18. Dean-León E, Parra-Vega V, Espinosa-Romero A, Fierro J (2004) Dynamical image-based PID uncalibrated visual servoing with fixed camera for tracking of planar robots with a heuristical predictor. In: Proc. 2nd IEEE Int. Conf. on Industrial Informatics 339–345

19. Shen Y, Sun D, Liu YH, Li KJ (2003) Asymptotic trajectory tracking of manipulators using uncalibrated visual feedback. IEEE/ASME Transactions on Mechatronics 8(1):87–9
20. Xiao D (2000) Sensor-hybrid position/force control of a robot manipulator in an uncalibrated enviroment. IEEE Transactions on Control System Technology 8(4):635–645
21. Nelson B, Khosla P (1996) Force and vision resolvability for assimilating disparate sensory feedback. IEEE Transactions on Robotics and Automation 12(5):714–731
22. Baeten J, De Schutter J (2001) Combined vision/force control at corners in planar robotic contour following. In: Proc. IEEE/ASME Int. Conf. on Advanced Intelligent Mechatronics 810–815
23. Hosoda K, Igarashi K, Asada M (1998) Adaptive hybrid control for visual and force servoing in an unknown environment. IEEE Robotics and Automation Magazine 5:39–43
24. Parra-Vega V, Arimoto S, Liu YH, Naniwa T (1994) Model-based adaptive hybrid control of robot manipulators under holonomic constraints. In: Proc. IFAC Symposium on Robot Control 475–480
25. Blajer W, Schiehleny W, Schirm W (1994) A projection criterion to the coordinate partitioning method for multi body dynamics. Archive of Applied Mechanics 84:86–98
26. García L, Dean-León E, Parra-Vega V, Espinosa-Romero A, Fierro J (2005) Experimental results of image-based adaptive visual servoing of 2D robots under jacobian and dynamic friction uncertainties. In: Proc. Int. Symposium on Computational Intelligence in Robotics and Automation
27. Parra-Vega V, Arimoto S, Liu YH, Hirzinger G, Akella P (2003) Dynamic sliding PID control for tracking of robot manipulators: theory and experiments. IEEE Transactions on Robotics and Automation 19:967–976

Passivity-Based Control of Multi-Agent Systems *

Nikhil Chopra and Mark W. Spong

Coordinated Science Laboratory
University of Illinois at Urbana-Champaign
1308 West Main Street
Urbana, IL 61801, USA
{nchopra,mspong}@uiuc.edu

Summary. In this paper we study passivity-based control for the problem of coordination and synchronization of multi-agent systems. We treat agents described by affine nonlinear systems that are input-output passive and that exchange information over a network described by an interconnection graph. We treat both linear interconnections on balanced, directed graphs and nonlinear interconnections on undirected graphs. We present synchronization results for both fixed and switching graphs. Finally, we treat the realistic case of time delay in the communication of information among agents. Our results unify several existing results from the literature on multi-agent systems.

As applications of our results we present a constructive methodology to solve the local exponential convergence problem for Kuramoto oscillators. We then apply our results to the general problem of synchronization of multiple Lagrangian systems. Using a network of simple pendula, the phenomena of oscillator death, synchronization, and anti-synchronization are all shown to be special cases of our results, depending on whether or not the natural frequencies of the pendula are identical or distinct. We also show that the general problem of multi-robot coordination can be handled using the results in this paper.

1 Introduction and Motivation

1.1 Energy Shaping and Passivity-Based Control

In this paper we present some recent results on passivity-based control of multi-agent systems on the occasion of the 70th birthday of Professor Suguru Arimoto. In our opinion, the fundamental work of Takegaki and Arimoto [34]

---

* Dedicated to the 70th birthday of Professor Suguro Arimoto. This research was partially supported by the Office of Naval Research under Grant N00014-02-1-0011, N00014-05-1-0186, and by the National Science Foundation under Grants ECS-0122412 and INT-0128656.

in 1981 marked the beginning of robot control as a distinct discipline. Prior to this work, most research on robot control relied on textbook linear methods based on first-order approximations to the nonlinear dynamics or on feedforward control. The Takegaki-Arimoto result introduced the notion of potential energy shaping as a way to stabilize an equilibrium set-point position of a robot manipulator. Potential-energy shaping is now a well-established method of control, not only for manipulators, but also for vehicles, bipedal robots, electric motors and many other systems. Indeed the lineage of many of the recent advanced Lagrangian and Hamiltonian methods in control can be traced back directly to the 1981 paper of Takegaki and Arimoto.

Potential-energy shaping is now recognized as a special case of the more general notion of *Passivity-Based Control*. The term passivity-based control was coined in 1988 by Ortega and Spong [25] to describe a class of adaptive control algorithms for manipulators. Thus, it is fitting that we discuss the application of passivity-based control in robotics as a tribute to the influence and legacy of Professor Arimoto.

1.2 Background

The problem of communication and control in multi-agent systems is important in numerous practical applications, such as sensor networks, unmanned aerial vehicles, and robot networks. Inspired by nature (schools of fish, flocks of birds) and driven by practical engineering considerations, the control problems are inherently distributed in nature. The goal then is to generate a desired collective behavior by local interaction among the agents. The pioneering model of boids [27], the discrete time flocking model of Vicsek et. al. [37] have been mathematically analyzed in [13]. Group coordination and formation stability problems have been recently addressed in [9, 18, 10, 20, 19, 35]. Agreement, group consensus and oscillator synchronization problems have been studied in [8, 14, 15, 17, 22, 26, 30, 28, 29, 32, 36]. The effect of communication delay on some of the aforementioned agreement protocols has been studied in [2, 28, 36, 38].

In this work we investigate a class of dynamic models for the networked agents that we believe unifies many of the results in the collective motion and oscillator synchronization literature. We treat the case where each agent in the network can be modeled as a nonlinear system that is affine in the control and input/output passive. The goal then is to drive the outputs of the agents to each other asymptotically. The assumption of passivity is a natural one for the types of problems considered. Many of the existing results in the literature model the agents as velocity-controlled particles; in other words, as first-order integrators, which are the simplest type of passive systems. On the other hand, quite complicated Lagrangian and Hamiltonian systems, such as n-degree-of-freedom robots, satisfy a natural and well-studied passivity property [31].

Passivity

To set the background and notation for what follows, consider a control affine nonlinear system of the form

$$\Sigma \begin{cases} \dot{x} = f(x) + g(x)u \\ y = h(x) \end{cases} \quad (1)$$

where $x \in R^n$, $u \in R^m$, and $y \in R^m$. The functions $f(.) \in R^n$, $g(.) \in R^{n \times m}$, and $h(.) \in R^m$ are assumed to be sufficiently smooth. The admissible inputs are taken to be piecewise continuous and locally square integrable and we note that the dimensions of the input and output are the same. We assume, for simplicity, that $f(0) = 0$ and $h(0) = 0$.

Definition 1. *The nonlinear system Σ is said to be passive if there exists a C^1 storage function $V(x) \geq 0$, $V(0) = 0$ and a function $S(x) \geq 0$ such that for all $t \geq 0$:*

$$V(x(t)) - V(x(0)) = \int_0^t u^T(s)y(s)ds - \int_0^t S(x(s))ds \quad (2)$$

The system Σ is strictly passive if $S(x) > 0$ and lossless if $S(x) = 0$.

An important characterization of passive systems is the following result, due to Moylan [23]:

Theorem 1. *The following statements are equivalent.*

(i) The system (1) is passive
(ii) There exists a C^1 scalar storage function $V : R^n \to R$ such that $V(x) \geq 0$, $V(0) = 0$, and $S(x) \geq 0$ such that

$$\begin{aligned} L_f V(x) &= -S(x) \\ L_g V(x) &= h^T(x) \end{aligned} \quad (3)$$

where $L_f V(x) = \frac{\partial V}{\partial x}^T f(x)$ and $L_g V(x) = \frac{\partial V}{\partial x}^T g(x)$.

Throughout this paper we assume that the ith agent under consideration can be represented as a passive system in the above form and possesses a radially unbounded C^2 positive definite storage function V_i.

Graph Theory and Communication Topology

Information exchange between agents can be represented as a graph. We give here some basic terminology and definitions from graph theory [11] sufficient to follow the subsequent development.

Definition 2. *By a* **graph** $\mathcal{G}$ *we mean a finite set* $\mathcal{V}(\mathcal{G}) = \{v_i, \ldots, v_N\}$, *whose elements are called* **nodes** *or* **vertices**, *together with set* $\mathcal{E}(\mathcal{G}) \subset \mathcal{V} \times \mathcal{V}$, *whose elements are called* **edges**. *An edge is therefore an ordered pair of distinct vertices.*

If, for all $(v_i, v_j) \in \mathcal{E}(\mathcal{G})$, *the edge* $(v_j, v_i) \in \mathcal{E}(\mathcal{G})$ *then the graph is said to be* **undirected**. *Otherwise, it is called a* **directed graph**.

An edge (v_i, v_j) *is said to be* **incoming with respect to** v_j *and* **outgoing with respect to** v_i *and can be represented as an arrow with vertex* v_i *as its tail and vertex* v_j *as its head.*

The **in-degree** *of a vertex* $v \in \mathcal{G}$ *is the number of edges that have this vertex as a head. Similarly, the* **out-degree** *of a vertex* $v \in \mathcal{G}$ *is the number of edges that have this vertex as the tail.*

If the in-degree equals the out-degree for all vertices $v \in \mathcal{V}(\mathcal{G})$, *then the graph is said to be* **balanced**.

A **path** *of length* r *in a directed graph is a sequence* $v_0, \ldots, v_r$ *of* $r+1$ *distinct vertices such that for every* $i \in \{0, \ldots, r-1\}$, (v_i, v_{i+1}) *is an edge.*

A **weak path** *is a sequence* $v_0, \ldots, v_r$ *of* $r+1$ *distinct vertices such that for each* $i \in \{0, \ldots, r-1\}$ *either* (v_i, v_{i+1}) *or* (v_{i+1}, v_i) *is an edge.*

A directed graph is **strongly connected** *if any two vertices can be joined by a path and is* **weakly connected** *if any two vertices can be joined by a weak path.*

We quote the following result from [11], which we will use in our subsequent analysis.

Lemma 1. *Let* $\mathcal{G}$ *be a directed graph and suppose it is balanced. Then* $\mathcal{G}$ *is strongly connected if and only if it is weakly connected.*

The remainder of this chapter is organized as follows. In section 2 we present results on output synchronization of networked passive systems in the case of linear coupling among agents. We demonstrate output synchronization for a system of N agents with possibly time-varying coupling gains, switching network topologies and time delays in the network. In Section 3 we treat the case of nonlinear coupling among the agents, with and without time delay. In Section 4 we present some important problems and examples in oscillator synchronization, general Lagrangian and Hamiltonian systems and multi-robot networks which can be addressed using our results.

2 Output Synchronization

The evolution of multi-agents networked passive systems depends fundamentally on their interconnection topology. We list below several assumptions regarding the interconnection topology that we will make in the sequel. The specific assumption(s) used in a given result below will be made clear in the statement of the theorem.

A1: The agents are weakly connected pointwise in time and form a balanced graph with respect to information exchange. Hence the information graph is also strongly connected from from Lemma 1.

A2: The agents form a balanced information graph pointwise in time, and there exists an infinite sequence of bounded, non-overlapping time intervals across which the agents are jointly connected.

A3: In addition to assumption A1, we assume that there exists a unique path between any two distinct nodes.

A4: The agents are weakly connected pointwise in time and form an undirected graph with respect to information exchange. As all undirected (bidirectional) graphs are also balanced, the agents are also strongly connected from Lemma 1.

A5: In addition to assumption A4, there exists a unique path between any two distinct nodes.

2.1 Output Synchronization with Linear Coupling

In this section we show that, when the information graph is balanced and weakly connected, linear coupling among agents is sufficient to achieve output synchronization. Recall that each agent's (passive) dynamics can be written for $i = 1, \ldots, N$ as

$$\begin{aligned} \dot{x}_i &= f_i(x_i) + g_i(x_i)u_i \\ y_i &= h_i(x_i) \end{aligned} \quad (4)$$

Definition 3. *Suppose we have a network of N agents as above. In the absence of communication delays, the agents are said to output synchronize if*

$$\lim_{t \to \infty} |y_i(t) - y_j(t)| = 0 \quad \forall i, j = 1, \ldots, N \quad (5)$$

We first analyze the case when the communication topology is fixed, i.e., the information graph does not change with time. Suppose that the agents are coupled together using the control

$$u_i = \sum_{j \in \mathcal{N}_i} K(y_j - y_i), \quad i = 1, \ldots, N \quad (6)$$

where K is a positive constant and $\mathcal{N}_i$ is the set of agents transmitting their outputs to the ith agent.

Theorem 2. *Consider the dynamical system described by (4) with the control (6). Then under assumption A1, the nonlinear system (4),(6) is globally stable and the agents output synchronize.*

Proof. Consider a positive definite Lyapunov function for the N agent system as

$$V = 2(V_1 + \cdots + V_N) \quad (7)$$

where V_i is the storage function for agent i. Using Theorem (1) and the control law (6), the derivative of this Lyapunov function along trajectories of the system is

$$\dot{V} = 2\sum_{i=1}^{N}(L_{f_i}V_i + L_{g_i}V_i u_i) = 2\sum_{i=1}^{N}(-S_i(x_i) + y_i^T u_i)$$

$$= -2\sum_{i=1}^{N} S_i(x_i) + 2\sum_{i=1}^{N}\sum_{j\in\mathcal{N}_i} y_i^T K(y_j - y_i) \tag{8}$$

$$= -2\sum_{i=1}^{N} S_i(x_i) - 2K\sum_{i=1}^{N}\sum_{j\in\mathcal{N}_i} y_i^T y_i + 2K\sum_{i=1}^{N}\sum_{j\in\mathcal{N}_i} y_j^T y_i$$

As the information exchange graph is balanced, we have

$$2K\sum_{i=1}^{N}\sum_{j\in\mathcal{N}_i} y_i^T y_i = K\sum_{i=1}^{N}\sum_{j\in\mathcal{N}_i} y_i^T y_i + K\sum_{i=1}^{N}\sum_{j\in\mathcal{N}_i} y_j^T y_j \tag{9}$$

and, therefore, it follows that

$$\dot{V} = -2\sum_{i=1}^{N} S_i(x_i) - K\sum_{i=1}^{N}\sum_{j\in\mathcal{N}_i}(y_i - y_j)^T(y_i - y_j) \leq 0 \tag{10}$$

Thus the system is globally stable and all signals are bounded. Consider the set $E = \{x_i \in R^{n\times 1},\ i = 1,\ldots,N \mid \dot{V} \equiv 0\}$. The set E is characterized by all trajectories such that $\{S_i(x_i) \equiv 0,\ (y_i - y_j)^T(y_i - y_j) \equiv 0\ \forall j \in \mathcal{N}_i,\ \forall i = 1,\ldots,N\}$. Lasalle's Invariance Principle [16] and strong connectivity of the network then implies output synchronization of (4).

Remark 1.

(i) In systems without drift, i.e. ($f_i(x_i) = 0$), it can be shown [5] that the outputs will converge to a common constant value.
(ii) Suppose that the dynamics of the N agents are given by identical linear systems of the form

$$\dot{x}_i = Ax_i + Bu_i$$
$$y_i = Cx_i \quad i = 1,\ldots,N \tag{11}$$

and let u_i be given by (6). Then, it is easily shown that, if the pair (C, A) is observable, output synchronization implies synchronization of the full state, i.e. $\lim_{t\to\infty}|x_i - x_j| = 0\ \forall i, j$.
(iii) In Theorem 2 the coupling gain K was taken to be a constant for simplicity. The result can be easily extended to the case where K is a positive definite matrix and thus, the linear consensus protocol of [28] can be admitted as a special case of Theorem 2.

The Case of Switching Topology

We next consider the case when the graph topology is not constant, such as in nearest neighbor scenarios. In the case of linear coupling, it turns out that output synchronization is still guaranteed for switching graphs under fairly mild assumptions.

Consider now the coupling control law

$$u_i(t) = \sum_{j \in \mathcal{N}_i(t)} K(t)(y_j - y_i), \quad i = 1, \ldots, N \tag{12}$$

In this equation we allow both the gain $K(t) > 0$ and the set $\mathcal{N}_i$ of neighbors of agent i to be time-dependent. We assume that the gain $K(t)$ is piecewise continuous and satisfies

$$K_l \leq K(t) \leq K_b, \quad K_l, K_b > 0 \quad \forall i, j \tag{13}$$

Theorem 3. *Consider the dynamical system described by (4), coupled together using the control law (12) together with the assumption A1. Suppose, in addition, that there are at most finitely many switches in the graph topology on any finite time interval. Then the system described is globally stable and the agents output synchronize.*

A more interesting scenario occurs when the agents are allowed to lose connectivity at every instant, but maintain connectivity in an average sense to be made precise. Let $t_{ij}(e), t_{ij}(d)$ denote the time instances at which the information link or the edge (i,j) is established and broken respectively. In the subsequent analysis we require that

$$t_{ij}(d) - t_{ij}(e) \geq \delta_{ij} > 0 \quad \forall i, j \in \mathcal{E} \tag{14}$$

The above assumption is similar to the dwell time assumption in Theorem 5 of [13], but is less stringent as it does not require the dwell time to be the same for every agent, but implies it to be uniformly bounded away from zero. The notion of joint connectivity was introduced in [13] (see also [36]), and has been used by several authors including [2, 22, 26, 30]. The agents are said to be jointly connected across the time interval $[t, t+T]$, $T > 0$ if the agents are weakly connected across the union $\cup_{\sigma \in [t, t+T]} \mathcal{E}(\mathcal{G}(\sigma))$. Note that the above assumption (14) implies that there are only finitely many distinct graphs in this union for each T. Our main result in this section is then

Theorem 4. *Consider the dynamical system described by (4), coupled together using the control described by (12) together with the assumption A2. Then the system is globally stable and the agents output synchronize.*

Proof. We note that a proof for this result implies Theorem 3 as a corollary. The proof basically proceeds by demonstrating that the Lyapunov function candidate

$$V = 2(V_1 + \ldots + V_N) \tag{15}$$

is a common Lyapunov function for the switching system. Using the pointwise balanced nature of the information graph, as in Theorem 2, the derivative of (15) can be written as

$$\dot{V}(t) = -2\sum_{i=1}^{N} S_i(x_i) - K(t)\sum_{i=1}^{N}\sum_{j \in \mathcal{N}_i(t)} (y_i - y_j)^T (y_i - y_j) \leq 0 \tag{16}$$

implying that the nonlinear system described by (4) and (12) is globally stable and every solution is bounded. Also, $\lim_{t \to \infty} V(t)$ exists and is finite since V is lower bounded by zero. Consider a infinite sequence $V(t_i)$ $i = 1, \ldots,$ where the times t_i approach infinity as i approaches infinity. Then, using Cauchy's convergence criteria, $\forall \epsilon \geq 0\ \exists M > 0$ s.t $\forall k > l \geq M$

$$|V(t_l) - V(t_k)| \leq \epsilon \tag{17}$$

As the above statement is true $\forall k, l > M$, in particular we choose k, l such that the time interval $[t_k, t_l]$ encompasses some time interval across which the agents are jointly connected. Existence of such a time interval is guaranteed by assumption A2. Therefore, from (16),

$$|\sum_{i=1}^{N} \sum_{j \in \mathcal{N}_i(t_k,t_l)} \int_{t_l}^{t_k} K(t)(y_i - y_j)^T (y_i - y_j) dt| \leq \epsilon$$

$$\Rightarrow |\sum_{j \in \mathcal{N}_i(t_k,t_l)} \int_{t_l}^{t_k} K(t)(y_i - y_j)^T (y_i - y_j) dt| \leq \epsilon \ \forall i$$

$$\Rightarrow |\sum_{j \in \mathcal{N}_i(t_k,t_l)} \int_{t_{ij}(e)}^{t_{ij}(d)} K(t)(y_i - y_j)^T (y_i - y_j) dt| \leq \epsilon \ \forall i$$

as ϵ is arbitrary this implies

$$\lim_{t \to \infty} \int_{t}^{t+\delta_{ij}} ||y_i - y_j||^2 dt = 0, \quad \forall j \in \mathcal{N}_i(t, t+T), \ \forall i$$

As $x_i \in \mathcal{L}_\infty\ \forall i$, $\dot{y}_i \in \mathcal{L}_\infty\ \forall i$, and hence the above limit implies that

$$\lim_{t \to \infty} ||y_i - y_j|| = 0 \ \forall j \in \mathcal{N}_i(t, t+T) \ \forall i$$

Weak connectivity of the network across the time interval $[t, t+T]$ then guarantees output synchronization of (4).

Remark 2.

(i) If the graph is undirected, the above results are still valid in the case that different gains (or weights) $K_{ij}(t)$ couple the agents, provided $K_{ij}(t) = K_{ji}(t)\ \forall i, j$.

2.2 Output Synchronization with Time Delay

In this section, we study the problem of output synchronization when there are communication delays in the network. The delays are assumed to be constant and bounded.

Definition 4. *The agents are said to output synchronize if*

$$\lim_{t \to \infty} |y_i(t - T_{ij}) - y_j(t)| = 0 \quad \forall i, j \quad j \neq i \tag{18}$$

where T_{ij} is the sum of the delays along the path from the ith agent to the jth agent.

It is to be noted that assumption A3 is needed to ensure uniqueness of $T_{ij} \; \forall j \neq i$. Let the agents be coupled together using the control

$$u_i(t) = \sum_{j \in \mathcal{N}_i} K(y_j(t - T_{ji}) - y_i) \quad i = 1, \ldots, N \tag{19}$$

where $K > 0$ is a constant and $\mathcal{N}_i$ is the set of agents transmitting their outputs to the ith agent.

Theorem 5. *Consider the dynamical system described by (4) with the control law (19). Then under the assumption A3, the nonlinear system (4), (19) is globally stable and the agents output synchronize in the sense of (18).*

Proof. Consider a positive definite Lyapunov Krasovskii functional for the N agent system as

$$V = K \sum_{i=1}^{N} \sum_{j \in \mathcal{N}_i} \int_{t-T_{ji}}^{t} y_j^T(\tau) y_j(\tau) d\tau + 2(V_1 + \ldots + V_N) \tag{20}$$

The derivative of this Lyapunov-functional along trajectories of the system is given as

$$\dot{V} = K \sum_{i=1}^{N} \sum_{j \in \mathcal{N}_i} (y_j^T y_j - y_j(t - T_{ji})^T y_j(t - T_{ji})) + 2 \sum_{i=1}^{N} (L_{f_i} V_i + L_{g_i} V_i u_i)$$

Using Theorem (1), the derivative reduces to

$$\dot{V} = K\sum_{i=1}^{N}\sum_{j\in\mathcal{N}_i}(y_j^T y_j - y_j(t-T_{ji})^T y_j(t-T_{ji})) + 2\sum_{i=1}^{N}(-S_i(x_i) + y_i^T u_i)$$

$$= K\sum_{i=1}^{N}\sum_{j\in\mathcal{N}_i}(y_j^T y_j - y_j(t-T_{ji})^T y_j(t-T_{ji}))$$

$$-2\sum_{i=1}^{N}S_i(x_i) + 2\sum_{i=1}^{N}\sum_{j\in\mathcal{N}_i}y_i^T K(y_j(t-T_{ji}) - y_i)$$

$$= K\sum_{i=1}^{N}\sum_{j\in\mathcal{N}_i}(y_j^T y_j - y_j(t-T_{ji})^T y_j(t-T_{ji}))$$

$$-2\sum_{i=1}^{N}S_i(x_i) - 2K\sum_{i=1}^{N}\sum_{j\in\mathcal{N}_i}y_i^T y_i + 2K\sum_{i=1}^{N}\sum_{j\in\mathcal{N}_i}y_j(t-T_{ji})^T y_i \qquad (21)$$

As the graph is balanced

$$K\sum_{i=1}^{N}\sum_{j\in\mathcal{N}_i}y_j^T y_j = K\sum_{i=1}^{N}\sum_{j\in\mathcal{N}_i}y_i^T y_i \qquad (22)$$

Using this in (21) yields,

$$\dot{V} = -K\sum_{i=1}^{N}\sum_{j\in\mathcal{N}_i}y_i^T y_i - K\sum_{i=1}^{N}\sum_{j\in\mathcal{N}_i}y_j(t-T_{ji})^T y_j(t-T_{ji})$$

$$+ 2K\sum_{i=1}^{N}\sum_{j\in\mathcal{N}_i}y_j(t-T_{ji})y_i - 2\sum_{i=1}^{N}S_i(x_i)$$

$$= -K\sum_{i=1}^{N}\sum_{j\in\mathcal{N}_i}(y_j(t-T_{ji}) - y_i)^T(y_j(t-T_{ji}) - y_i) - 2\sum_{i=1}^{N}S_i(x_i) \leq 0$$

Therefore the system is globally stable, every solution is bounded, $\lim_{t\to\infty}V(t)$ exists and is finite. Differentiating $\dot{V}$ we obtain

$$\ddot{V} = -2\sum_{i=1}^{N}\frac{\partial S_i(x_i)}{\partial x_i}\dot{x}_i - 2K\sum_{i=1}^{N}\sum_{j\in\mathcal{N}_i}\left(\left(\frac{\partial y_i}{\partial x_i} - \frac{\partial y_j(t-T_{ji})}{\partial x_j}\right)(\dot{x}_i - \dot{x}_j(t-T_{ji}))\right)^T$$

$$\cdot(y_j(t-T_{ji}) - y_i) \qquad (23)$$

By assumption, all partial derivatives are defined and are continuous. Using the fact that $x_i \in \mathcal{L}_\infty$ $\forall i$, we have $\frac{\partial S_i(x_i)}{\partial x_i}, \frac{\partial y_i}{\partial x_i}, \dot{x}_i \in \mathcal{L}_\infty$ $\forall i$ (as $x_i \in \mathcal{L}_\infty$, R.H.S of (4) is bounded) which implies that $\ddot{V} \in \mathcal{L}_\infty$ and hence, the function $\dot{V}$ is uniformly continuous. Invoking Barbalat's Lemma [16] we conclude that $\lim_{t\to\infty}\dot{V} = 0$. Thus,

$$\lim_{t\to\infty} |y_j(t - T_{ji}) - y_i| = 0 \quad \forall j \in \mathcal{N}_i \quad i = 1, \ldots, N$$

Strong connectivity of the interconnection graph then implies output synchronization of (4) in the sense of (18).

2.3 Nonlinear Coupling Among Agents

We next consider the case of nonlinear coupling. The agents are assumed to be coupled together using any nonlinear passive control strategy. We demonstrate that under such a framework the multi-agent system is globally stable and output synchronizes for a class of such controls.

Let $\chi(z) : R \to R$ be a nonlinear function satisfying the following properties

(i) $\chi(z)$ is continuous and locally Lipschitz
(ii) $\chi(-z) = -\chi(z) \ \forall z \neq 0$

If the argument $z \in R^n$, then by $\chi(z) \in R^{n \times 1}$ we mean the function $\chi(.)$ acting componentwise on the vector z. Let the agents be coupled together using the control

$$u_i = \sum_{(i,j) \in \mathcal{E}(\mathcal{G})} \chi(y_j - y_i) \quad i = 1, \ldots, N \quad (24)$$

which takes as argument the difference of each agent's output with that of its neighbors. In the sequel, for the sake of brevity $\chi(y_j - y_i) \equiv \chi_{ji}$ and similarly for all other controls and scalar functions unless mentioned otherwise.

The controls $\chi_{ji} \ \forall i,j$ are assumed to be passive, therefore they satisfy the differential version of the passivity property outlined earlier,

$$\dot{V}_{ji} + S_{ji} = (y_j - y_i)^T \chi_{ji} \quad (25)$$

Let the Lyapunov function candidate for the system be given as

$$V = \sum_{i=1}^{N} V_i + \sum_{(i,j) \in \mathcal{E}(\mathcal{G})} V_{ji} \quad (26)$$

Our main result in this section is the following:

Theorem 6. *Consider the system (4) together with the control law given by (24). Then under assumption A4, the nonlinear system (4), (24) is globally stable and all trajectories converge to the largest invariant set where $\dot{V} = 0$.*

Proof. Using Theorem (1), the control law (24) and (25), the derivative of the Lyapunov function (26) along trajectories of the system is given as

$$\dot{V} = \sum_{i=1}^{N}(L_{f_i}V_i + L_{g_i}V_i u_i) + \sum_{(i,j)\in\mathcal{E}(\mathcal{G})} \dot{V}_{ji}$$

$$= \sum_{i=1}^{N}(-S_i(x_i) + y_i^T u_i) + \sum_{(i,j)\in\mathcal{E}(\mathcal{G})}(-S_{ji} + (y_j - y_i)^T \chi_{ji})$$

$$= -\sum_{i=1}^{N}S_i(x_i) + \sum_{i=1}^{N}\sum_{(i,j)\in\mathcal{E}(\mathcal{G})} y_i^T \chi_{ji} + \sum_{(i,j)\in\mathcal{E}(\mathcal{G})}(-S_{ji} + (y_j - y_i)^T \chi_{ji})$$

Let us examine the term

$$\sum_{i=1}^{N}\sum_{(i,j)\in\mathcal{E}(\mathcal{G})} y_i^T \chi_{ji} \qquad (27)$$

in the above expression for $\dot{V}$. As the information exchange structure is bidirectional, it is easy to see that corresponding to every term $y_i^T \chi_{ji}$ for some i,j there is a corresponding term $y_j^T \chi_{ij}$ in the summation (27). Then, using Property 2 of the function $\chi(.)$, we can rewrite this term as $-y_j^T \chi_{ji}$. Thus

$$\sum_{i=1}^{N}\sum_{(i,j)\in\mathcal{E}(\mathcal{G})} y_i^T \chi_{ji} = -\sum_{(i,j)\in\mathcal{E}(\mathcal{G})} (y_j - y_i)^T \chi_{ji} \qquad (28)$$

It follows that

$$\dot{V} = -\sum_{i=1}^{N} S_i(x_i) - \sum_{(i,j)\in\mathcal{E}(\mathcal{G})} S_{ji} \leq 0$$

Therefore the system is globally stable, $\lim_{t\to\infty} V(t)$ exists and is finite. Consider the set $E = \{x_i \in R^{n\times 1},\ i = 1,\ldots,N\ |\ \dot{V}{\equiv}0\}$. The set E is characterized by all trajectories such that $\{S_i(x_i){\equiv}0,\ S_{ji}{\equiv}0\ \forall \in (i,j) \in \mathcal{E}(\mathcal{G}),\ \forall i = 1,\ldots,N\}$. Using Lasalle's Invariance Principle [16], all solutions of the dynamical system given by (4) and (24) converge to M as $t \to \infty$, where M is the largest invariant set contained in E.

Asymptotic properties of the signals, and in specific output synchronization, depend on the scalar functions $S_i, S_{ji}\ \forall i,j$, and consequently on the specific choice of the control action $\chi(.)$. It is important to note that by suitably choosing the controller dissipation functions $S_{ij}\ \forall i,j$, we can steer the difference of the agents' outputs to a desired set. If $S_{ij} \equiv 0 \Leftrightarrow (y_i - y_j) \equiv 0\ \forall i,j$, then output synchronization results. We next investigate a class of controls which ensure output synchronization, namely the sector nonlinearities.

Let $\phi(z) : R \to R$ be a nonlinear function satisfying the following properties

(i) $\phi(z)$ is continuous and locally Lipschitz
(ii) $\phi(z) = 0 \Leftrightarrow z = 0$
(iii) $\phi(-z) = -\phi(z) \ \forall z \neq 0$
(iv) $\alpha z^2 \leq \phi(z)z \leq \beta z^2$

where $\alpha, \beta > 0$. As before, if the argument $z \in R^n$, then by $\phi(z)$ we mean the function $\phi(.)$ acting componentwise on the vector z. Suppose that the control law u_i is given by

$$u_i = \sum_{(i,j) \in \mathcal{E}(\mathcal{G})} \phi_{ji} \quad i = 1, \ldots, N \tag{29}$$

Then we can show the following

Theorem 7. *Consider the system (4) together with the control law (29). Then, under assumption A4, the nonlinear system (4), (29) is globally stable and the agents output synchronize in the sense of (5).*

Proof. It can be easily computed that for the control $\phi(.)$

$$V_{ji} = 0 \ ; \ S_{ji} = (y_j - y_i)^T \phi_{ji}$$

Therefore using Theorem 6 the closed loop system given by (4),(29) is globally stable and all trajectories converge to the largest invariant set where $S_{ji} \equiv 0$. Property 2 of the control $\phi(.)$ together with strong connectivity of the network then implies output synchronization of (4).

Remark 3.

(i) Theorem 7 can also be taken as the generalized version of the nonlinear consensus protocol of [28].
(ii) A proof for synchronization of Kuramoto oscillators, for the case when the oscillators have identical natural frequencies [14, 22], follows from the application of Theorem 7.

3 Nonlinear Coupling with Time-Delay

In this section we demonstrate delay independent stability (and in some cases output synchronization) of the multiagent system using any nonlinear passive control strategy. The key tool we use is the scattering transformation which was earlier used in the problem of bilateral teleoperation [1], [24] to guarantee delay independent stability. In the new setting (with the scattering transformation) the agents transmit the so called scattering variables instead of their outputs to their neighbors. The reader is referred to [1], [24] and the references within to get a background on the scattering transformation.

Let the agents be coupled together using the control

$$u_i = \sum_{(i,j)\in\mathcal{E}(\mathcal{G})} \chi(y_{js} - y_i) \quad i = 1,\ldots,N \tag{30}$$

where as before, the controls $\chi(.) \in R \to R$ act componentwise on the vector argument $(y_{js} - y_i)$. In the sequel, for the sake of brevity $\chi(y_{js} - y_i) \equiv \chi_{jsi}$ and similarly for all other controls and scalar functions unless mentioned otherwise. The reference variables y_{js}, y_{is} are derived out of the scattering transformation. The scattering transformation is given as

$$\begin{aligned} s_{ij}^+ &= \tfrac{1}{\sqrt{2b}}(-\chi_{jsi} + by_{js}) \; ; \; s_{ij}^- = \tfrac{1}{\sqrt{2b}}(-\chi_{jsi} - by_{js}) \\ s_{ji}^+ &= \tfrac{1}{\sqrt{2b}}(\chi_{isj} + by_{is}) \; ; \; s_{ji}^- = \tfrac{1}{\sqrt{2b}}(\chi_{isj} - by_{is}) \end{aligned} \tag{31}$$

where $b > 0$ is a constant. The superscript $+, -$ for the scattering variables is a convention for the direction of power flow. As mentioned before, the agents transmit the scattering variables instead of their outputs. This is illustrated schematically in Figure 1. The ith agent transmits the scattering variable s_{ij}^+ to the jth agent who receives it as the scattering variable s_{ji}^+. The jth agent then uses the the control χ_{isj} to extract the signal y_{is} out of the incoming scattering variable s_{ji}^+. A similar procedure is used to obtain the reference signal y_{js} by the ith agent. It is important to observe from (30) that the ith agent is participating in n_i closed loops as the one demonstrated in Figure 1, where n_i is the number of neighbors of the ith agent. As the scattering variables

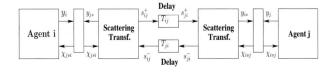

Fig. 1. The Scattering Transformation

are transmitted across the network, they satisfy the following relation

$$\begin{aligned} s_{ji}^+ &= s_{ij}^+(t - T_{ij}) \\ s_{ij}^- &= s_{ji}^-(t - T_{ji}) \end{aligned} \tag{32}$$

The scattering transformation renders the communication channel passive and the storage function for the communication channel is given as

$$\begin{aligned} V_{channel}^{ij} &= \frac{1}{2}\int_0^t (\|s_{ij}^+\|^2 - \|s_{ji}^+\|^2 + \|s_{ji}^-\|^2 - \|s_{ij}^-\|^2)d\tau \\ &= \frac{1}{2}\int_0^t (\|s_{ij}^+\|^2 - \|s_{ij}^+(\tau - T_{ij})\|^2 + \|s_{ji}^-\|^2 - \|s_{ji}^-(\tau - T_{ji})\|^2)d\tau \\ &= \frac{1}{2}\int_{t-T_{ij}}^t \|s_{ij}^+\|^2 d\tau + \frac{1}{2}\int_{t-T_{ji}}^t \|s_{ji}^-\|^2 d\tau \; \geq 0 \end{aligned}$$

where the notation $||.||$ denotes the Euclidean norm of the enclosed signal. It is easy to verify from (31) that

$$V^{ij}_{channel} = \frac{1}{2}\int_0^t (||s^+_{ij}||^2 - ||s^+_{ji}||^2 + ||s^-_{ji}||^2 - ||s^-_{ij}||^2)d\tau$$
$$= \int_0^t (-\chi^T_{jsi}y_{js} - \chi^T_{isj}y_{is})d\tau \geq 0 \tag{33}$$

The controls $\chi_{jsi}\ \forall i,j$ are assumed to be passive, therefore they satisfy the differential version of the passivity property outlined earlier,

$$\dot{V}_{jsi} + S_{jsi} = (y_{js} - y_i)^T \chi_{jsi} \tag{34}$$

where symbols have the previously defined interpretation. Our main result in this section is:

Theorem 8. *Consider the system (4) together with the control law (30) and the scattering transformation (31). Then under assumption A5, the nonlinear system (4), (30), (31) is globally stable.*

Proof. Consider the following Lyapunov function candidate for the system as following Lyapunov function

$$V = \sum_{i=1}^N V_i + \sum_{(i,j)\in\mathcal{E}(\mathcal{G})} V^{ij}_{channel} + \sum_{i=1}^N \sum_{(i,j)\in\mathcal{E}(\mathcal{G})} V_{jsi} \tag{35}$$

The derivative of V along trajectories of the system can be computed using Theorem 1, (33), (34) as

$$\dot{V} = -\sum_{i=1}^N S_i(x_i) + \sum_{i=1}^N \sum_{(i,j)\in\mathcal{E}(\mathcal{G})} y_i^T \chi_{jsi} + \sum_{(i,j)\in\mathcal{E}(\mathcal{G})} (-\chi^T_{jsi}y_{js} - \chi^T_{isj}y_{is})$$
$$+ \sum_{i=1}^N \sum_{(i,j)\in\mathcal{E}(\mathcal{G})} -S_{jsi} + (y_{js} - y_i)^T \chi_{jsi}$$

It can be verified that

$$\sum_{i=1}^N \sum_{(i,j)\in\mathcal{E}(\mathcal{G})} y_i^T \chi_{jsi} + (y_{js} - y_i)^T \chi_{jsi} + \sum_{(i,j)\in\mathcal{E}(\mathcal{G})} (-\chi^T_{jsi}y_{js} - \chi^T_{isj}y_{is}) = 0$$

Therefore the derivative of the Lyapunov function reduces to

$$\dot{V} = -\sum_{i=1}^N S_i(x_i) - \sum_{i=1}^N \sum_{(i,j)\in\mathcal{E}(\mathcal{G})} S_{jsi} \leq 0$$

Thus the system is globally stable, every solution is bounded and $\lim_{t\to\infty} V(t)$ exists and is finite.

As was the case in Theorem 6, asymptotic properties of the signals depend on the scalar functions S_i, S_{jsi} $\forall i,j$, and consequently on the specific choice of the control action $\chi(.)$. It seems likely that Theorem 8 can be extended to ensure that the difference of agents' outputs reach a desired limit set by suitably choosing the controller dissipation functions S_{ij} $\forall i,j$. The corresponding analysis tool for the same would be the extension of Lasalle's invariance principle for autonomous time-delay systems [12]. This shall be a subject of our future research.

Next, we study case where the control $\chi(.) \equiv \phi(.)$ where $\phi(.)$ retains the properties described in Section 2.3. The control law u_i for the ith agent is then given by

$$u_i = \sum_{(i,j) \in \mathcal{E}(\mathcal{G})} \phi_{jsi} \quad i = 1, \ldots, N \tag{36}$$

Then we can show the following

Theorem 9. *Consider the system (4) together with the control law (36) and the scattering transformation (31). Assuming A5, the nonlinear system (4), (36), (31) is globally stable and the agents output synchronize in the sense of (18).*

Proof. It is to be noted that for the control $\phi(.)$

$$V_{jsi} = 0 \; ; \; S_{jsi} = (y_{js} - y_i)^T \phi_{jsi}$$

Consider the following Lyapunov function function

$$V = \sum_{i=1}^{N} V_i + \sum_{(i,j) \in \mathcal{E}(\mathcal{G})} V_{channel}^{ij} \tag{37}$$

Following Theorem 8, the derivative of (37) along trajectories is given as

$$\dot{V} = -\sum_{i=1}^{N} S_i(x_i) - \sum_{i=1}^{N} \sum_{(i,j) \in \mathcal{E}(\mathcal{G})} (y_{js} - y_i)^T \phi_{jsi} \leq 0$$

Therefore the system is globally stable, $\lim_{t \to \infty} V(t)$ exists and is finite. As $x_i \in \mathcal{L}_\infty$ $\forall i$, we have that $\ddot{V} \in \mathcal{L}_\infty$, and hence the function $\dot{V}$ is uniformly continuous. Invoking Barbalat's Lemma we conclude that $\lim_{t \to \infty} \dot{V} = 0$. Thus, $\lim_{t \to \infty} (y_{js} - y_i)^T \phi_{jsi} = \lim_{t \to \infty} (y_{is} - y_j)^T \phi_{isj} = 0$ $(i,j) \in \mathcal{E}(\mathcal{G})$. This implies that

$$\lim_{t \to \infty} (y_{js} - y_i) = \lim_{t \to \infty} (y_{is} - y_j) = 0 \; (i,j) \in \mathcal{E}(\mathcal{G}) \tag{38}$$

To establish output synchronization further analysis is needed. The transmission equations (32) can be rewritten as

$$y_{is} = y_{js}(t - T_{ij}) - \frac{1}{b}(\phi_{isj} + \phi_{jsi}(t - T_{ij}))$$

$$y_{js} = y_{is}(t - T_{ij}) - \frac{1}{b}(\phi_{jsi} + \phi_{isj}(t - T_{ji}))$$

From the above two equations we have

$$\lim_{t \to \infty} y_{is} = \lim_{t \to \infty} y_{js}(t - T_{ij}) - \lim_{t \to \infty} \frac{1}{b}(\phi_{isj} + \phi_{jsi}(t - T_{ij}))$$

$$\lim_{t \to \infty} y_{js} = \lim_{t \to \infty} y_{is}(t - T_{ji}) - \lim_{t \to \infty} \frac{1}{b}(\phi_{jsi} + \phi_{isj}(t - T_{ji}))$$

As $\lim_{t \to \infty} \phi_{isj} = \lim_{t \to \infty} \phi_{jsi} = 0$, we have

$$\lim_{t \to \infty} y_{is} = \lim_{t \to \infty} y_{js}(t - T_{ij})$$

$$\lim_{t \to \infty} y_{js} = \lim_{t \to \infty} y_{is}(t - T_{ji})$$

Therefore using (38) we have that

$$\lim_{t \to \infty} y_i = \lim_{t \to \infty} y_j(t - T_{ji})$$

$$\lim_{t \to \infty} y_j = \lim_{t \to \infty} y_i(t - T_{ij}) \quad (i,j) \in \mathcal{E}(\mathcal{G})$$

Strong connectivity of the network then implies output synchronization.

Remark 4. It is interesting to note that output synchronization in the sense of (18) implies existence of equations of the kind

$$\lim_{t \to \infty} |y_i - y_i(t - T_1)| = \ldots = \lim_{t \to \infty} |y_i - y_i(t - T_m)| = 0$$

where m is the number of closed communication loops in which the ith agent participates, and $T_s \quad s = 1, \ldots, m$ is the closed loop delay in the sth loop. Therefore, the results in this section hint at the existence of delay induced periodicity in the outputs [39].

4 Examples

Synchronization of Kuramoto Oscillators

The Kuramoto model consists of a population of N oscillators whose dynamics are governed by the following equations

$$\dot{\theta}_i = \omega_i + \frac{K}{N} \sum_{j=1}^{N} \sin(\theta_j - \theta_i), \quad i = 1, \ldots, N \tag{39}$$

where $\theta_i \in S^1$ is the phase of the ith oscillator, $\omega_i \in R$ is its natural frequency and $K > 0$ is the coupling gain. It is to be noted that the coupling between the oscillators is *all-to-all* and hence bidirectional. The problem then is to characterize the coupling gain K so that the oscillators phase velocities synchronize, i.e.

$$\lim_{t \to \infty} |\dot{\theta}_i - \dot{\theta}_j| = 0 \quad \forall i, j \tag{40}$$

To get a nice intuition about the problem, the oscillators may also be thought of as points moving on a unit circle. Imagining these oscillators as points on a circle, the points then become frequency locked, or in other words they start moving with the same angular frequency asymptotically. The reader is referred to [32] for an excellent survey of the problem and the subsequent efforts to address it.

Kuramoto demonstrated that in the continuum limit where $N \to \infty$, when the natural frequencies are distributed according to a unimodal probability density, then for a large enough coupling gain K the oscillator population splits into two groups, one of which starts synchronizing and eventually start moving with the mean natural frequency of the group. Numerical simulations hint that the oscillators synchronize exponentially. To date there is no analysis which shows that the oscillators in the original Kuramoto model (where the oscillators have different natural frequencies) synchronize exponentially (even locally) and quoting Strogatz [32] "Nobody has even touched the problems of global stability and convergence". In this example we demonstrate that Theorem 3 and Theorem 7 provide a constructive tool to address the synchronization problem for the Kuramoto oscillators.

We analyze the Kuramoto model for the case where

- There are finite number N of oscillators.
- The natural frequencies ω_i $\forall i$ are picked arbitrarily from the set of reals.

First, consider the case when $|\theta_i| \leq \frac{\pi}{4}$ and the natural frequencies are same for all oscillators and are given by ω_s. Transforming the system to a rotating frame by the transformation $\Phi_i \to \theta_i - \omega_s t$ $\forall i$, the system (39) can be written

$$\dot{\Phi}_i = u \qquad u = \frac{K}{N} \sum_{j=1}^{N} \sin(\Phi_j - \Phi_i)$$

$$y_i = \Phi_i \quad i = 1, \ldots, N \tag{41}$$

This above system is passive with storage function $V = \frac{1}{2}\Phi^T \Phi$, where $\Phi = [\Phi_1 \ldots \Phi_N]^T$. From Theorem 7, it immediately follows that

$$\lim_{t \to \infty} |\Phi_j - \Phi_i| = 0 \quad \forall i, j = 1, \ldots, N \tag{42}$$

Reverting back to the inertial frame we have

$$\lim_{t \to \infty} |\dot{\theta}_i - \omega_s| = 0 \quad \forall i \tag{43}$$

and, hence the oscillators synchronize to the common frequency ω_s for all $K \geq 0$.

The more interesting scenario is when all oscillators have different natural frequencies. In the subsequent analysis we additionally assume that

- All oscillators at $t = 0$ are contained in the set $\mathcal{D}$ where $\mathcal{D} = \{\theta_i, \theta_j \mid |\theta_i - \theta_j| < \frac{\pi}{2} \quad \forall i, j = 1, \ldots, N\}$.

As we are interested in dynamics of the angular frequencies, we differentiate the Kuramoto model (39) to get

$$\ddot{\theta}_i = \frac{K}{N} \sum_{j=1}^{N} \cos(\theta_j - \theta_i)(\dot{\theta}_j - \dot{\theta}_i), \quad i = 1, \ldots, N \tag{44}$$

Choosing $\dot{\theta}_i = x_i$ and $\cos(\theta_j - \theta_i) = g_{ji}(t)$, we can construct a state space representation for the Kuramoto model as

$$\dot{x}_i = u \; ; \; u = \frac{K}{N} \sum_{j=1}^{N} g_{ji}(t)(x_j - x_i)$$

$$y_i = x_i \tag{45}$$

At this point let us recall a result from [8].

Theorem 10. *Consider the previously described system. Let all initial phase differences at t=0 be contained in the compact set* $\mathcal{D} = \{\theta_i, \theta_j \mid |\theta_i - \theta_j| \leq \frac{\pi}{2} - 2\epsilon \quad \forall i, j = 1, \ldots, N\}$ *where* $\epsilon < \frac{\pi}{4}$ *is an arbitrary nonnegative number. Then there exists a coupling gain* $K_{inv} > 0$ *such that* $(\theta_i - \theta_j) \in \mathcal{D} \quad \forall t > 0$.

This result ensures that for a sufficiently large coupling gain K_{inv}, the phase differences are positively invariant with respect to $\mathcal{D}$. Selecting $K > K_{inv}$, $(\theta_j - \theta_i) \in D \quad \forall t \geq 0 \Rightarrow \cos(\theta_j - \theta_i) = g_{ji}(t) > 0$. As $\cos(.)$ is an even function, $g_{ij}(t) = g_{ji}(t) \quad \forall i, j$.

The dynamics given by (45) represent a first order integrator that is passive with $V = \frac{1}{2} x^T x$ as the storage function, where $x = [x_1 \ldots x_N]^T$ is the vector representing the angular rates of the oscillators. Choosing $\phi(z) = x_j - x_i$, noting that $K_{ij}(t) = g_{ij}(t)$, and invoking the nonlinear version of Theorem 3 (see Remarks 2, 3) we have

$$\lim_{t \to \infty} |x_i - x_j| = 0 \quad \forall i, j \text{ or}$$

$$\lim_{t \to \infty} |\dot{\theta}_i - \dot{\theta}_j| = 0 \quad \forall i, j \tag{46}$$

and, hence the oscillators synchronize in the sense of (40). To the best of authors' knowledge, it was demonstrated for the first time in [8] that the oscillators locally exponentially synchronize to the mean frequency.

Mechanical Systems and Multi-Robot Networks

Following [31], the Euler-Lagrange equations of motion for N mechanical systems (under the gravitational potential field) are given as

$$M_i(q_i)\ddot{q}_i + C_i(q_i,\dot{q}_i)\dot{q}_i + g_i(q_i) = \tau_i$$
$$y_i = \dot{q}_i \quad i = 1,\ldots,N \tag{47}$$

where $q \in R^n$ are the generalized configuration coordinates for the system with n degree of freedom, $\tau \in R^n$ is the vector of generalized forces acting on the system, $M(q)$ is the $n \times n$ symmetric, positive definite inertia (generalized mass) matrix, $C(q,\dot{q})$ is the $n \times n$ vector of Centripetal and Coriolis torques and $g(q)$ is the $n \times 1$ vector of gravitational torques. Although the above equations of motion are coupled and non-linear, they exhibit certain fundamental properties due to their Lagrangian dynamic structure.

Property 1 The inertia matrix $M(q)$ is symmetric positive definite and there exists positive constants m and M such that

$$mI \leq M(q) \leq MI \tag{48}$$

Property 2 Under an appropriate definition of the matrix C, the matrix $\dot{M} - 2C$ is skew symmetric.

The dynamics (47) are passive with

$$V_i(q_i,\dot{q}_i) = \frac{1}{2}\dot{q}_i^T M_i(q_i)\dot{q}_i + G_i(q_i) \tag{49}$$

where $G(q_i)$ is the gravitational potential energy and $g_i(q_i) = \frac{\partial G_i}{\partial q_i}$. It can be verified (using Property 2) that

$$V_i(t) - V_i(0) = \int_0^t \tau_i(s)^T \dot{q}_i(s) ds \tag{50}$$

and hence the dynamics are passive with $(\tau_i, \dot{q}_i)$ as the input-output pair.

It is easily seen that $V_i \ \forall i$ is positive definite. Let the interconnection structure between the N systems satisfy assumption A1, and the system are coupled together using the control

$$\tau_i = \sum_{j \in \mathcal{N}_i} K(\dot{q}_j - \dot{q}_i), \quad i = 1,\ldots,N \tag{51}$$

Then invoking Theorem 2 with

$$V = 2(V_1 + \ldots + V_N) \tag{52}$$

as the Lyapunov function candidate, the systems output synchronize and

$$\lim_{t \to \infty} |\dot{q}_i(t) - \dot{q}_j(t)| = 0 \quad \forall i, j = 1, \ldots, N \tag{53}$$

Similarly, the agents output synchronize in the presence of communication delays using Theorem 5.

Example 1: Consider a system constituting point masses, hence the matrix $M(q) = m$ is a constant, and $P(q) = 0$. The equations of motion for a N body system reduce to

$$m_i \ddot{q}_i = \tau_i$$
$$y_i = \dot{q}_i \quad \forall i \tag{54}$$

Let the systems be coupled together using the control (51). It is easily seen that this system is passive with $V = \sum_{i=1}^{N} 2V_i = \sum_{i=1}^{N} m_i \dot{q}_i^T \dot{q}_i$ as the positive semi-definite storage function. If the coupling torques are given by (51), then using the calculations in Theorem 2 and Barbalat's Lemma it can be shown that $\lim_{t \to \infty} \tau_i = 0$ which implies velocity synchronization. Furthermore, it is easily seen that $\sum_{i=1}^{N} \tau_i = 0$, and thus from (54) $\sum_{i=1}^{N} m_i \ddot{q}_i = 0$. Hence the total momentum $\sum_{i=1}^{N} p_i = \sum_{i=1}^{N} m_i \dot{q}_i$ of the system is invariant, and

$$\lim_{t \to \infty} \dot{q}_i = \frac{m_1 \dot{q}_1(0) + \ldots + m_N \dot{q}_N(0)}{m_1 + \ldots + m_N} \tag{55}$$

where $\dot{q}_i(0)$ is the velocity of the ith particle at time $t = 0$.

Consider a network of four agents, with a ring like communication structure. The agent dynamics are given by

$$m_1 \ddot{q}_1 = K(\dot{q}_2 - \dot{q}_1)$$
$$m_2 \ddot{q}_2 = K(\dot{q}_3 - \dot{q}_2)$$
$$m_3 \ddot{q}_3 = K(\dot{q}_4 - \dot{q}_3)$$
$$m_4 \ddot{q}_4 = K(\dot{q}_1 - \dot{q}_4) \tag{56}$$

Let $m_1 = 1, m_2 = 2, m_3 = 4, m_4 = 4.5$ and $\dot{q}_1(0) = 3, \dot{q}_2(0) = -4, \dot{q}_3(0) = -1, \dot{q}_4(0) = 2$, so that $\sum_{i=1}^{4} m_i \dot{q}_i(0) = 0$. As seen in Figure 2, the agent velocities converge to the origin as the initial mean momentum of the system was zero.

Example 2: The next example of two simple pendula is used to bring out some interesting phenomena in oscillator synchronization. Consider two pendula of lengths L_1 and L_2 with the following dynamics

$$\ddot{q}_1 + \frac{g}{L_1} \sin(q_1) = u_1$$
$$\ddot{q}_2 + \frac{g}{L_2} \sin(q_2) = u_2 \tag{57}$$

and choose $u_1 = -u_2 = K(\dot{q}_2 - \dot{q}_1)$. The simple pendulum is passive from $u_i \to \dot{q}_i$ $i = 1, 2$ with $V_i = \frac{1}{2}\dot{q}_i^2 + \frac{g}{L_i}(1 - \cos(q_i))$ as the positive definite

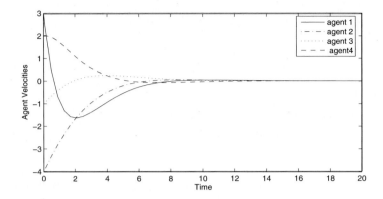

Fig. 2. The agent velocities converges to the origin.

storage function. Invoking Theorem 2, all trajectories converge to the largest invariant set where $\dot{q}_1 \equiv \dot{q}_2$. From (57), this implies

$$\frac{g}{L_1}\sin(q_1) \equiv \frac{g}{L_2}\sin(q_2) \tag{58}$$

If the pendulum lengths are different, the largest invariant set is given as $q_i \equiv 0 \Rightarrow \dot{q}_i \equiv 0$, and hence the oscillations die out asymptotically as seen in Figure 3.

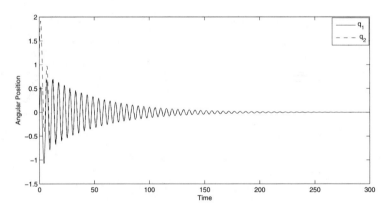

Fig. 3. The oscillations die out when the pendula have different lengths.

If the two pendulums are identical, the largest invariant set is of the kind $q_1 \equiv q_2$, and thus the oscillators synchronize as seen in Figure 4.

Passivity from $u_i \to \dot{q}_i$ guarantees passivity from $-u_i \to -\dot{q}_i$. Exploiting this fact we may chose the control $u_1 = u_2 = -K(\dot{q}_2 - (-\dot{q}_1))$. Using Theorem 2 all trajectories approach the largest invariant set where $\dot{q}_1 \equiv -\dot{q}_2$.

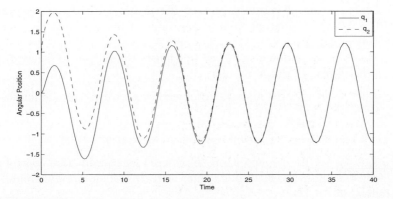

Fig. 4. The pendula move synchronously when they have identical lengths

From (57) we get

$$\frac{g}{L_1}\sin(q_1) \equiv -\frac{g}{L_2}\sin(q_2) \tag{59}$$

As before, for non-identical pendula the oscillations die out, for identical pendula the largest invariant set is of the kind $q_1 \equiv -q_2$, and the pendula move asynchronously (Figure 5).

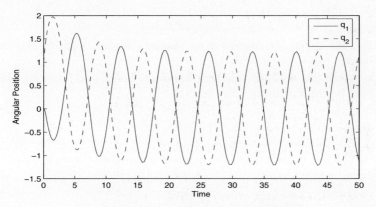

Fig. 5. Identical pendula move out of phase

Next we address the tracking problem for multiple robots when there are delays in the network. In such a scenario the goal is to coordinate the motion of multiple robots and ensure that every robot tracks the position (and not only velocity using (51)) of every other robot asymptotically. It is evident that to develop an effective coordination strategy within the passivity framework, the following goals need to be accomplished

- A feedback control law that renders the manipulator dynamics passive with respect to an output from which the joint displacement information is observable.
- A passive coordination control law which uses this output to *kinematically lock* the motion of the N robots.

In order to achieve the first design objective, we choose torque for the ith robot as

$$\tau_i = -M_i(q_i)\lambda \dot{q}_i - C_i(q_i,\dot{q}_i)\lambda q_i + g_i(q_i) + \bar{\tau}_i \qquad (60)$$

where $\bar{\tau}_i$ is the additional motor torque required for coordination control and λ is a constant positive definite diagonal matrix. The system dynamics (47) now reduce to

$$\dot{q}_i = -\lambda q_i + r_i$$
$$M_i \dot{r}_i + C_i r_i = \bar{\tau}_i \qquad (61)$$

where

$$r_i = \dot{q}_i + \lambda q_i \qquad (62)$$

is the new output of the ith robot. Consider $V_i = \frac{1}{2}r_i^T M_i r_i$ as the positive semi-definite storage function for each robot. It can be verified (using Property 2) that

$$V_i(t) - V_i(0) = \int_0^t \bar{\tau}_i^T(z) r_i(z) dz \quad i = 1, \ldots, N \quad \forall t > 0$$

Therefore the dynamics described by (61) are lossless with $(\bar{\tau}_i, r_i)$ as the input-output pair.

The robots are assumed to satisfy assumption A3, and following (19), let the robots be coupled using a the control

$$\bar{\tau}_i = \sum_{j \in \mathcal{N}_i} K(r_j(t - T_{ji}) - r_i) \quad i = 1, \ldots, N \qquad (63)$$

where $K > 0$ is a constant. Let the positive semi-definite storage function for the system be given by

$$V = K \sum_{i=1}^{N} \sum_{j \in \mathcal{N}_i} \int_{t-T_{ji}}^{t} r_j^T(\tau) r_j(\tau) d\tau + 2(V_1 + \ldots + V_N)$$

Following the calculations in Theorem 5 (and noting that $S_i(r_i) = 0 \;\; \forall i$)

$$\dot{V} = -K \sum_{i=1}^{N} \sum_{j \in \mathcal{N}_i} (r_j(t - T_{ji}) - r_i)^T (r_j(t - T_{ji}) - r_i)$$

$$= -K^{-1} \sum_{i=1}^{N} \bar{\tau}_i^T \bar{\tau}_i \;\; \leq 0$$

Thus $r_i \in \mathcal{L}_\infty \Rightarrow q_i, \dot{q}_i \in \mathcal{L}_\infty$ $\forall i$ (from (61)). It can seen from the above equation that the signals $r_j(t - T_{ji}) - r_i \in \mathcal{L}_2$ $\forall j \in \mathcal{N}_i$, $i = 1, \ldots, N$. Using (62)

$$r_j(t - T_{ji}) - r_i = \left(\dot{q}_j(t - T_{ji}) - \dot{q}_i\right) + \lambda\left(q_j(t - T_{ji}) - q_i\right)$$
$$= \dot{e}_{ji} + \lambda e_{ji}$$

where $e_{ji} = q_j(t - T_{ji}) - q_i$ represents the position tracking error between the jth robot and the ith robot. Therefore the signal $\dot{e}_{ji} + \lambda e_{ji} \in \mathcal{L}_2$ which guarantees asymptotic convergence of the tracking error e_{ji} to the origin. Strong connectivity of the network then implies

$$\lim_{t \to \infty} |q_i(t - T_{ij}) - q_j| = 0 \quad \forall i, j$$

and thus the robots track each other asymptotically. The above results were used in [7] to solve the bilateral teleoperation problem without using the scattering transformation approach [1], and thus was the first result that guaranteed delay independent asymptotic position and force tracking in a nonlinear bilateral teleoperation system.

Example 3: Consider four point masses with the the previously described dynamics (54). It was shown recently in [18] that the flocking and consensus problem for a second order drift free system is nontrivial, and mimicking the linear consensus protocols in [28] can potentially make the system unstable for certain choices of the coupling gains. We apply the proposed methodology to ensure asymptotic convergence of the position and velocity tracking errors to the origin. Following (60), choosing $\lambda = 1$ and applying the preliminary feedback

$$\tau_i = -m_i \dot{q}_i + \bar{\tau}_i \quad \forall i$$

the dynamics reduce to

$$\dot{q}_i = -q_i + r_i$$
$$m_i \dot{r}_i = \bar{\tau}_i \quad \forall i$$

where $r_i = \dot{q}_i + q_i$. Let the communication structure among the agents be described by a ring topology. Then the system dynamics (using (63)) are given as

$$m_1 \dot{r}_1 = K(r_2(t - T_{21}) - r_1)$$
$$m_2 \dot{r}_2 = K(r_3(t - T_{32}) - r_2)$$
$$m_3 \dot{r}_3 = K(r_4(t - T_{43}) - r_3)$$
$$m_4 \dot{r}_4 = K(r_1(t - T_{14}) - r_4)$$

where $K = 1$, m_i $\forall i$ are chosen same as before, T_{ij} denotes the delay from the ith agent to the jth agent. In this simulation $T_{21} = .1s, T_{32} = .2s, T_{43} = 1s, T_{14} = 1.5s$ and as seen in Figure 6, the agents synchronize asymptotically.

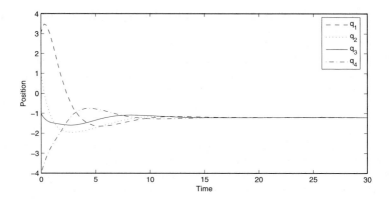

Fig. 6. The four agents synchronize in spite of the communication delay

5 Conclusions

In this paper a new formulation was developed which views the multi-agent coordination and control problem from an input-output perspective. This approach has been shown to unify a lot of results in the literature where the information graph is restricted to be balanced and connected over uniformly bounded time intervals. Given that agents have nonlinear control affine, input/output passive, open loop stable dynamics, control laws were developed to achieve output synchronization. In the specific case where the coupling control is linear in the outputs, it was shown that output synchronization is robust to arbitrary constant (bounded) time delays. Finally, the proposed results were used as constructive tools in solving many problems such as the open problem of exponential synchronization of Kuramoto oscillators, output synchronization of Lagrangian systems and the multi-robot coordination problem.

Systems which are minimum phase and relative degree one, are feedback equivalent to a passive system. Hence the proposed methodology is also applicable to any nonlinear system satisfying the above two properties. It will be interesting to see the extension of the formulation to a system with lower triangular structure via backstepping. Future research involves extension of the proposed approach to rooted communication graphs.

References

1. R. J. Anderson and M. W. Spong, "Bilateral control of teleoperators with time delay," *IEEE Transactions on Automatic Control*, Vol. 34, No. 5, pp. 494-501, May 1989.
2. D. Angeli and P.-A. Bliman, "Stability of leaderless multi-agent systems. extension of a result by Moreau,". preprint:*math.OC/0411338*, (http://arxiv.org/list/math.oc/0411).

3. V. D. Blondel, J. M. Hendrickx, A. Olshevsky, and J. N. Tsitsiklis, "Convergence in multiagent coordination, consensus, and flocking," *IEEE Conference on Decision and Control*, Spain, Dec. 2005.
4. N. Chopra and M. W. Spong, "Output synchronization of nonlinear systems with time delay in communication," *IEEE Conference on Decision and Control*, to appear, San Diego, Dec. 2006.
5. N. Chopra and M. W. Spong, "Output synchronization of networked passive systems," *IEEE Transactions on Automatic Control*, submitted for publication, Dec. 2005.
6. N. Chopra, M. W. Spong, R. Ortega, and N. E. Barabanov, "On tracking performance in bilateral teleoperation," *IEEE Transactions on Robotics*, to appear.
7. N. Chopra and M. W. Spong, "On synchronization of networked passive systems with time delays and application to bilateral teleoperation," In *Annual Conference of Society of Instrument and Control Engineers of Japan*, Okayama, Japan, Aug. 6-10, 2005.
8. N. Chopra and M. W. Spong, "On synchronization of Kuramoto oscillators," *IEEE Conference on Decision and Control*, Spain, Dec. 2005.
9. J. Cortés, S. Martínez, T. Karatas, and F. Bullo, "Coverage control for mobile sensing networks," *IEEE Transactions on Robotics and Automation*, Vol. 20, No. 2, pp. 243-255, April 2004.
10. J. A. Fax and R. M. Murray, "Information flow and cooperative control of vehicle formations," *IEEE Transactions on Automatic Control*, Vol. 49, No. 9, pp. 1465-1476, 2004.
11. C. Godsil and G. Royle, *Algebraic Graph Theory*, Springer Graduate Texts in Mathematics No. 207, New York, 2001.
12. J. K. Hale and S. M. Verduyn, *Introduction to Functional Differential Equations*, Applied Mathematical Sciences, Springer-Verlag, New York, 1993.
13. A. Jadbabaie, J. Lin, and A. S. Morse, "Coordination of groups of mobile autonomous agents using nearest neighbor rules," *IEEE Transactions on Automatic Control*, Vol. 48, pp. 988-1001, June 2003.
14. A. Jadbabaie, N. Motee, and M. Barahona, "On the stability of the Kuramoto model of coupled nonlinear oscillators," *American Control Conference*, pp. 4296 - 4301, 2004.
15. E. W. Justh and P. S. Krishnaprasad, "A simple control law for UAV formation flying," Technical Report 2002-38, Institute for Systems Research, 2002.
16. H. K. Khalil, *Nonlinear Systems*, Prentice Hall, Upper Saddle River, New Jersey, 2002.
17. Y. Kuramoto, *In International Symposium on Mathematical Problems in Theoretical Physics, Lecture Notes in Physics,*, Vol. 39, Springer, New York, 1975.
18. D. J. Lee and M. W. Spong, "Stable flocking of multiple inertial agents on balanced graphs," *IEEE Transactions on Automatic Control*, Submitted.
19. N. E. Leonard and E. Fiorelli, "Virtual leaders, artificial potentials and coordinated control of groups," *IEEE Conference on Decision and Control*, pp. 2968-2973, Orlando, Florida, 2001.
20. J. A. Marshall, M. E. Broucke, and B. A. Francis, "Formation of vehicles in cyclic pursuit," *IEEE Transactions on Automatic Control*, Vol. 49, No. 11, pp. 1963-1974, Nov. 2004.
21. T. Matiakis, S. Hirche, and M. Buss, "The scattering transformation for networked control systems," *IEEE Conference on Control Applications*, pp. 705-710, Toronto, CA, Aug. 2005.

22. L. Moreau, "Stability of multi-agent systems with time-dependent communication links," *IEEE Transactions on Automatic Control*, Vol. 50, No. 2, pp. 169–182, Feb. 2005.
23. P. J. Moylan, "Implications of passivity for a class of nonlinear systems," *IEEE Transactions on Automatic Control*, Vol. 19, pp. 373-381, Aug. 1974.
24. G. Niemeyer and J.J.-E., Slotine, "Stable adaptive teleoperation," *IEEE Journal of Oceanic Engineering*, Vol. 16, No. 1, pp. 152-162, Jan. 1991.
25. R. Ortega and M. W. Spong, "Adaptive motion control of rigid robots: a tutorial," *IEEE Conference on Decision and Control*, Vol. 2, pp. 1575-1584, San Antonio, TX, Dec. 1988.
26. W. Ren and R. W. Beard, "Consensus seeking in multi-agent systems using dynamically changing interaction topologies," *IEEE Transactions on Automatic Control*, Vol. 5, No. 5, pp. 655-661, May 2005.
27. C. W. Reynolds, "Flocks, herds, and schools: A distributed behavioural model," *Computer Graphics*, 21 (4), 25-34, 1987.
28. R. Olfati-Saber and R. M. Murray, "Consensus problems in networks of dynamic agents with switching topology and time-delays," *IEEE Transactions on Automatic Control*, Vol. 49, pp. 1520 - 1533, Sept. 2004.
29. R. Sepulchre, D. Paley, and N. Leonard, "Collection motion and oscillator synchronization," *Cooperative Control, Lecture Notes in Control and Information Sciences*, Vol. 309, Springer Verlag, 2004.
30. J.-J. E. Slotine and W. Wang, "A study of synchronization and group cooperation using partial contraction theory," *Cooperative Control, Lecture Notes in Control and Information Sciences*, Vol. 309, Springer Verlag, 2004.
31. M. W. Spong, S. Hutchinson, and M. Vidyasagar, *Robot Modeling and Control*, John Wiley & Sons, Inc., New York, 2006.
32. S. H. Strogatz, "From Kuramoto to Crawford: exploring the onset of synchronization in populations of coupled oscillators," *Physica D*, 143:1-20, 2000.
33. S. H. Strogatz, *SYNC: The Emerging Science of Spontaneous Order*, Hyperion Press, New York, 2003.
34. M. Takegaki and S. Arimoto, "A new feedback method for dynamic control of manipulators," *Trans. of the ASME, Journal of Dynamic Systems, Measurement, and Control*, Vol. 103, No.2, pp. 119-125 (1981).
35. H. G. Tanner, G. J. Pappas, and V. Kumar, "Leader-to-formation stability," *IEEE Transactions on Robotics and Automation*, Vol. 20, No. 3, pp. 443-455, 2004.
36. J. N. Tsitsiklis, D. P. Bertsekas and M. Athans, "Distributed Asynchronous Deterministic and Stochastic Gradient Optimization Algorithms," *IEEE Transactions on Automatic Control*, Vol. 31, No. 9, pp. 803-812, 1986.
37. T. Vicsek, A. Czirck, E. Ben-Jacob, I. Cohen, and O. Schochet, "Novel type of phase transitions in a system of self-driven particles," *Physical Review Letters*, Vol. 75, No. 6, pp. 1226-1229, Aug. 1995.
38. W. Wang and J.-J. E. Slotine, "Contraction analysis of time-delayed communications and group cooperation," *IEEE Transactions on Automatic Control*, Vol. 51, No. 4, pp. 712-717, April 2006.
39. S. Yanchuk, "Discretization of frequencies in delay coupled oscillators," *Physical Review E*, 72, 036205 2005.

Navigation Functions for Dynamical, Nonholonomically Constrained Mechanical Systems

Gabriel A. D. Lopes[1] and Daniel E. Koditschek[2]

[1] Electrical Engineering and Computer Science Department, University of Michigan, Ann Arbor, MI, USA glopes@umich.edu
[2] Electrical and Systems Engineering Department, University of Pennsylvania, 200 33rd st., Philadelphia, 19103, PA, USA kod@seas.upenn.edu

Summary. In this review we explore the possibility of adapting first order hybrid feedback controllers for nonholonomically constrained systems to their dynamical counterparts. For specific instances of first order models of such systems, we have developed gradient based hybrid controllers that use Navigation functions to reach point goals while avoiding obstacle sets along the way. Just as gradient controllers for standard quasi-static mechanical systems give rise to generalized "PD-style" controllers for dynamical versions of those standard systems, so we believe it will be possible to construct similar "lifts" in the presence of non-holonomic constraints notwithstanding the necessary absence of point attractors.

1 Introduction

The use of total energy as a Lyapunov function for mechanical systems has a long history [1] stretching back to Lord Kelvin [2]. Unquestionably, Arimoto [3] represents the earliest exponent of this idea within the modern robotics literature, and, in tribute to his long and important influence, we explore in this paper its extension into the realm of nonholonomically constrained mechanical systems.

The notion of total energy presupposes the presence of potential forces arising from the gradient of a scalar valued function over the configuration space. We focus our interest on "artificial cost functions" introduced by a designer to encode some desired behavior as originally proposed by Khatib [4, 5]. However, we take a global view of the task, presuming a designated set of prohibited configurations — the "obstacles" — and a designated set of selected configurations — the "goal," which we restrict in this paper to be an isolated single point. We achieve the global specification through the introduction of a *Navigation Function* (NF) [6] — an artificial potential function that attains a maximum value of unity on the entire boundary of the obstacle set, and its only local minimum, at zero, exactly on the isolated goal point. Such functions are guaranteed to exist over any configuration space of relevance to physical mechanical systems [7], and constructive examples have been furnished for a variety of task domains [6, 8, 9, 10].

NF-generated controls applied to completely actuated mechanical systems force convergence to the goal from almost every initial condition and guarantee that no motions will intersect the obstacle set along the way. In the dynamical setting, where the role of kinetic energy is important, they achieve a pattern of behavior analogous to that of similarly controlled corresponding quasi-static dynamics. For example, in the one degree of freedom case, the dynamical setting is represented by the familiar spring-mass-damper system

$$m\ddot{q} + c\dot{q} + kq = 0 \tag{1}$$

and the corresponding quasi-static model arises through a neglect of the inertial forces, $m \to 0$ in (1), yielding

$$c\dot{q} + kq = 0 \tag{2}$$

To illustrate the nature of NF-gradient-based controllers in this simple setting, take the configuration space to be $\mathcal{Q} := \{q \in \mathbb{R} : |q| \leq 1\}$ with navigation function $\varphi(q) := \frac{1}{2}kq^2$, implying that $\{0\} = \varphi^{-1}[0]$ is the goal and $\{-1, 1\} = \varphi^{-1}[1]$ the obstacle set. We imagine that both systems, (1), (2), arise from application of the NF-gradient control law, $u := -\nabla\varphi$, to the respective open loop,

$$u = m\ddot{q} + c\dot{q}$$

or

$$u = c\dot{q}.$$

We observe that φ is a global Lyapunov function for (2) guaranteeing that all initial conditions give rise to motions that avoid the obstacle set while converging asymptotically on the goal set. Analogously, the total energy, $\mu := \frac{1}{2}\dot{q}^2 + \varphi(q)$ is a Lyapunov function for the velocity-limited extension of $\mathcal{Q}$, $\mathcal{X} := \mu^{-1}[0, 1] = \{(q, \dot{q}) \in \mathbb{R}^2 : \mu(q, \dot{q}) \leq 1\}$. This guarantees that all initial conditions in $\mathcal{X}$ give rise to motions that avoid the obstacle (and, in fact, are repelled from the entire boundary, $\mu^{-1}[1]$) while converging asymptotically on the zero velocity goal set, $\mu^{-1}[0] = \{(0,0)\}$. A more general global version of Arimoto's [3] adaptation of Lord Kelvin's observations has been presented in greater detail in [11].

In contrast, for incompletely actuated mechanical systems, when the degrees of freedom exceed the number of independently controlled actuators, the applicability of NF-gradient-based controllers to either dynamical or quasi-static mechanical systems remains largely unexplored. One particularly important class of such systems arises in the presence of nonholonomic constraints — systems with intrinsically unavailable velocities whose absence cannot be expressed in terms of configuration space obstacles [12]. In such settings, there is an inherent degree of underactuation, since no amount of input work can be exerted in the forbidden directions. In an echo of Arimoto's [3] "lift" of first order gradient dynamics (2) to damped second order mechanical dynamics (1), this paper explores the relationship between quasi-static and fully dynamical NF-gradient controllers for a class of nonholonomic systems.

One very important general observation about nonlinear systems that throws a shadow on every aspect of this exploration was made two decades ago by Brockett [13] who pointed out that nonlinear systems may be completely controllable while failing to be smoothly stabilizable. Nonholonomically constrained systems suffer this defect, so that no smooth feedback controller, NF-gradient or otherwise, could ever stabilize a single point goal. However, switching controllers incur no such limitation.

In the next section we will introduce a switching controller for a broad class of systems that alternately runs "down" and then "across" the gradient slope to bring all motions arbitrarily close to the goal without hitting any obstacles. For these classes we can prove this analytically. The next section addresses the same class of systems, now cast in the dynamical setting. We show how to recast the hybrid controller to give the analogous result for these second order systems. Finally, as an illustrative example, we present numerical simulations of the rolling disk defined in a configuration space with a simple sensory model that imposes a particular configuration space topology.

2 Hybrid Controller for Nonholonomic Kinematic Systems

We start by presenting a class of controllers defined in $\mathbb{R}^3$ for nonholonomic kinematic systems. For an in depth exposure please see [14]. Consider the class of smooth and piecewise analytic, three degree of freedom, drift-free control systems

$$\dot{q} = B(q)u, \quad q \in \mathcal{Q} \subset \mathbb{R}^3; \quad u \in \mathbb{R}^2, \tag{3}$$

where $\mathcal{Q}$ is a smooth and piecewise analytic, compact, connected three dimensional manifold with a boundary, $\partial \mathcal{Q}$ (that separates the acceptable from the forbidden configurations of $\mathbb{R}^3$), possessing a distinguished interior *goal point*, $q^* \in \mathcal{Q}$. In this section we will impose very general assumptions on B and construct a hybrid controller that guarantees local convergence to an arbitrarily small neighborhood of the goal state while avoiding any forbidden configurations along the way.[3]

We find it convenient to write (3) using the *nonholonomic projection matrix* [15], H into the image of B:

---

[3] In the next section, we will introduce more specialized assumptions that extend the basin of attraction to include almost every initial configuration in $\mathcal{Q}$.

$$H(q) = B(q)B(q)^\dagger = B(q)\left(B(q)^T B(q)\right)^{-1} B(q)^T \tag{4}$$
$$\dot{q} = H(q)v, \quad q \in \mathcal{Q} \subset \mathbb{R}^3; \quad v \in \mathbb{R}^3 \tag{5}$$

2.1 Two Controllers and Their Associated Closed Loop Dynamics

It is useful to compare the unconstrained system $\dot{q} = v$ with the constrained version (5). Let φ be a navigation function defined in $\mathcal{Q}$. For the input $v = -\nabla\varphi$ the unconstrained system is globally asymptotically stable at the origin. Using φ as a control Lyapunov function yields $\dot{\varphi} = -\|\nabla\varphi\|^2$. Given this result, a naive approach to attempt stabilizing system (5) is to use the same input $v = -\nabla\varphi$. Define the vector field $f_1 : \mathcal{Q} \to T\mathcal{Q}$ such that $f_1(q) := -H(q)\nabla\varphi(q)$ and the system

$$\dot{q} = f_1(q) = -H(q)\nabla\varphi(q) \tag{6}$$

Since H has a 1-dimensional kernel it follows that (6) has a 1 dimensional center manifold

$$\mathcal{W}^c := \{q \in \mathcal{Q} : H(q)\nabla\varphi(q) = 0\},$$

as corroborated by explicitly computing the Jacobian of f_1 at q^*:

$$Df_1|_{q^*} = -(\underbrace{\nabla\varphi|_{q^*}}_{=0} \otimes I)DH^S - HD^2\varphi = -\left.HD^2\varphi\right|_{q^*} \tag{7}$$

Using φ as a control Lyapunov function, La Salle's invariance theorem states that system (6) has its limit set in $\mathcal{W}^c$:

$$\dot{\varphi} = -\nabla\varphi^T H \nabla\varphi$$
$$= -\|H\nabla\varphi\|^2 \begin{cases} = 0 \text{ if } q \in \mathcal{W}^c \\ < 0 \text{ if } q \notin \mathcal{W}^c \end{cases} \tag{8}$$

Figure 1 illustrates the topology associated with (6): the projection H imposes a co-dimension 1 foliation complementary to the center manifold. The *stable manifold*, $\mathcal{W}^s$, is the leaf containing the goal, q^*. The input

$$u_1 := B(q)^\dagger \nabla\varphi(q) \tag{9}$$

alone cannot stabilize system (6) at the origin, since no smooth time invariant feedback controller has a closed loop system with an asymptotically stable equilibrium point [12]. Nevertheless, for any initial condition outside $\mathcal{W}^c$ an infinitesimal motion in the direction of f_1 reduces the energy φ. If there can be found a second controller that "escapes" $\mathcal{W}^c$ without increasing φ then it is reasonable to imagine that iterating the successive application of these two controllers might well lead eventually to the goal. We now pursue this idea by introducing the following controller,

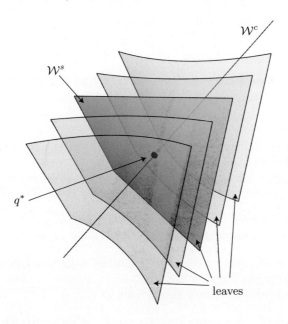

Fig. 1. Conceptual illustration of the flow associated with (6). Each leaf is an invariant manifold with all trajectories collapsing into $\mathcal{W}^c$.

$$u_2 := B(q)^\dagger \left[A(q) \times \nabla\varphi(q) \right], \tag{10}$$

leading to the closed loop vector field

$$\dot{q} = f_2(q)$$
$$f_2(q) := A(q) \times \nabla\varphi(q) \tag{11}$$

where $A(q) := \times B(q)$ is the cross product of the columns of $B(q)$. [4] Note that the nonholonomic constraint expressed in (3) can be represented by the implicit equation $A^T(q)\dot{q} = 0$. Since the derivative of φ in the direction of f_2 is

$$L_{f_2}\varphi = \nabla\varphi(q)^T \cdot (A(q) \times \nabla\varphi(q)) = 0, \tag{12}$$

it follows that f_2 is φ-invariant — i.e. the energy, φ, is constant along its motion. Moreover $0 = A^T(A \times \nabla\varphi) = A^T f_2$, verifying that f_2 indeed satisfies the constraint (3).

2.2 Assumptions, a Strategy, and Preliminary Analysis

Given the previous two vector fields — one which is energy decreasing; the other energy conserving — we now sketch a strategy that brings initial conditions of system (3) to within an arbitrarily small neighborhood ϵ of the goal,

---

[4] We will assume in A3 that B has rank two at each point.

by way of motivating the subsequent definitions and claims that arise in the formal proofs to follow. Let $f_1^t(q)$ and $f_2^t(q)$ denote the flows of f_1 and f_2 respectively.

(1). If $q_0 \in \mathcal{W}^c$ then follow a direction in $\text{im}(H)$ for a finite amount of time t_0 such that $f_1^{t_0}(q_0) \notin \mathcal{W}^c$ and $\varphi \circ f_1^{t_0}(q_0) < 1$ for all $t \in (0, t_0)$.

(2). If $q_0 \notin \mathcal{W}^c$ and $\varphi(q_0) > \epsilon$

    2.1) Use a scaled version of f_2 for time τ_2 to escape a δ-neighborhood of $\mathcal{W}^c$, keeping the energy φ constant.

    2.2) Use controller f_1, for time τ_1, to decrease the energy φ, stopping at a γ-neighborhood of $\mathcal{W}^c$ such that $f_1^{\tau_1}(q) \notin \mathcal{W}^c$ and $\gamma < \delta$.

We now introduce a number of assumptions, definitions and their consequences that will allow us to formalize each of the previous steps:

A1 $\mathcal{Q}$ is a smooth compact connected manifold with boundary.
A2 φ is a navigation function in $\mathcal{Q}$.
A3 H has rank two, uniformly throughout $\mathcal{Q}$.

Assumption A1 gives the proper setting for the existence of a navigation function in the configuration space. Assumption A3 assures the foliation sketched in figure 1.

Define the *local surround* of the goal to be the closed "hollow sphere", $\mathcal{Q}_s := \varphi^{-1}[\Phi_s]$, with $\Phi_s := [\epsilon, \varphi_s]$ whose missing inner "core" is the arbitrarily small open neighborhood, $\mathcal{Q}_\epsilon := \varphi^{-1}[\Phi_\epsilon]$; $\Phi_\epsilon := [0, \epsilon)$, and whose outer "shell", $\mathcal{Q}_1 := \varphi^{-1}[\Phi_1]$, with $\Phi_1 := (\varphi_s, 1]$, includes the remainder of the free configuration space. φ_s is defined to be the largest level such that all the smaller levels, $\varphi_0 \in (0, \varphi_s)$ are homeomorphic to the sphere, S^2, and are all free of critical points, $\|\nabla \varphi\|^{-1}[0] \cap \varphi^{-1}[(0, \varphi_s)] = \emptyset$.

The restriction to φ-invariant topological spheres precludes limit sets of f_2 more complex than simple equilibria in the local surround. However, has in the examples section of [14], one can provide more specialized conditions resulting in the guarantee that the algorithm brings almost every initial condition in the "outer" levels, $\mathcal{Q}_1$ into the local surround, $\mathcal{Q}_s$ and, thence, into the goal set $\mathcal{Q}_\epsilon$.

Lemma 1 ([14]). *Given the previous assumptions*

$$f_1^{-1}[0] \cap \mathcal{Q}_s \equiv f_2^{-1}[0] \cap \mathcal{Q}_s \equiv \mathcal{W}^c \cap \mathcal{Q}_s. \tag{13}$$

To formally express the "δ-neighborhood" described in the stabilization strategy we start by defining the function $\xi : \mathcal{Q} - \{q^*\} \to [0, 1]$:

$$\xi(q) := \frac{\|H(q)\nabla\varphi(q)\|^2}{\|\nabla\varphi(q)\|^2} \tag{14}$$

The quantity $\|H(q)\nabla\varphi(q)\|^2$ evaluates to zero only in $\mathcal{W}^c$. Therefore in a small neighborhood of $\mathcal{W}^c$ the level sets of $\|H(q)\nabla\varphi(q)\|^2$ define a "tube" around $\mathcal{W}^c$. The denominator of (14) normalizes ξ such that $0 \leq \xi \leq 1$. Moreover it produces a "pinching" of the tube at the goal q^*.

Lemma 2 ([14]). *For all $\varphi_0 \in \Phi_s$, $\varphi^{-1}[\varphi_0]$ intersects the unit level set of ξ, i.e., $\xi^{-1}[1] \cap \varphi^{-1}[\varphi_0] \neq \emptyset$.*

Corollary 1 ([14]). *For all $\varphi_0 \in \Phi_s$ the level set $\varphi^{-1}[\varphi_0]$ intersects every level set of ξ, i.e., $\xi^{-1}[\alpha] \cap \varphi^{-1}[\varphi_0] \neq \emptyset$ for all $\alpha \in [0,1]$.*

Lemma 3 ([14]). *A sufficient condition for the Jacobian of $f_2(q)$ evaluated at $\mathcal{W}^c - \|\nabla\varphi\|^{-1}[0]$ to have at least one eigenvalue with non-zero real part is that the control Lie algebra on B spans $\mathbb{R}^3$.*

Lemma 4 ([14]). *The Jacobian of $f_2(q)$ evaluated at $\mathcal{W}^c \cap \mathcal{Q}_s$ has two non-zero real part eigenvalues with the same sign.*

Now consider the implicit equation,

$$\xi(q) = \xi^* \Leftrightarrow \|H(q)\nabla\varphi(q)\|^2 = \xi^* \|\nabla\varphi(q)\|^2 \qquad (15)$$

At the goal any ξ^* satisfies (15). Although ξ is not defined at q^* all of its level sets intersect at q^*. Finally, define the parameterized cone $\mathcal{C}_\gamma$ around $\mathcal{W}^c$, and its complement $\mathcal{C}_\gamma^c := \mathcal{Q} - \mathcal{C}_\gamma - \{q^*\}$, by:

$$\mathcal{C}_\gamma = \{q \in \mathcal{Q} - \{q^*\} : \xi(q) \leq \gamma\} \qquad (16)$$

We follow by imposing conditions on H and A such that the vector field f_2 can afford the needed "escape" from $\mathcal{W}^c$.

Lemma 5 ([14]). *Suppose system (3) satisfies assumptions A1-A3 and, hence, the previous lemmas. Then, there exists a function $\sigma : \mathcal{Q} \to \mathbb{R}$ that renders the system*

$$\dot{q} = \sigma(q)A(q) \times \nabla\varphi(q) = \bar{f}_2(q) \qquad (17)$$

unstable at $\mathcal{W}^c \cap \mathcal{Q}_s$.

Corollary 2 ([14]). *Under the conditions of the previous lemma, there can be found a $\tau \in (0, \infty)$ such that for all $q_0 \in \xi^{-1}[\delta/2]$ we have $\xi \circ \bar{f}_2^\tau(q_0) \geq \delta$.*

Figure 2 illustrates the steps used in the previous proof. Trajectories starting inside $\mathcal{N} - \mathcal{C}_\gamma^c$ will traverse $\partial\mathcal{C}_\gamma$ and $\partial\mathcal{C}_\delta$ in finite time.

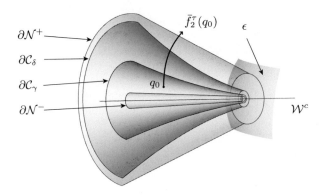

Fig. 2. Illustration of the construction used in the proof of corollary 2.

2.3 A Hybrid Controller and Proof of its Local Convergence

Given the previous result define the time variables τ_1, τ_2 and the scalars $\gamma < \delta$ such that:

$$\tau_1(q,\gamma) := \begin{cases} \min\{t > 0 \mid \xi(f_1^t(q)) = \gamma\} & \text{if } q \in \mathcal{C}_\gamma^c \\ 0 & \text{otherwise} \end{cases}$$

$$\tau_2(q,\delta) := \begin{cases} \min\{t > 0 \mid \xi(\bar{f}_2^t(q)) = \delta\} & \text{if } q \in \mathcal{C}_\delta - \mathcal{W}^c \\ 0 & \text{otherwise} \end{cases}$$

I.e., τ_1 is the time to reach the γ neighborhood of $\mathcal{W}^c$ using vector field f_1 and τ_2 is the time to escape from a γ neighborhood to a δ neighborhood of $\mathcal{W}^c$ using vector field $\bar{f}_2$.

This results in the following maps:

$$f_1^{\tau_1} : \overline{\mathcal{C}_\gamma^c} \to \partial \mathcal{C}_\gamma \tag{18}$$

$$\bar{f}_2^{\tau_2} : \mathcal{Q}_s - \mathcal{W}^c \to \overline{\mathcal{C}_\delta^c} \subset \overline{\mathcal{C}_\gamma^c}, \tag{19}$$

where $\overline{\mathcal{C}}$ is the closure of $\mathcal{C}$. With $\delta = 2\gamma$ define the map $P : \mathcal{Q}_s - \mathcal{W}^c \to \partial \mathcal{C}_\gamma$

$$P(q) = f_1^{\tau_1(\cdot, \gamma)} \circ \bar{f}_2^{\tau_2(q, 2\gamma)}(q) \tag{20}$$

and consider the recursive equation:

$$q_{k+1} = P(q_k). \tag{21}$$

The set $\partial \mathcal{C}_\gamma$ can be interpreted as a Poincaré section for the discrete system (21). We are now ready to present the following result:

Theorem 1 ([14]). *There exists an iteration number, $N : \mathcal{Q}_s \to \mathbb{N}$ such that the iterated hybrid dynamics, P^N brings $\mathcal{Q}_s$ to $\mathcal{Q}_\epsilon$.*

Proof. Define

$$N := \min\{n \in \mathbb{N} | 0 \leq N \leq N_\epsilon | \varphi \circ P^n(q_0) \leq \epsilon\},$$

and $\Delta\varphi(q) := \varphi \circ P(q) - \varphi(q)$. Since $\mathcal{Q}_s$ is a compact set it follows that $|\Delta\varphi|$ achieves its minimum value, Δ_ϵ, on that set, hence at most $N_\epsilon := ceiling(\varphi_s - \epsilon)/\Delta_\epsilon$ iterations are required before reaching $\mathcal{Q}_\epsilon$.

Note that all initial conditions in the pre-image of the "local surround", $\mathcal{R} := \bigcup_{t>0} f_1^{-t}(\mathcal{Q}_s - \mathcal{W}^c)$ are easily included in the basin of the goal, $\mathcal{Q}_s$, by an initial application of the controller u_1. While it is difficult to make any general formal statements about the size of $\mathcal{R}$, we show in the next section that for all the examples we have tried, the "missing" initial conditions, $\mathcal{Q} - \mathcal{R} = \mathcal{Z}$, comprise a set of empty interior (in all but one case $\mathcal{Z}$ is actually empty) because all of $\mathcal{W}^c$, excepting at most a set of measure zero, is included in $\mathcal{Q}_s$. In configuration spaces with more complicated topology, there is no reason to believe that this pleasant situation would prevail. To summarize, the following algorithm is guaranteed to bring all initial configurations in $\mathcal{R}$ to the goal set, $\mathcal{Q}_s$:

(1). $\forall q_0 \in \mathcal{Q}_s - \mathcal{W}^c$, follow successive applications of (21), i.e. use the inputs to equation (3):

$$u_1(q) := B^\dagger(q)\nabla\varphi(q) \tag{22}$$
$$u_2(q) := \sigma(q)B^\dagger(q)J(A(q))\nabla\varphi(q) \tag{23}$$

(2). $\forall q_0 \in \mathcal{W}^c$ use the input

$$u_3 := \begin{bmatrix} \alpha_1 \\ \alpha_2 \end{bmatrix}, \tag{24}$$

for a small amount of time t_3 such that $\varphi \circ f_3^{t_3}(q_0) < 1$, with $f_3(q) := B(q)u_3$.

(3). $\forall q_0 \in \mathcal{R} - \mathcal{Q}_s$, use the input u_1 for time t until $f_1^t(q_0) \in \mathcal{Q}_s$.

2.4 Other Considerations

- **Limit cycles in the level sets of** φ. In many practical applications switching between controllers f_1 and f_2 using a small δ-neighborhood is far too conservative. It may be possible to escape $\mathcal{W}^c$ by more than just the small collar $\xi^{-1}[\delta]$. If we could recognize the passage into $\mathcal{W}^s$ and switch off controller u_2 (i.e. turn $\mathcal{W}^s$ into an attractor of a suitable modified form of f_2) then a final application of controller u_1 is guaranteed to achieve the goal state, q^*. The hope of reworking the form of u_2 so that the resulting closed loop vector field, f_2, has its forward limit set solely in $\mathcal{W}^s$ thus raises the question of when there exists limit cycles in the level sets of φ

for the flow of f_2. More importantly, we seek a condition that guarantees that every trajectory of f_2 starting in a small neighborhood of $\mathcal{W}^c$ can intersect $\mathcal{W}^s$ either by forward or inverse time integration of system (11). Note that f_2 generates a planar flow, making the Bendixon's criteria a natural candidate for such condition. Several authors [16, 17, 18, 19] have developed extensions to Bendixson's criteria for higher dimensional spaces, obtaining in general conditions that preclude invariant sub-manifolds on some set. For systems with first integrals, such as some classes of systems that result from nonholonomic constraints, the conditions simplify to a divergence style test. Feckan's theorem (see [16]) states that in open subsets where $\operatorname{div} f_2 \neq 0$ there can exist no invariant submanifolds of any level precluding cyclic orbits. The divergence measure can be used this way to detect limit cycles. Note however, that the previous result does not preclude quasi-periodic orbits.

- **Computational heuristic substitutes for** σ. The σ function introduced in Lemma 5 modifies the flow of f_2 rendering the center manifold unstable. Having that property is sufficient for stabilization, but more can be accomplished. By careful craft of σ one can minimize the number of switches between controllers f_1 and $\bar{f}_2$ necessary to reach the desired neighborhood of the goal. If the stable manifold $\mathcal{W}^s$ is contained in the zero set of σ and $\mathcal{W}^s$ is made attractive by $\bar{f}_2$ for any point in $\mathcal{Q}_s$ then one gets $f_1^\infty \circ \bar{f}_2^\infty(\mathcal{Q}_s) = q^*$, i.e., only 2 steps are necessary to reach the goal. Different methods for approximating σ are presented in [14]. Specifically the function σ is replaced by the divergence of f_2 in a neighborhood of $\mathcal{W}^c$; by maximizing ξ, since that implies escaping $\mathcal{W}^c$ in some measure; or replacing σ by and implicit polynomial stable manifold approximation (please see [20, 14] for invariant manifold computations).

3 Hybrid Controller for Nonholonomic Dynamic Systems

In this section we look into the "lift" of the algorithm proposed in the previous section to nonholonomically constrained dynamical systems. The resulting corollaries arise naturally from the ideas introduced in [21]. Let (25) and (26) be the system of equations for unconstrained systems [22] and nonholonomically constrained systems [23] with $q, u \in \mathbb{R}^n$ and $v \in \mathbb{R}^m, m < n$:

$$M(q)\ddot{q} + c(q,\dot{q}) = u \tag{25}$$

$$M(q)\ddot{q} + c(q,\dot{q}) = A(q)^T \lambda + B(q)v \tag{26}$$
$$A(q)\dot{q} = 0$$

where M is the mass matrix, c the Coriolis term, A and B represent the actuation constraints defined in section 2 and λ is a vector of Lagrange multipliers.

We start by recalling some notation and lemmas required for the subsequent proofs. Using the "stack-kronecker notation" [24, 25, 26] consider the following linear map:

$$\acute{M}_q : x \mapsto [x \otimes I]^T D_q M^S \tag{27}$$

and the skew-symmetric value operator:

$$J_q(x) := \acute{M}_q(x) - \acute{M}_q^T(x) \tag{28}$$

Lemma 6 ([21]). *For any curve, $q : \mathbb{R} \to \mathcal{Q}$, and any vector, $x \in T_{q(t_0)}\mathcal{Q}$,*

$$\dot{M}_q|_{t_0} x = \acute{M}_{q(t_0)}(x) \dot{q}|_{t_0} \tag{29}$$

Lemma 7 ([21]). *Given a Lagrangian with kinetic energy, κ, with no potential forces present, and with an external torque or force actuating at every degree of freedom as specified by the vector, τ, the equations of motion may be written in the form:*

$$M(q)\ddot{q} + c(q, \dot{q}) = \tau \tag{30}$$

where

$$c(q, x) = C(q, x)x \tag{31}$$

and

$$C(q, x) := \frac{1}{2}\acute{M}(q_x) - \frac{1}{2}J_q(x) \tag{32}$$

Notice that the representation of the Coriolis and centripetal forces in terms of the bilinear operator valued map C only coincide at $\dot{q}$ with the quadratic expression $c(q, \dot{q})$. In general they are not the same.

Corollary 3 ([21]). *For any motion $q : \mathbb{R} \to \mathcal{Q}$, and any tangent vector, $x \in T_{\mathcal{Q}}q(t)$,*

$$x^T \left[\frac{1}{2}\dot{M}(q) - C(q, \dot{q}) \right] x \equiv 0 \tag{33}$$

Proof. From the previous lemma,

$$x^T \left[\frac{1}{2}\dot{M}(q) - C(q, \dot{q}) \right] x = -\frac{1}{2} x^T J_q(\dot{q}) x = 0$$

3.1 Embedding the Limit Behavior of Gradient Dynamics

Controller $f_1(q) = -H(q) \cdot \nabla \varphi(q)$, introduced in Section 2.1, aims to reach a fixed point in the center manifold $\mathcal{W}^c$. In order to lift the controller into a 2nd order system, theorem 2, concerning limit sets of gradient dynamics, is complemented with corollary 4. Let the state variables p_1, p_2 represent $q, \dot{q}$ respectively and let $\mathcal{P} = \mathcal{TQ}$ be the tangent bundle of $\mathcal{Q}$ for system (25).

Theorem 2 (Koditschek[21]). Let φ be a Morse function on $\mathcal{Q}$ which is exterior directed on the boundary $\partial \mathcal{Q}$, surpasses de value $\mu > 0$ on the boundary, and has a local minima at the points $\mathcal{G} := \{q_i\}_{i=1}^n \subset \mathcal{Q}$. Let $K_2 > 0$ denote some positive definite symmetric matrix. Consider the set of "bounded total energy" states

$$\mathcal{P}^\mu := \left\{ \begin{bmatrix} p_1 \\ p_2 \end{bmatrix} \in \mathcal{P} : \varphi(p_1) + \frac{1}{2} p_2^T M p_2 \leq \mu \right\} \tag{34}$$

Under the feedback algorithm

$$u := -K_2 p_2 - D\varphi^T(p_1) \tag{35}$$

$\mathcal{P}^\mu$ is a positive invariant set of the closed loop dynamical system within which all initial conditions excluding a set of measure zero take $\mathcal{G}$ as their positive limit set.

Let H be the nonholonomic projection matrix. Define $Q(q) := I - H(q)$ to be the nonholonomic converse projection matrix. Notice that $ker(A) = ker(Q)$ and therefore $Q(q)\dot{q} = 0$. Let $\mathcal{W}_0^c := \{q \in \mathcal{W}^c \wedge \dot{q} = 0\}$. As shown in Section 2.1, $\mathcal{W}^c$ is the center manifold of the system $\dot{q} = -H(q) \cdot \nabla \varphi(q)$. Rewriting equation (26) with a new input $v := B(q)^\dagger u$ we get:

$$M(q)\ddot{q} + c(q, \dot{q}) = A(q)^T \lambda + H(q)u$$
$$A(q)\dot{q} = 0 \tag{36}$$

Corollary 4. Let $K_2 = \bar{K}_2 H(q)$ with $\bar{K}_2 > 0$ denoting a positive definite symmetric matrix. Under the conditions of theorem 2 all the initial conditions of the system (36), excluding a set of measure zero, take $\mathcal{W}_0^c$ as their positive limit set.

Proof. Let $V = \varphi(q) + \frac{1}{2} \dot{q}^T M(q) \dot{q}$ be a Lyapunov function for (5). Then

$$\dot{V} = D\varphi \dot{q} + \frac{1}{2} \dot{q}^T \dot{M} \dot{q} - \dot{q}^T \left(H K_2 \dot{q} + H D\varphi^T + H C \dot{q} \right) + \underbrace{\dot{q}^T A^T \lambda}_{=0}$$

$$= \underbrace{D\varphi Q \dot{q}}_{=0} + \underbrace{\dot{q}^T J_q \dot{q}}_{=0} - \dot{q}^T H K_2 \dot{q}$$

$$= -\dot{q}^T H^T \bar{K}_2 H \dot{q}$$

$H^T \bar{K}_2 H$ is a semi-definite positive matrix. $\dot{V}$ is null when either $\dot{q} = 0$ or $H\dot{q} = 0$. Since $A(q)\dot{q} = 0 \Rightarrow H\dot{q} \neq 0$ then the largest invariant set is the interception of the previous sets with $\mathcal{W}_0^c$ resulting in $\mathcal{W}_0^c$. La Salle's theorem guarantees that (5) with input (35) takes $\mathcal{W}_0^c$ as the forward limit.

3.2 Embedding of More General Dynamics

We now seek to lift the controller $f_2(q)$ defined in Section 2.1 to a 2nd order system. We do so by adding once again a level regulator term, so that the reference dynamics attracts to a particular level set. First recall the embedding of general reference dynamics: let f be a reference vector field with Lyapunov function μ, and let $F(p) := p_2 - f(p_1)$. Consider the control algorithm,

$$u = -K_2 F - D\mu^T + MDfp_2 + Cf \quad (37)$$

which applied to the mechanical system (25) yields a closed loop form, $\dot{p} = h(p)$,

$$h(p) := \begin{bmatrix} p_2 \\ Dfp_2 - M^{-1}\left[K_2 F + CF + D\mu^T\right] \end{bmatrix}$$

Theorem 3 (Koditschek[21]). *If μ is a strict Lyapunov function for f on $\mathcal{Q}$, then*

$$V := \mu + \frac{1}{2} F^T M F$$

is a strict Lyapunov function for h on $\mathcal{P}$.

In system (5) the set of images of H for each point on $\mathcal{Q}$ is the tangent bundle of $\mathcal{Q}$. Therefore, since H is a projection operator then $\forall x \in \mathcal{P}$ we have $H(p_x).x = x$. Define $\bar{H} = M^{-1}HM$ and $\bar{Q} = M^{-1}QM$. For system (36) with input (37) the closed loop is written in the following way:

$$h(p) := \begin{bmatrix} p_2 \\ \bar{H}Dfp_2 - M^{-1}\left[HK_2 F + HCF + HD\mu^T + QCp_2 - A^T\lambda\right] \end{bmatrix}$$

Corollary 5. *Suppose $Hf = f$, i.e. the reference vector field respects the nonholonomic constraints. If μ is a strict Lyapunov function for f on $\mathcal{Q}$ for system (36), then*

$$V := \mu + \frac{1}{2} F^T M F$$

is a strict Lyapunov function for h on $\mathcal{P}$.

Proof. First note that $Q.f = 0$; $Q.p_2 = 0$; $A.f = 0$.

$$\dot{V} = D\mu \, p_2 + \frac{1}{2} F^T \dot{M} F + F^T M \bar{Q} D f p_2 +$$
$$- F^T \left(H K_2 F + H C F + H D \mu^T + Q C p_2 \right) + F^T A^T \lambda$$

$$= D\mu H f - F^T H K_2 F - \underbrace{f^T Q (M D f \dot{q} - C f)}_{=0} + \underbrace{f^T A^T \lambda}_{=0}$$

$$= D\mu f - F^T K_2 F$$

According to the hypothesis, the first term is negative except on the largest invariant set of $F = 0$. The second term is always negative except in $F^{-1}[0]$. The interception of the two results in the limit set of $\dot{q} = f(q)$.

The previous result show that as long as the reference dynamics respects the nonholonomic constraints we can apply Theorem 3 directly. Notice that Corollary 5 also applies to controller $-H(q).\nabla\varphi(q)$. In general such restrictive dynamics are not necessary for that controller, so using Corollary 4 gives a better tool since we are only interested on the limit set.

4 Simulations

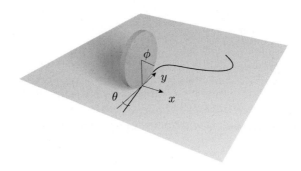

Fig. 3. The vertical rolling disk.

In this section we present simulation examples for the unicycle or vertical rolling disk depicted in Figure 3. The unicycle is commonly defined in the SE(2) configuration space with constraint equation $\dot{x}\sin(\theta) - \dot{y}\cos(\theta) = 0$. Since we are interested in simulations in a dynamic setting we will follow instead Bloch's vertical rolling disk [27], defined in the configuration space $\mathcal{Q} = \mathbb{R}^2 \times S^1 \times S^1 = \mathsf{SE}(2) \times S^1$ with coordinates $q = (x, x, \theta, \phi)$. The equations of motion for the vertical rolling disk are:

$$(mR^2 + I)\ddot{\phi} = u_1$$
$$J\ddot{\theta} = u_2, \qquad (38)$$

with the constraint equations:

$$\dot{x} = R\cos(x)\dot{\phi}$$
$$\dot{y} = R\sin(x)\dot{\phi}. \qquad (39)$$

Differentiating (39) in time and replacing $\ddot{\phi}$ from (38), one obtains a complete set of equations of motion that verifies the nonholonomic constraints for initial conditions that also verify (39):

$$\ddot{x} = -R\sin(\theta)\dot{\theta}\dot{\phi} + \frac{R\cos(\theta)}{mR^2 + I}u_1$$
$$\ddot{y} = R\cos(\theta)\dot{\theta}\dot{\phi} + \frac{R\sin(\theta)}{mR^2 + I}u_1 \qquad (40)$$

This system is now written in the form $M(q)\ddot{q} + c(q,\dot{q}) = B(q)u$ for which corollaries 4, 5 apply directly. The $A(q)$ and $B(q)$ matrices are:

$$A := \begin{bmatrix} -\frac{I+mR^2}{R} & -\sin(\theta) \\ 0 & \cos(\theta) \\ 0 & 0 \\ \cos(\theta) & 0 \end{bmatrix} \;;\; B := \begin{bmatrix} \frac{R\cos(\theta)}{mR^2+I} & 0 \\ \frac{R\sin(\theta)}{mR^2+I} & 0 \\ 0 & 1 \\ 1 & 0 \end{bmatrix} \qquad (41)$$

Although the algorithms presented in section 2 are defined only in $\mathbb{R}^3$, by close inspection of A and B one realizes that, for this particular example, by choosing a $\mathbb{R}^4$ navigation function defined only by the first three parameters of q we will obtain the "same" controller has in $\mathbb{R}^3$. Let φ be a navigation function such that $\nabla\varphi = [\varphi_x, \varphi_y, \varphi_\theta, 0]^T$. Next compute the $\mathbb{R}^4$ cross product of A and $\nabla\varphi$:

$$\times(A, \nabla\varphi) = \sum_{i,j,k,l=0}^{4} \epsilon_{ijkl}\nabla\varphi_j A_{k1} A_{l2} \hat{e}_i \qquad (42)$$

$$= \frac{-1}{\cos(\theta)} \begin{bmatrix} \varphi_\theta \cos(\theta) \\ \varphi_\theta \sin(\theta) \\ -\varphi_x \cos(\theta) - \varphi_y \sin(\theta) \\ -\frac{I+mR^2}{R}\theta \end{bmatrix}, \qquad (43)$$

where ϵ_{ijkl} denotes the permutation tensor, $\hat{e}_i$ are the canonical basis vectors and $A_{i,j}$ is the ith row, jth column of A. We now compare with the function

f_2, defined in equation (11) for the $\mathbb{R}^3$ unicycle with $A_3 = [-\sin(\theta), \cos(\theta), 0]^T$ and $\nabla \varphi_3 := [\varphi_x, \varphi_y, \varphi_\theta]^T$:

$$f_2 = A_3 \times \nabla \varphi_3 = \tag{44}$$

$$= \begin{bmatrix} \varphi_\theta \cos(\theta) \\ \varphi_\theta \sin(\theta) \\ -\varphi_x \cos(\theta) - \varphi_y \sin(\theta) \end{bmatrix} \tag{45}$$

The two previous computations produce, in effect, the same behavior for the variables x, y and θ. For each fixed coordinate ϕ in $\mathcal{Q} \subset \mathbb{R}^4$ one obtains a copy of the topology of $\mathsf{SE}(2)$. Therefore, from here on, although the configuration space is defined in $\mathbb{R}^4$, we will only be interested in x, y and θ.

4.1 Navigation Function

Kantor and Rizzi [28] solved the problem of positioning a robot in relation to a single engineered beacon by using the notion of Sequential Composition of Controllers [29]. The final approach to the goal is implemented using Ikeda's Variable Constraint Control. We recast the problem with a NF-encoding according to the approach described in the previous sections, to recover in simulation behavior comparable to that obtained in [28].

Let h be a change of coordinates from $\mathsf{SE}(2) \times S^1$ to double polar coordinates times S^1 that we denote by $\mathcal{P}$ with coordinates $p = [\eta, \mu, d, \phi]^T$:

$$\begin{bmatrix} \eta \\ \mu \\ d \\ \phi \end{bmatrix} = h(x, y, \theta, \phi) = \begin{bmatrix} \arctan(y/x) \\ \theta - \arctan(y/x) \\ \sqrt{x^2 + y^2} \\ \phi \end{bmatrix} \tag{46}$$

Obstacles are introduced on the field of view so that the robot maintains a range of distances to the beacon and keeps facing it:

$$\mu_m < \mu < \mu_M; \qquad d_m < d < d_M \tag{47}$$

Consider the following potential function:

$$\bar{\varphi}(p) := \frac{\left(2 - \cos(\eta - \eta^*) - \cos(\mu - \mu^*) + (d - d^*)^2\right)^k}{(1 - \cos(\mu - \mu_m))(1 - \cos(\mu - \mu_M))} \cdot \frac{1}{(d - d_m)(d_M - d)}, \tag{48}$$

and its "squashed" navigation function version $\check{\varphi} : \mathcal{P} \to [0, 1]$:

$$\check{\varphi}(p) := \frac{\bar{\varphi}(p)^l}{\epsilon + \bar{\varphi}(p)^l} \tag{49}$$

The navigation function written in the $\mathcal{Q}$ coordinates is $\varphi(q) = \check{\varphi} \circ h(q)$ and its derivative:

$$\nabla \varphi(q) = Dh^T(q) \cdot \nabla \check{\varphi} \circ h(q) \qquad (50)$$

We choose to present the Kantor-Rizzi example as the canonical illustration of our ideas due to the interesting topology of the configuration space. Since $\mathcal{Q}$ is not simply connected the level sets of φ change from topological spheres close to the goal q^* to topological tori close to the boundary of $\mathcal{Q}$. Initial conditions stating in the tori will generate quasi-periodic orbits when f_2 is used. In the dynamical setting this provides a good example of the applicability of corollary 5, resulting in the generation of reference dynamics that attract to a particular level set.

4.2 Kinematic Rolling Disk

We first simulate the previously described system in a kinematic setting by solving the system:

$$\dot{q} = B(q)u, \qquad (51)$$

and using the control functions f_1, f_2 defined in section 2:

$$u_1(q) := f_1(q) = -H\nabla\varphi \qquad (52)$$
$$u_2(q) := \sigma(q)f_2(q) = \times(A, \nabla\varphi)\sigma \qquad (53)$$

Figure 4 illustrates the resulting simulation where the initial condition is $q_0 = [1, 1, -\frac{3\pi}{4}, 0]^T$, the desired goal is $q_0 = [0, -2, \frac{\pi}{2}, 0]^T$, the body parameters are $I = J = m = R = 1$, the obstacles are $\mu_m = -\frac{\pi}{4}; \mu_M = \frac{\pi}{4}; d_m = 1; d_M = 3$ and $\sigma(q) = x$. The manifold $\{q \in \mathcal{Q} : x = 0\}$ is a good local approximation for the stable manifold $\mathcal{W}^s$ of the system $\dot{q} = f_1(q)$. One can observe that from the initial time to t_s the controller f_2 keeps the energy constant while moving exactly in the level set $\varphi^{-1}[\varphi^*]$, with $\varphi^* = 0.98$. At time t_s we switch to controller f_1 and the resulting final position is very close to the goal. Looking at ϕ in the "positions" graphic one observes that the robot does a back and forward motion, necessary to the parallel park maneuver. This comes as a natural consequence of moving in the surface of the torus shown in the "trajectories" plot.

4.3 Dynamic Rolling Disk

For the dynamic setting we solve the system defined by equations (38) and (40):

$$M(q)\ddot{q} + c(q, \dot{q}) = H(q)u, \qquad (54)$$

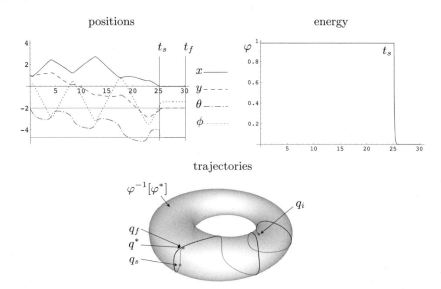

Fig. 4. Kinematic simulation of the vertical rolling disk.

with control functions (35) and (37):

$$u_1(q, \dot{q}) := -K_2 \dot{q} - \nabla \varphi \tag{55}$$
$$u_2(q, \dot{q}) := -K_2 F - D\mu^T + MDf_2\dot{q} + Cf_2, \tag{56}$$

where $F(q, \dot{q}) := \dot{q} - f_2(q)$ and $\mu := \alpha(\varphi - \varphi^*)^2$.

The first simulation, depicted in Figure 5, uses a high gain $\alpha = 5000$ in the function μ to track the level set as close as possible while the controller f_2 is in use. That results in a good tracking but very jerky steering motion, visible in the first part of the "velocities" and "trajectories" plots. The damping matrix K_2 is set to the identity matrix, resulting in low damping, as observed in the intervals $[t_s, t_f]$ of the "positions" and "velocities" plots.

For the second simulation in the dynamic setting, depicted in Figure 6, the parameter $\alpha = 250$ provokes a less accurate tracking of the desired level set φ^*, when using f_2, as one can observe in the "energy" and "trajectories" plots. However, the resulting motion is smoother then the previous simulation. For the controller f_1, the damping matrix $K_2 = 10I$ slows down the approach to the desired goal, elimination any oscillations as seen in the "energy" plot.

The damping matrix K_2, and the Lyapunov function μ are the design parameters for the control of equation (54).

5 Conclusions

This exploratory discussion paper addresses the reuse of navigation functions developed for fully actuated bodies in the setting of nonholonomic constrained

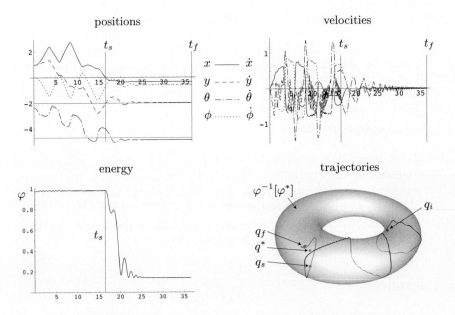

Fig. 5. Dynamic simulation of the vertical rolling disk.

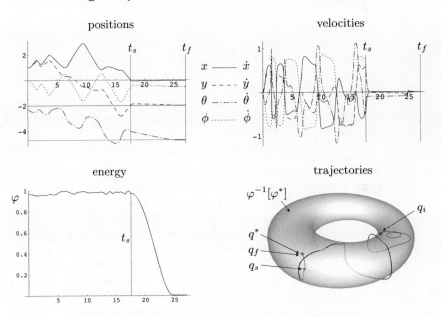

Fig. 6. Dynamic simulation of the vertical rolling disk.

systems for both kinematic and dynamic versions of the model. We suggest how the vector fields developed for kinematic systems can be lifted to the dynamic setting with the introduction of damping and proportional gain type constants. The simulations suggest this lifting can be readily realized in real applications, by proper choice of the damping and gain.

6 Acknowledgments

We thank John Guckenheimer, George Kantor, Al Rizzi and Anthony Bloch for a number of stimulating and helpful conversations bearing on the problem addressed in this text. This research was supported by the The University of Michigan, by DARPA/ONR N00014-98-1-0747, by DARPA/SPAWAR N660011-03-C-8045 and by Fundação para a Ciência e Tecnologia - Portugal, with the fellowship PRAXIS XXI/BD/18148/98.

References

1. Koditschek DE (1989) Robot planning and control via potential functions. In Khatib O, Craig J, Lozano-Pérez T, eds.: The Robotics Review. MIT Press 349–368
2. Thompson SW, Tait PG (Cambridge) Treatise on Natural Philosophy. University of Cambridge Press, 1886
3. Arimoto S, Miyazaki F (1984) Stability and robustness of PID feedback control for robot manipulators of sensory capability. In: Robotics Research, First International Symposium, Cambridge, MA, MIT Press
4. Khatib O, Maitre JFL (1978) Dynamic control of manipulators operating in a complex environment. In: Proceedings Third International CISM-IFToMM Symposium, Udine, Italy 267–282
5. Khatib O (1986) Real time obstacle avoidance for manipulators and mobile robots. 5(1):90–99
6. Rimon E, Koditschek DE (1992) Exact robot navigation using artificial potential fields. IEEE Transactions on Robotics and Automation 8(5):501–518
7. Koditschek DE, Rimon E (1990) Robot navigation functions on manifolds with boundary. Advances in Applied Mathematics 11:412–442
8. Cowan NJ, Weingarten JD, Koditschek DE (2002) Visual servoing via navigation functions. IEEE Trans. Rob. Aut. 18:521–533
9. Koditschek DE, Bozma HI (2001) Assembly as a noncooperative game of its pieces: analysis of 1d sphrere assemblies. Robotica 19(1):93–108
10. Karagoz C, Bozma H, Koditschek D (2004) Feedback-based event-driven parts moving. IEEE Transactions on Robotics and Automation 20(6):1012–1018
11. Koditschek DE (1991) Applications of natural motion control. ASME Journal of Dynamic Systems, Measurement, and Control 113(4):552–557
12. Bloch A (2003) Nonholonomic Mechanics and Control. Interdisciplinary Applied Mathematics. Springer-Verlag NY
13. Brockett RW (1981) New Directions in Applied Mathematics. In: Control theory and singular riemannian geometry. Springer 11–27

14. Lopes GAD, Koditschek DE (2006) Visual servoing for nonholonomically constrained three degree of freedom kinematic systems. submitted to IEEE Int. Journal of Computer Vision, Special Issue: Vision and Robotics
15. Lopes GAD, Koditschek DE (2004) Level sets and stable manifold approximations for perceptually driven nonholonomically constrained navigation. In: IEEE/RSJ International Conference on Intelligent Robots and Systems, Sendai, Japan
16. Fečkan M (2001) A generalization of Bendixson's criterion. In: Proceedings of the American Mathematical Society. Volume 129. 3395–3399
17. Fečkan M (2001) Criteria on the nonexistence of invariant lipschitz submanifolds for dynamical systems. Journal of Differential Equations (174):392–419
18. Li MY (200) Dynamics of differential equations on invariant manifolds. Journal of Differential Equations (168):295–320
19. Giraldo L, Gascón FG (2000) New proof and generalizations of the demidowitsch-schneider criterion. Journal of Mathematical Physics 41(9)
20. Guckenheimer J, Holmes P (1983) Nonlinear Oscillations, Dynamica Systems, and Bifurcations of Vector Fields. Verlag, New York
21. Koditschek DE (1987) Adaptive techniques for mechanical systems. In: Fifth Yale Workshop on Applications of Adaptive Systems Theory, New Haven, CT, Center for Systems Science, Yale University 259–265
22. Goldstein H (1950) Classical Mechanics. Addison-Wesley, Reading Mass.
23. Bloch A, Reyhanoglu M, McClamroch H (1992) Control and stabilization of nonholonomic dynamical systems. IEEE Trans. Aut. Control 37:1746–1757
24. (MacDuffee CC) The Theory of Matrices. (reprint of first edition) edn. Chelsea Publishing Company, New York
25. Bellman R (1965) Introduction to Matrix Analysis. McGraw Hill, New York
26. Koditschek DE (1987) Robot Control Systems. In: Encyclopedia of Artificial Inteligence. John Wiley and Sons, Inc. 902–923
27. Bloch A, Krishnaprasad P, Marsden J, Murray R (1996) Nonholonomic mechanical systems with symmetry. Arch. Rat. Mech. An. 136:21–99
28. Kantor G, Rizzi A (2003) Sequential composition for control of underactuated systems. Technical Report CMU-RI-TR-03-23, Robotics Institute, Carnegie Mellon
29. Burridge RR, Rizzi AA, Koditschek DE (1999) Sequential composition of dynamically dexterous robot behaviors. The International Journal of Robotics Research 18(6):534–555

Planning and Control of Robot Motion Based on Time-Scale Transformation

Sadao Kawamura[1] and Norimitsu Sakagami[2]

[1] Ritsumeikan University, Department of Robotics,
  1-1-1 Nojihigashi, Kusatsu-shi, Shiga 525-8577, Japan
  kawamura@se.ritsumei.ac.jp
[2] Tokai University, Department of Naval Architecture and Ocean Engineering,
  3-20-1 Orido, Shimizu-ku, Shizuoka-shi, 424-8610, Japan
  sakagami@scc.u-tokai.ac.jp

Summary. In this paper, we discuss control scheme design methods of a complex system that comprises many elements. To control such a system, we assert the importance of a control scheme design based on the structure, which is physically characterized. As examples of the control scheme design method, iterative learning control and time-scale transformation are introduced. This paper particularly describes that the dynamics of a robot and a contact environment are physically characterized by time-scale transformation. The effectiveness of the time-sale transformation is shown in cases where a robot moves in air and in water. It is difficult to model the hydrodynamic effect in the neighborhood of the robot when it moves in water. It is claimed that the difficulty on modeling is overcome if time-scale transformation is applied. Finally, the usefulness of time-scale transformation is demonstrated through examination of some experimental results.

1 Introduction

1.1 Design Methods of Control Schemes for a Complex System

We consider design methods of control schemes for a complex system that comprises many elements. A way of resolving the system into simple elements is generally useful at first to control such a complex system. Next, the element is investigated and modeled. Finally, each element's model parameters are identified. It might be natural to consider that the control of the whole system is realized by combining each identified element. Such a technique can be called a control scheme design method by an inverse process of element resolution of a system. In this way, excellent control performance is achievable in some cases. However, for cases in which the modeling and the parameter estimation of the element are incomplete, an incomplete part is accumulated by each element. Consequently, the required performance is often unachievable.

On the other hand, another control scheme design method uses the input-output relation of the system without considering each element inside the system. In memory-based control, which is represented by neural nets and so on, it is not necessary to identify each element of the system minutely as the model. We might need to devote attention only to the relation between input and output. That is, this method of control scheme design accumulates the input-output relation. However, this control scheme is not compatible with the physical model. Therefore, even if a suitable performance is obtainable through the input-output relation, it is often difficult to physically understand the control performance and to apply the obtained input-output relation data to another objective system.

Therefore, we consider a design method that is different from the two above. First, the method resolves an objective system into elements and it models each element. Here, it is extremely important to note that the inverse process of resolving to the elements is not necessary. Instead of the inverse process of the element resolution, the structure of each element that is physically characterized is considered in this paper. We assert the importance of the control scheme design method based on the structure, which is physically characterized. It is expected that the control performance can be physically expressed and that the obtained results are applicable to another objective system because the physical structure of each element is clear. Moreover, it is presumed that the more complex the object system is, the more useful this control system design method can be.

1.2 Inverse Dynamics and Iterative Learning Control

Here we consider an inverse dynamics problem for mechanical systems such as robots. That is, the problem is to obtain a desired torque pattern that produces a given desired motion pattern. For analyses of multivariable nonlinear robot dynamics, each element can correspond to each link of the robot. It is considered that each element is integrated into the overall robot dynamics. To identify each element, the physical parameters of the link length, the mass, the moment of inertia and so on are required. It is possible to calculate the desired input torque based on those parameter values and the desired motion (position, velocity and acceleration) if all parameters are identified perfectly. It is the inverse process of resolving to elements and is well known as the computed torque method. However, if modeling errors and parameter identification errors occur, then this method cannot make the desired input torque. Furthermore, the neural network method or memory-based control method was proposed. Those methods are very useful when the modeling of the objective system is very difficult. However, because the physical structure is unknown, the acquired data cannot be used easily when conditions such as the contact condition to the environment are changed. Moreover, in some cases, an enormous number of test motions is necessary.

To solve such problems, iterative learning control was proposed [1]-[4]. In iterative learning control, we devote attention to the input and output of the system and the trial motion is repeated using the actual robot system. At each trial motion, the input pattern is modified by the difference between the desired motion pattern and the actual robot motion pattern. The modified input torque pattern is used at the next trial motion. The input-output data at each sampling time are put into memory in a computer. Therefore, it might be called a kind of memory-based control. However, the convergence of the actual motion to the desired one is mathematically proven based on the physical structure of robot dynamics, which is called passivity. In iterative learning control, it is not necessary to identify the parameters of robot dynamics exactly. Moreover, even though repeatable disturbances exist, the desired motion is obtainable. The practical usefulness was demonstrated through experimental results [5],[6].

1.3 Planning and Control of Robot Motion Based on Time-Scale Transformation

The desired motion is obtained using the iterative learning control even if the parameters of the system are not known. The subsequent problem was whether or not it is possible to synthesize a new input torque pattern that produces a new desired motion pattern using a combination of input torque patterns that have been already obtained through iterative learning control. It was proven that a new desired input torque pattern for a different speed motion pattern is formed by some combination of the desired input patterns [7]. Because the motion trajectory is same in space but different in time, the changing of the motion speed means that the time-scale is changed in robot dynamics. We clarified the physical structure with the time-scale transformation on nonlinear robot dynamics in order to obtain a new desired input pattern without additional iteration [7]. In [7], the linear time-scale transformation was investigated. Results showed that nonlinear robot dynamics can be classified as the zero-th, first, and second-order terms of the coefficient with time-scale changing. The same characterization with time-scale transformation on robot dynamics had already been pointed out by Hollerbach [8]. However, the purpose of the paper was the characterization of robot dynamics and was not the use of the input torque patterns obtained through iterative learning control. According to our results, it is possible to produce another speed pattern without additional iterative operations if the robot has once learned three different speed motion patterns.

Our result in [7] treated the linear time-scale transformation or the coefficient of the time-scale function as constant. Next, we proposed nonlinear time-scale transformation in which the time-scale function can have nonlinearity with some conditions [9]. From this result, we were able to set arbitrary time-scale functions for robot motions. Nonlinear time-scale transformation became possible. Therefore, optimal control such as minimum-time-control

and so on were resolved as a planning problem with the time-scale function
[10]. Generally speaking, an optimal control input is computed based on the
estimated parameter values of the system. Therefore, if estimation errors are
apparent, then the computed control input cannot become optimal. On the
other hand, in time-scale transformation, parameter estimation is not necessary. As a result, the optimal control input is found only from the input torque
patterns obtained through iterative learning control.

Next, we consider the case in which a robot contacts on an environment.
Modeling and the identification of the contact environment are generally difficult. A robot itself has multivariable nonlinear dynamics. Furthermore, the
contact environment of the robot is added and the complexity of the total dynamics increases. In such a case, it is possible to form the desired input pattern
and to plan an optimal motion easily by clarifying the physical characteristics through time-scale transformation of the complex dynamics including the
contact environment. In the following sections, we introduce cases in which a
robot mechanically contacts a rigid surface and a robot moves in water. The
modeling and identification of hydrodynamic terms in the total dynamics are
extremely difficult in practice when a robot moves in water. On the other
hand, if the physical characteristics of the robot's dynamics of and the water
environment are clarified in the sense of the time-scale transformation, the
desired input patterns can be formed easily. An optimal control input is obtainable even though neither the parameter identification nor the iteration is
utilized. In the following parts of this paper, the effectiveness of the time-scale
transformation is explained theoretically and the usefulness of this method is
demonstrated using some experimental results.

2 Characterization of Robot and Environment Dynamics

2.1 Robot and Environment Dynamics

Suppose that a robot moves in contact with an environment and that the
dynamics of the robot and the environment are represented as

$$\boldsymbol{d}_r(\boldsymbol{q}, \dot{\boldsymbol{q}}, \ddot{\boldsymbol{q}}) + \boldsymbol{d}_e = \boldsymbol{u}, \tag{1}$$

where $\boldsymbol{q}$, $\dot{\boldsymbol{q}}$, and $\ddot{\boldsymbol{q}}$ respectively represent the joint angle, velocity, and acceleration vectors, $\boldsymbol{d}_r$ denotes robot dynamics, $\boldsymbol{d}_e$ means environment dynamics,
and $\boldsymbol{u}$ denotes control input.

Figure 1 shows the cases in which a robot mechanically contacts a rigid
surface and a robot moves in water. When $\boldsymbol{d}_e = 0$, Eq. (1) represents that a
robot manipulator moves in free space without contacts.

2.2 Passivity

For iterative learning control, passivity of the robot and the environment dynamics plays an important role in acquiring desired motion patterns $\boldsymbol{q}_d(t)$ ($t \in$

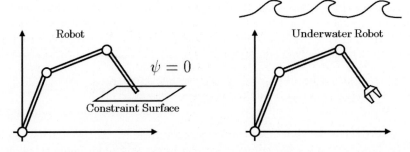

Fig. 1. A robot and its environment

$[0, T]$). Robot dynamics are known to be physically characterized by passivity [4]. Convergence of the robot motion to the desired motion in the learning process is guaranteed based on the passivity condition. As described in [4], when a robot moves in free space, the passivity condition is satisfied and the robot can learn the desired motion in the learning process. In the case that a robot moves in contact with an environment, we might investigate whether the robot and the environment dynamics given by Eq. (1) can satisfy the passivity condition or not. When a robot moves with mechanical constraints, the passivity condition is satisfied under some assumptions [4]. On the other hand, the authors theoretically and experimentally investigated the case of underwater robots in [12]. Unfortunately, the passivity condition with underwater robot dynamics has not been checked rigorously yet, but several sets of experimental results show that the motion of the underwater robot manipulator converges to the desired one with some conditions even if the hydrodynamic forces and moments influence the iterative learning process.

2.3 Classification of Time-Scale Transformation

Time-scale transformation is a useful tool to classify nonlinear dynamics of robots and environments. We explain the characteristics of time-scale transformation and classify nonlinear dynamics from the viewpoint of time-scale transformation.

If we consider a vector function $\boldsymbol{x}(t)$, a time-scale is transformed easily by $\boldsymbol{x}(r(t))$, where $r(t)$ is a time-scale function. Mathematically, the relationship between $\boldsymbol{x}(t)$ and $\boldsymbol{x}(r(t))$ is as follows:

$$\boldsymbol{x}(t) = \boldsymbol{x}(r(t)). \tag{2}$$

In the case of a function $d\boldsymbol{x}(t)/dt$, the corresponding transformation is expressed as

$$\frac{d\boldsymbol{x}(t)}{dt} = \frac{dr}{dt}\frac{d\boldsymbol{x}(r)}{dr}. \tag{3}$$

Namely, the derivative of the time-scale function $r(t)$ must be multiplied. In the same manner as Eq. (3), a function $\mathrm{d}^2 x(t)/\mathrm{d}t^2$ generates a square term of $\mathrm{d}r/\mathrm{d}t$. Here, it is important to classify nonlinear dynamics from the viewpoint of the time-scale transformation as follows:

D^0 class : zero order of $\mathrm{d}r/\mathrm{d}t$ (ex. gravitational term),
D^1 class : first order of $\mathrm{d}r/\mathrm{d}t$ (ex. viscous friction term),
D^2 class : second order of $\mathrm{d}r/\mathrm{d}t$ (ex. inertia, centrifugal and Coriolis terms).

Therefore, the robot dynamics can be characterized as D^0, D^1 and D^2 classes in the sense of time-scale transformation. Regarding environmental dynamics, the structure of the environmental dynamics should be investigated. Of course, a higher order of $\mathrm{d}r/\mathrm{d}t$ might be considered if necessary. However, it is presumed that D^0, D^1, and D^2 classes are sufficient for many cases because motion equations such as the Newton-Euler equation belong to D^0, D^1, and D^2 classes.

3 Planning and Control of Robot Motion under an Endpoint Constraint

3.1 Dynamics and Iterative Learning Control of Robots under an Endpoint Constraint

Here, we consider the case in which an endpoint of a robot manipulator is constrained on a rigid and fixed surface. Assume that the dynamics of a robot with 3 degrees of freedom (D.O.F.) in contact with a constraint surface $\psi = 0$, as shown on the left side of Fig. 1, are described as

$$\boldsymbol{R}\left(\boldsymbol{q}(t)\right)\frac{\mathrm{d}^2 \boldsymbol{q}(t)}{\mathrm{d}t^2} + \frac{\mathrm{d}\boldsymbol{q}^T(t)}{\mathrm{d}t}\boldsymbol{S}\left(\boldsymbol{q}(t)\right)\frac{\mathrm{d}\boldsymbol{q}(t)}{\mathrm{d}t} + \boldsymbol{B}_q\frac{\mathrm{d}\boldsymbol{q}(t)}{\mathrm{d}t} + \boldsymbol{g}\left(\boldsymbol{q}(t)\right)$$
$$= \boldsymbol{u}(t) - \boldsymbol{\gamma}_\psi^T\left(\boldsymbol{q}(t)\right)\lambda(t) - \boldsymbol{J}^T\left(\boldsymbol{q}(t)\right)\boldsymbol{B}_p\frac{\mathrm{d}\boldsymbol{p}(t)}{\mathrm{d}t}. \quad (4)$$

The matrices and the vectors in the above equation are given as follows:

$\boldsymbol{R}(\boldsymbol{q}) \in \mathcal{R}^{3 \times 3}$: inertia matrix,
$\boldsymbol{S}(\boldsymbol{q}) \in \mathcal{R}^{3 \times 3 \times 3}$: position dependent tensor in formulation of Coriolis and centrifugal forces,
$\boldsymbol{B}_q \in \mathcal{R}^{3 \times 3}$: coefficient matrix of viscous friction at joint parts,
$\boldsymbol{g}(\boldsymbol{q}) \in \mathcal{R}^{3 \times 1}$: position-dependent vector of gravity force,
$\boldsymbol{p}(t) \in \mathcal{R}^{3 \times 1}$: task-oriented coordinates given by $[x(t), y(t), z(t)]^T$,
$\boldsymbol{J}(\boldsymbol{q}) \in \mathcal{R}^{3 \times 1}$: Jacobian which is defined as $\boldsymbol{J} = \partial \boldsymbol{p}/\partial \boldsymbol{q}$,
$\boldsymbol{\gamma}_\psi(\boldsymbol{q}) \in \mathcal{R}^{1 \times 3}$: vector defined as $\boldsymbol{\gamma}_\psi(\boldsymbol{q}) = \frac{\partial \psi}{\partial \boldsymbol{p}}\frac{\partial \boldsymbol{p}}{\partial \boldsymbol{q}} = \frac{\partial \psi}{\partial \boldsymbol{p}}\boldsymbol{J}(\boldsymbol{q})$

$\lambda(t)$: contact force with the normal direction to the constraint surface,
$\boldsymbol{B}_p \in \mathcal{R}^{3\times 3}$: coefficient matrix of viscous friction at the contact point on the surface,

For simplicity, Eq. (4) is rewritten as

$$\boldsymbol{R}\left(\boldsymbol{q}(t)\right)\frac{\mathrm{d}^2\boldsymbol{q}(t)}{\mathrm{d}t^2} + \frac{\mathrm{d}\boldsymbol{q}^T(t)}{\mathrm{d}t}\boldsymbol{S}\left(\boldsymbol{q}(t)\right)\frac{\mathrm{d}\boldsymbol{q}(t)}{\mathrm{d}t} + \boldsymbol{B}\left(\boldsymbol{q}(t)\right)\frac{\mathrm{d}\boldsymbol{q}(t)}{\mathrm{d}t} + \boldsymbol{g}\left(\boldsymbol{q}(t)\right)$$
$$= \boldsymbol{u}(t) - \boldsymbol{\gamma}_\psi^T\left(\boldsymbol{q}(t)\right)\lambda(t) \tag{5}$$

where

$$\boldsymbol{B}\left(\boldsymbol{q}(t)\right) = \boldsymbol{B}_q + \boldsymbol{J}^T(\boldsymbol{q}(t))\boldsymbol{B}_p\boldsymbol{J}(\boldsymbol{q}(t)).$$

Iterative Learning Control of Robot Manipulators with Endpoint Constraints

In the case where the endpoint of a robot manipulator is constrained on a rigid and fixed surface, iterative learning control guarantees that the robot motion and the contact force converge to the desired motion and the desired force as the trial number $k \to \infty$. The convergence of this iteration was proven mathematically by Arimoto based on passivity and orthogonality in [4].

3.2 Time-Scale Transformation

Classification of Robot Dynamics on Time-Scale Transformation

Presuming that the iterative learning control realizes four desired motions $\boldsymbol{q}_i(r_i)$ ($i = 1, 2, 3, 4$), the spatial paths on these motions are identical, but the time trajectories are different. In [7], we used three desired motions to form a new desired motion. In that case in [7], the time-scale functions were linear with a standard time-scale. Nonlinear time-scale functions must be considered to treat general cases. For this purpose, we introduce the fourth desired motion. The role of the fourth motion will be explained in the following section.

All motions have the same spatial path. Therefore, the same relation as that in Eq. (2) is obtained as

$$\boldsymbol{q}_1\left(r_1(t)\right) = \boldsymbol{q}_2\left(r_2(t)\right) = \boldsymbol{q}_3\left(r_3(t)\right) = \boldsymbol{q}_4\left(r_4(t)\right), \tag{6}$$

where $r_i(t)$ ($i = 1, 2, 3, 4$) are appropriate time-scale functions.

Presume cases in which the contact forces λ_i ($i = 1, 2, 3, 4$) have an identical pattern, but have different speeds. That is, the contact forces λ_i ($i = 1, 2, 3, 4$) have the following relationships:

$$\lambda_1\left(r_1(t)\right) = \lambda_2\left(r_2(t)\right) = \lambda_3\left(r_3(t)\right) = \lambda_4\left(r_4(t)\right). \tag{7}$$

Regarding the velocity and acceleration, as shown in Eq. (3) a linear term and a square term of dr/dt are obtained, respectively. In Eq. (5), we know that the robot dynamics with mechanical constraints are characterized by the time-scale transformation. We classify the robot dynamics with mechanical constraints as

D^0 class : gravitational, contact force terms,

D^1 class : viscous friction terms,

D^2 class : inertia, centrifugal and Coriolis terms.

Conditions for The Time-Scale Transformation

Without loss of generality, the standard time-scale t is given as $r_1(t) = t$. Moreover, the following conditions are set for the time-scale functions.

I. For $r_i(t)$ $(i = 0, 1, 2, 3, 4)$:
  (i) $r_i(0) = 0$, $r_i(T_1) = T_i$ (T_i : Terminal time),
  (ii) $r_i(t) \in C^2$ $t \in [0, T_i]$,
  (iii) $0 < dr_i(t)/dt < \infty$ $t \in [0, T_i]$.

II. For $r_2(t)$ and $r_3(t)$:
  $r_2(t) = k_2 t$, $r_3(t) = k_3 t$
  where k_2 and k_3 are positive constant scalars such that $k_2 \neq 1$, $k_3 \neq 1$, and $k_2 \neq k_3$.

III. For $r_4(t)$:
  $1/\frac{dr_4(t)}{dt} \neq 0$, $\frac{d^2 r_4(t)}{dt^2} \neq 0$ $t \in [0, T_4]$

In condition I, (ii) and (iii) mean that the functions $dr_i(t)/dt$ are continuously differentiable and $r_i(t)$ are monotonically increasing functions. Here, it is noteworthy that the motion $q_i(r_i)$ becomes more rapid if $dr_i(t)/dt < 1$. It slows if $dr_i(t)/dt > 1$, as compared with $q_1(t)$. Condition II means that the time-scale functions of the desired motions q_1, q_2 and q_3 must have a linear relation. Condition III plays an important role to treat the nonlinear time-scale function.

Preliminary for Formation of a New Desired Torque Pattern

The four input torque patterns $u_i(r_i)$ $(i = 1, 2, 3, 4)$ to generate the four desired motion patterns exactly $q_i(r_i)$ and the four desired contact force patterns $\lambda_i(r_i)$ $(i = 1, 2, 3, 4)$ are acquired through iterative learning control, meaning that those input patterns can satisfy

$$R(q_i(r_i)) \frac{d^2 q_i(r_i)}{dr_i^2} + \frac{dq_i^T(r_i)}{dr_i} S(q_i(r_i)) \frac{dq_i(r_i)}{dr_i} + B(q_i(r_i)) \frac{dq_i(r_i)}{dr_i}$$
$$+ g(q_i(r_i)) = u_i(r_i) - \gamma_\psi^T(q_i(r_i)) \lambda_i(r_i) \quad (i = 1, 2, 3, 4). \tag{8}$$

Now, the following equations are satisfied from the characteristics of the time-scale functions.

$$\frac{d\boldsymbol{q}_1(t)}{dt} = \frac{dr_i}{dt}\frac{d\boldsymbol{q}_i(r_i)}{dr_i},$$

$$\frac{d^2\boldsymbol{q}_1(t)}{dt^2} = \frac{d^2 r_i}{dt^2}\frac{d\boldsymbol{q}_i(r_i)}{dr_i} + \left(\frac{dr_i}{dt}\right)^2 \frac{d^2\boldsymbol{q}_i(r_i)}{dr_i^2}. \quad (9)$$

Therefore, we obtain:

$$\frac{d\boldsymbol{q}_i(r_i)}{dr_i} = a_i \frac{d\boldsymbol{q}_1(t)}{dt},$$

$$\frac{d^2\boldsymbol{q}_i(r_i)}{dr_i^2} = a_i^2 \frac{d^2\boldsymbol{q}_1(t)}{dt^2} - a_i^3 b_i \frac{d\boldsymbol{q}_1(t)}{dt} \quad (10)$$

where

$$a_i \triangleq 1/\frac{dr_i}{dt}, \quad b_i \triangleq \frac{d^2 r_i}{dt^2}. \quad (11)$$

Substituting Eqs. (6), (7) and (10) into Eq. (8) yields

$$-a_i^3 b_i \boldsymbol{f}_{14}(t) + a_i^2 \boldsymbol{f}_{11}(t) + a_i \boldsymbol{f}_{12}(t) + \boldsymbol{f}_{13}(t) = \boldsymbol{u}_i(r_i) \quad (i=1,2,3,4) \quad (12)$$

where

$$\boldsymbol{f}_{11}(t) = \boldsymbol{R}(\boldsymbol{q}_1(t))\frac{d^2\boldsymbol{q}_1(t)}{dt^2} + \frac{d\boldsymbol{q}_1^T(t)}{dt}\boldsymbol{S}(\boldsymbol{q}_1(t))\frac{d\boldsymbol{q}_1(t)}{dt}, \quad (13)$$

$$\boldsymbol{f}_{12}(t) = \boldsymbol{B}(\boldsymbol{q}_1(t))\frac{d\boldsymbol{q}_1(t)}{dt}, \quad (14)$$

$$\boldsymbol{f}_{13}(t) = \boldsymbol{g}(\boldsymbol{q}_1(t)) + \boldsymbol{\gamma}_\psi^T(\boldsymbol{q}_1(t))\boldsymbol{\lambda}_1(t), \quad (15)$$

$$\boldsymbol{f}_{14}(t) = \boldsymbol{R}(\boldsymbol{q}_1(t))\frac{d\boldsymbol{q}_1(t)}{dt}. \quad (16)$$

Here, it is crucial to know whether the input patterns $\boldsymbol{u}_i$ in Eq. (12) contain the vector $\boldsymbol{f}_{14}(t)$ in Eq. (16), even though the vector $\boldsymbol{f}_{14}(t)$ does not exist in the original robot dynamics. In other words, vector $\boldsymbol{f}_{14}(t)$ is produced from the nonlinearity of the time-scale functions. In the following, it will be shown that the vector $\boldsymbol{f}_{14}(t)$ is obtained under conditions I, II, and III with the time-scale functions $r_i(t)$.

First, using matrix form, Eq. (12) is rewritten as

$$\boldsymbol{A}\boldsymbol{f} = \boldsymbol{u}, \quad (17)$$

where the following pertain.

$$\boldsymbol{A} \triangleq \begin{bmatrix} a_1^2 \boldsymbol{I} & a_1 \boldsymbol{I} & \boldsymbol{I} & -a_1^3 b_1 \boldsymbol{I} \\ a_2^2 \boldsymbol{I} & a_2 \boldsymbol{I} & \boldsymbol{I} & -a_2^3 b_2 \boldsymbol{I} \\ a_3^2 \boldsymbol{I} & a_3 \boldsymbol{I} & \boldsymbol{I} & -a_3^3 b_3 \boldsymbol{I} \\ a_4^2 \boldsymbol{I} & a_4 \boldsymbol{I} & \boldsymbol{I} & -a_4^3 b_4 \boldsymbol{I} \end{bmatrix}, \quad \boldsymbol{f} \triangleq \begin{bmatrix} \boldsymbol{f}_{11}(r_1) \\ \boldsymbol{f}_{12}(r_2) \\ \boldsymbol{f}_{13}(r_3) \\ \boldsymbol{f}_{14}(r_4) \end{bmatrix}, \quad \boldsymbol{u} \triangleq \begin{bmatrix} \boldsymbol{u}_1(t) \\ \boldsymbol{u}_2(k_2 t) \\ \boldsymbol{u}_3(k_3 t) \\ \boldsymbol{u}_4(r_4) \end{bmatrix}. \quad (18)$$

From $r_1(t) = t$ and condition II of the time-scale functions, matrix A can be rewritten as

$$A = \begin{bmatrix} & & & 0 \\ & B & & 0 \\ & & & 0 \\ a_4^2 I & a_4 I & I & -a_4^3 b_4 I \end{bmatrix}, \quad B = \begin{bmatrix} a_1^2 I & a_1 I & I \\ a_2^2 I & a_2 I & I \\ a_3^2 I & a_3 I & I \end{bmatrix}. \quad (19)$$

Because matrix B is a block Vandermonde matrix and neither a_4 nor b_4 becomes zero from Eq. (11) and condition III, matrix A is always non-singular. Therefore, we obtain

$$f = [f_{11}(t), f_{12}(t), f_{13}(t), f_{14}(t)]^T = A^{-1} u, \quad (20)$$

$$f_{14} = [0\ 0\ 0\ I]\, f = [0\ 0\ 0\ I]\, A^{-1} u. \quad (21)$$

Formation of the Desired Input Torque Pattern

Here, the input torque pattern u_0 to realize an arbitrary desired motion q_0 is formed from u_1, u_2 and u_3 and the already obtained compensation vector f_{14}.

It is considered that r_i ($i = 1, 2, 3, 4$) is a function of r_0. Furthermore, we define

$$\alpha_i \triangleq 1/\frac{d r_i}{d r_0}, \qquad \beta_i \triangleq 1/\frac{d^2 r_i}{d^2 r_0}. \quad (22)$$

In the same manner as Eq. (10), the following equations

$$\frac{d q_i(r_i)}{d r_i} = \alpha_i \frac{d q_0(r_0)}{d r_0},$$
$$\frac{d^2 q_i(r_i)}{d r_i^2} = \alpha_i^2 \frac{d^2 q_0(r_0)}{d r_0^2} - \alpha_i^3 \beta_i \frac{d q_0(r_0)}{d r_0} \quad (23)$$

hold, and we obtain the dynamics of

$$\alpha_i^2 f_{01}(r_0) + \alpha_i f_{02}(r_0) + f_{01}(r_0) = u_i(r_i) + \alpha_i^3 \beta_i f_{04}(r_0), \quad (i = 1, 2, 3)\quad(24)$$

where

$$f_{01}(r_0) = R(q_0(r_0))\frac{d^2 q_0(r_0)}{d r_0^2} + \frac{d q_0^T(r_0)}{d r_0} S(q_0(r_0)) \frac{d q_0(r_0)}{d r_0}, \quad (25)$$

$$f_{02}(r_0) = B(q_0(r_0)) \frac{d q_0(r_0)}{d r_0}, \quad (26)$$

$$f_{03}(r_0) = g(q_0(r_0)) + \gamma_\psi^T(q_0(r_0)) \lambda_0(r_0), \quad (27)$$

$$f_{04}(r_0) = R(q_0(r_0)) \frac{d q_0(r_0)}{d r_0}. \quad (28)$$

Because

$$\alpha_2 = \mathrm{d}r_0/\mathrm{d}(r_2 t) = \alpha_1/k_2, \qquad \alpha_3 = \mathrm{d}r_0/\mathrm{d}(r_3 t) = \alpha_1/k_3, \qquad (29)$$
$$\beta_2 = \mathrm{d}^2(r_2 t)/\mathrm{d}r_0^2 = k_2 \beta_1, \qquad \beta_3 = \mathrm{d}^2(r_3 t)/\mathrm{d}r_0^2 = k_3 \beta_1, \qquad (30)$$

and vector $\boldsymbol{f}_{04}$ is obtained from $\boldsymbol{f}_{14}$ in the following manner

$$\boldsymbol{f}_{04} = \boldsymbol{R}(\boldsymbol{q}_0(r_0)) \frac{\mathrm{d}\boldsymbol{q}_0(r_0)}{\mathrm{d}r_0} = \boldsymbol{R}(\boldsymbol{q}_0(r_0)) \frac{\mathrm{d}\boldsymbol{q}_1(t)}{\mathrm{d}t} \frac{\mathrm{d}t}{\mathrm{d}r_0} = \frac{\boldsymbol{f}_{14}(t)}{\alpha_1}, \qquad (31)$$

we can rewrite Eq. (24) as

$$\boldsymbol{A}^* \boldsymbol{f}^* = \boldsymbol{u}^* \qquad (32)$$

where

$$\boldsymbol{A}^* \triangleq \begin{bmatrix} \alpha_1^2 \boldsymbol{I} & \alpha_1 \boldsymbol{I} & \boldsymbol{I} \\ (\alpha_1/k_2)^2 \boldsymbol{I} & (\alpha_1/k_2)\boldsymbol{I} & \boldsymbol{I} \\ (\alpha_1/k_3)^2 \boldsymbol{I} & (\alpha_1/k_3)\boldsymbol{I} & \boldsymbol{I} \end{bmatrix}, \quad \boldsymbol{f}^* \triangleq \begin{bmatrix} \boldsymbol{f}_{01}(r_0) \\ \boldsymbol{f}_{02}(r_0) \\ \boldsymbol{f}_{03}(r_0) \end{bmatrix},$$

$$\boldsymbol{u}^* \triangleq \begin{bmatrix} \boldsymbol{u}_1(t) + \alpha_1^2 \beta_1 \boldsymbol{f}_{14}(t) \\ \boldsymbol{u}_2(k_2 t) + (\alpha_1^2 \beta_1/k_2^2)\boldsymbol{f}_{14}(t) \\ \boldsymbol{u}_3(k_3 t) + (\alpha_1^2 \beta_1/k_3^2)\boldsymbol{f}_{14}(t) \end{bmatrix}. \qquad (33)$$

Here, matrix $\boldsymbol{A}^*$ is in Vandermonde form and is non-singular. Therefore, we obtain

$$\begin{aligned} [\boldsymbol{I}\ \boldsymbol{I}\ \boldsymbol{I}]\, \boldsymbol{A}^{*-1} \boldsymbol{u}^* &= \boldsymbol{f}_{01}(r_0) + \boldsymbol{f}_{02}(r_0) + \boldsymbol{f}_{03}(r_0) \\ &= \boldsymbol{R}(\boldsymbol{q}_0(r_0)) \frac{\mathrm{d}^2 \boldsymbol{q}_0(r_0)}{\mathrm{d}r_0^2} + \frac{\mathrm{d}\boldsymbol{q}_0^T(r_0)}{\mathrm{d}r_0} \boldsymbol{S}(\boldsymbol{q}_0(r_0)) \frac{\mathrm{d}\boldsymbol{q}_0(r_0)}{\mathrm{d}r_0} \\ &+ \boldsymbol{B}(\boldsymbol{q}_0(r_0)) \frac{\mathrm{d}\boldsymbol{q}_0(r_0)}{\mathrm{d}r_0} + \boldsymbol{g}(\boldsymbol{q}_0(r_0)) + \boldsymbol{\gamma}_\psi^T(\boldsymbol{q}_0(r_0)) \lambda_0(r_0) \\ &= \boldsymbol{u}_0(r_0). \end{aligned} \qquad (34)$$

The desired input torque pattern $\boldsymbol{u}_0(r_0)$ for a different speed motion pattern is obtainable through some combination of input patterns $\boldsymbol{u}_i(r_i)$ ($i = 1, 2, 3, 4$).

3.3 Time-Optimal Control Method

Here, we propose a motion planning method that realizes a minimum-time motion without parameter estimation of robot dynamics. In the proposed method, limitations of input torque and velocity can be incorporated easily into the planning process. To simplify the explanation, we treat only the limitations on the input torque, which are represented as

$$u_{min}^j \leq u_{opt}^j(t) \leq u_{max}^j \qquad (j = 1, \cdots, n), \qquad (35)$$

where u_{min}^j and u_{max}^j are the minimum and the maximum bounds on the j-th element of the input torque.

Considering definitions of $\boldsymbol{A}^*$ and $\boldsymbol{u}^*$, Eq. (34) is rewritten as

$$\begin{aligned}\boldsymbol{u}_0(r_0) &= [\boldsymbol{I}\ \boldsymbol{I}\ \boldsymbol{I}]\,\boldsymbol{A}^{*-1}\boldsymbol{u}^* \\ &= -\frac{\boldsymbol{f}_{14}(t)}{\alpha_1^3}(-\alpha_1^3\beta_1) + \frac{(\alpha_1 - k_2)(\alpha_1 - k_3)}{(1 - k_2)(1 - k_3)\alpha_1^2}\boldsymbol{u}_1(t) \\ &\quad - \frac{(\alpha_1 - k_3)(\alpha_1 - 1)}{(k_2 - k_3)(k_2 - 1)\alpha_1^2}\boldsymbol{u}_2(k_2 t) + \frac{(\alpha_1 - 1)(\alpha_1 - k_2)}{(k_3 - 1)(k_3 - k_2)\alpha_1^2}\boldsymbol{u}_3(k_3 t)\end{aligned} \quad (36)$$

Now, scalars k_2, k_3 and vectors $\boldsymbol{u}_1, \boldsymbol{u}_2, \boldsymbol{u}_3, \boldsymbol{f}_{14}$ are already obtained from the motions $\boldsymbol{q}_i(r_i)$. Moreover, β_1 is rewritten as

$$\beta_1 = \frac{d^2 t}{dr_0^2} = \frac{dt}{dr_0}\frac{d(dt/dr_0)}{dt} = \frac{1}{\alpha_1}\frac{d\alpha_1^{-1}}{dt} = -\frac{1}{\alpha_1^3}\frac{d\alpha_1}{dt}. \quad (37)$$

Therefore, substituting Eq. (37) into Eq. (36), we obtain

$$\boldsymbol{u}_0(r_0) = \boldsymbol{\chi}_1(t, \alpha_1)\frac{d\alpha_1}{dt} + \boldsymbol{\chi}_2(t, \alpha_1), \quad (38)$$

where

$$\begin{aligned}\boldsymbol{\chi}_1(t, \alpha_1) &= -\frac{\boldsymbol{f}_{14}(t)}{\alpha_1^3} \\ \boldsymbol{\chi}_2(t, \alpha_1) &= \frac{(\alpha_1 - k_2)(\alpha_1 - k_3)}{(1 - k_2)(1 - k_3)\alpha_1^2}\boldsymbol{u}_1(t) - \frac{(\alpha_1 - k_3)(\alpha_1 - 1)}{(k_2 - k_3)(k_2 - 1)\alpha_1^2}\boldsymbol{u}_2(k_2 t) \\ &\quad + \frac{(\alpha_1 - 1)(\alpha_1 - k_2)}{(k_3 - 1)(k_3 - k_2)\alpha_1^2}\boldsymbol{u}_3(k_3 t).\end{aligned} \quad (39)$$

That is to say, each element of the input torque pattern $\boldsymbol{u}_0(r_0)$ has a linear function of $d\alpha_1/dt$.

Now, $\boldsymbol{u}_{opt}(r_{opt})$ is given by expanding or shrinking the time-scale r_0. Therefore, the time-scale function r_0 is expanded or shrunk so that the input torque pattern $\boldsymbol{u}_0(r_0)$ satisfies the limits of the input torque as

$$u_{min}^j \leq \chi_1^j(t, \alpha_1)\frac{d\alpha_1}{dt} + \chi_2^j(t, \alpha_1) \leq u_{max}^j, \quad (j = 1, \cdots, n). \quad (40)$$

Therein, $\chi_1^j(t, \alpha_1)$ and $\chi_2^j(t, \alpha_1)$ respectively denote elements of j-th joint for $\boldsymbol{\chi}_1(t, \alpha_1)$ and $\boldsymbol{\chi}_2(t, \alpha_1)$.

We analyze the paths of $d\alpha_1/dt$ in Eq. (40) on the $t - \alpha_1$ phase plane to obtain the optimal motion. From Eq. (40), by defining

$$L^j(t,\alpha_1) = \begin{cases} (u^j_{min} - \chi_2^j)/\chi_1^j & (\chi_1^j > 0), \\ (u^j_{max} - \chi_2^j)/\chi_1^j & (\chi_1^j < 0) \end{cases}$$

$$H^j(t,\alpha_1) = \begin{cases} (u^j_{max} - \chi_2^j)/\chi_1^j & (\chi_1^j > 0), \\ (u^j_{min} - \chi_2^j)/\chi_1^j & (\chi_1^j < 0) \end{cases} \quad (41)$$

we have

$$L^j(t,\alpha_1) \le \frac{d\alpha_1}{dt} \le H^j(t,\alpha_1) \quad (j=1,\cdots,n). \quad (42)$$

These inequalities represent the upper limits and the lower limits of $d\alpha_1/dt$ at the j-th joint. In the case $\chi_1^j = 0$, Eq. (42) is satisfied over the whole range of $d\alpha_1/dt$ when $u^j_{min} \le \chi_2^j \le u^j_{max}$ $(j=1,\cdots,n)$. On the other hand, the inequalities are not satisfied over the whole range of $d\alpha_1/dt$ when $\chi_2^j < u^j_{min}$ or $u^j_{max} < \chi_2^j$ $(j=1,\cdots,n)$.

Let us define the minimum value of $L^j(t,\alpha_1)$ as $ML(t,\alpha_1)$ and the maximum value of $H^j(t,\alpha_1)$ as $MH(t,\alpha_1)$ among all the joints. When $MH(t,\alpha_1)$ is greater than $ML(t,\alpha_1)$, we have

$$ML(t,\alpha_1) \le \frac{d\alpha_1}{dt} \le MH(t,\alpha_1). \quad (43)$$

This inequality represents an enabled bound of direction at any arbitrary point on the $t - \alpha_1$ phase plane. In contrast, if the above inequality does not hold on the $t - \alpha$ phase plane, no direction is enabled.

Because of the conditions for the time-scale function $r_0(t)$, $\alpha_1(t)$ must satisfy $\alpha_1 \in C^1$ and $\alpha_1 > 0$. The time-optimal motion $\boldsymbol{q}_{opt}(r_{opt})$ is determined from a path of α_1, which surrounds a minimum region on the phase plane, because

$$\int_0^t \alpha_1 d\tau = \int_0^t \frac{dr_0(\tau)}{d\tau} d\tau = r_0(t) - r_0(0) = r_0(t) = r_{opt}(t). \quad (44)$$

Because the time-scale function $r_{opt}(t)$ is obtained using the above equation, the time-optimal motion $\boldsymbol{q}_{opt}(r_{opt})$ can also be obtained.

4 Planning and Control of the Underwater Robot's Motion

4.1 Dynamics and Iterative Learning Control of Underwater Robot Manipulators

For underwater robots, the environment dynamics $\boldsymbol{d}_e$ in Eq. (1) represent hydrodynamic terms such as added-mass, drag, and buoyancy terms. Several researchers, e.g., [14], [15], [16], have studied the models of the dynamics of underwater manipulators.

Here, we use a strip-theory approach to model the hydrodynamic terms. First, each link of a manipulator as shown in Fig. 2 is divided into m strips. If the manipulator has n links, the number of all strips is $n \times m$. Let us define a characteristic velocity $v_{j,k}$ for a jk-th strip as a perpendicular flow velocity. Then, the added-mass force $\Delta f_{Aj,k}$ and the drag force $\Delta f_{Dj,k}$ acting on the jk-th strip can be given respectively as

$$\Delta f_{Aj,k} = C_{A(j,k)} \frac{\mathrm{d}v_{j,k}}{\mathrm{d}t}, \tag{45}$$

$$\Delta f_{Dj,k} = C_{D1(j,k)} v_{j,k} + C_{D2(j,k)} |v_{j,k}| v_{j,k} \tag{46}$$

where $C_{A(j,k)}$, $C_{D1(j,k)}$ and $C_{D2(j,k)}$ are parameters that consist of the hydrodynamic coefficients, the density of the fluid, the geometric shape of the manipulator and so on. Here, we suppose that the parameters on a strip are constants.

Through the summation of $\Delta f_{Aj,k}$ and $\Delta f_{Dj,k}$ taken over the j-th link, the hydrodynamic forces acting on the j-th link can be computed. Consequently, the robot dynamics including hydrodynamic terms are described as follows:

$$\boldsymbol{R}\left(\boldsymbol{q}(t)\right) \frac{\mathrm{d}^2 \boldsymbol{q}(t)}{\mathrm{d}t^2} + \boldsymbol{R}_A\left(\boldsymbol{q}(t)\right) \frac{\mathrm{d}\boldsymbol{v}(t)}{\mathrm{d}t} + \frac{\mathrm{d}\boldsymbol{q}^T(t)}{\mathrm{d}t} \boldsymbol{S}\left(\boldsymbol{q}(t)\right) \frac{\mathrm{d}\boldsymbol{q}(t)}{\mathrm{d}t}$$

$$+ \boldsymbol{D}_1\left(\boldsymbol{q}(t)\right) \boldsymbol{v}(t) + |\boldsymbol{v}(t)|^T \boldsymbol{D}_2\left(\boldsymbol{q}(t)\right) \boldsymbol{v}(t) + \boldsymbol{B}_q \frac{\mathrm{d}\boldsymbol{q}(t)}{\mathrm{d}t}$$

$$+ \boldsymbol{g}\left(\boldsymbol{q}(t)\right) + \boldsymbol{b}\left(\boldsymbol{q}(t)\right) = \boldsymbol{u}(t). \tag{47}$$

The matrices and the vectors in Eq. (47) are defined as
$\boldsymbol{v} \in \mathcal{R}^{nm \times 1}$: flow velocity vector
$\boldsymbol{v} = [v_{1,1}, v_{1,2}, \cdots, v_{n,m}]^T$,
$|\boldsymbol{v}| = [|v_{1,1}|, |v_{1,2}|, \cdots, |v_{n,m}|]^T$,
$\boldsymbol{R}_A(\boldsymbol{q}) \in \mathcal{R}^{n \times nm}$: added inertia matrix,
$\boldsymbol{D}_1(\boldsymbol{q}) \in \mathcal{R}^{n \times nm}$: matrix of linear drag force,
$\boldsymbol{D}_2(\boldsymbol{q}) \in \mathcal{R}^{nm \times n \times nm}$: tensor of quadratic drag force,
$\boldsymbol{b}(\boldsymbol{q}) \in \mathcal{R}^{n \times 1}$: vector of buoyancy force.

Iterative Learning Control of Underwater Robot Manipulators

We theoretically and experimentally investigated the performance of the iterative learning control for underwater robot manipulators in [12]. Some assumptions for the passivity condition in the case of underwater robot manipulators were evaluated theoretically. The effectiveness of the iterative learning control was demonstrated through experiments using a 3-D.O.F. robot manipulator. Experimental results showed that learning control realized the desired motions in water. Moreover, the robustness of the learning control was investigated through some additional experiments.

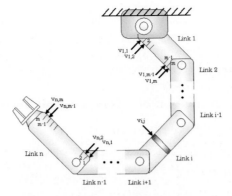

Fig. 2. Robot manipulator and a flow-velocity model

4.2 Time-Scale Transformation

Classification of Time-Scale Transformation

Suppose that learning control realizes four desired motions $\boldsymbol{q}_i(r_i)$ ($i = 1, 2, 3, 4$). The time-scale functions $r_i(t)$ ($i = 1, 2, 3, 4$) should satisfy conditions I, II, and III, which are given in the previous section. The four motions have an identical spatial path. Therefore, we obtain

$$\boldsymbol{q}_1(r_1(t)) = \boldsymbol{q}_2(r_2(t)) = \boldsymbol{q}_3(r_3(t)) = \boldsymbol{q}_4(r_4(t)). \tag{48}$$

Here, we assume that the flow velocities around the manipulator satisfy

$$v_{j,k}(r_1) = \frac{\mathrm{d}r_i}{\mathrm{d}t} v_{j,k}(r_i), \quad (i = 2, 3, 4) \tag{49}$$

when the time-scale is changed. It is presumed that the flow velocities $v_{j,k}$ around the manipulator depend largely on the time-scale r_i of the motions $\boldsymbol{q}_i$ when the manipulator moves in still water. In such a case, it is expected that the relation represented by Eq. (49) can hold. In fact, our experimental results [13] show that this assumption is appropriate.

From such points, we classify the hydrodynamic term $\boldsymbol{d}_e$ from the viewpoint of time-scale transformation as:

D^0 class : buoyancy term,

D^1 class : linear drag term,

D^2 class : added inertia, quadratic drag terms.

Here, we regard the linear and quadratic drag terms respectively as D^1 and D^2 class.

Conditions for Time-Scale Transformation

Control and planning of underwater robot manipulators presented here also require the four time-scale functions $r_i(t)$ ($i = 1, 2, 3, 4$). Therefore, we also set conditions I, II and III.

Preliminary for Formation of a New Desired Torque Pattern

In place of Eq. (8), we have the following.

$$\boldsymbol{R}(\boldsymbol{q}_i(r_i))\frac{\mathrm{d}^2\boldsymbol{q}_i(r_i)}{\mathrm{d}r_i^2} + \boldsymbol{R}_A(\boldsymbol{q}_i(r_i))\frac{\mathrm{d}\boldsymbol{v}_i(r_i)}{\mathrm{d}r_i} + \frac{\mathrm{d}\boldsymbol{q}_i^T(r_i)}{\mathrm{d}r_i}\boldsymbol{S}(\boldsymbol{q}_i(r_i))\frac{\mathrm{d}\boldsymbol{q}_i(r_i)}{\mathrm{d}r_i}$$
$$+\boldsymbol{D}_1(\boldsymbol{q}_i(r_i))\boldsymbol{v}_i(r_i) + |\boldsymbol{v}_i(r_i)|^T \boldsymbol{D}_2(\boldsymbol{q}_i(r_i))\boldsymbol{v}_i(r_i) + \boldsymbol{B}_q\frac{\mathrm{d}\boldsymbol{q}_i(r_i)}{\mathrm{d}r_i}$$
$$+\boldsymbol{g}(\boldsymbol{q}_i(r_i)) + \boldsymbol{b}(\boldsymbol{q}_i(r_i)) = \boldsymbol{u}_i(r_i), \quad (i = 1, 2, 3, 4). \tag{50}$$

The following equations in reference to Eq. (49),

$$\boldsymbol{v}_1(t) = \frac{\mathrm{d}r_i}{\mathrm{d}t}\boldsymbol{v}_i(r_i),$$

$$\frac{\mathrm{d}\boldsymbol{v}_1(t)}{\mathrm{d}t} = \frac{\mathrm{d}^2 r_i}{\mathrm{d}t^2}\boldsymbol{v}_i(r_i) + \left(\frac{\mathrm{d}r_i}{\mathrm{d}t}\right)^2 \frac{\mathrm{d}\boldsymbol{v}_i(r_i)}{\mathrm{d}r_i}, \tag{51}$$

are satisfied. Therefore, we obtain

$$\boldsymbol{v}_i(r_i) = a_i \boldsymbol{v}_1(t),$$
$$\frac{\mathrm{d}\boldsymbol{v}_i(r_i)}{\mathrm{d}r_i} = a_i^2 \frac{\mathrm{d}\boldsymbol{v}_1(t)}{\mathrm{d}t} - a_i^3 b_i \boldsymbol{v}_1(t), \tag{52}$$

where

$$a_i \triangleq 1/\frac{\mathrm{d}r_i}{\mathrm{d}t}, \quad b_i \triangleq \frac{\mathrm{d}^2 r_i}{\mathrm{d}t^2}. \tag{53}$$

In place of Eq. (12), substituting Eqs. (10), (48) and (52) into Eq. (50) yields

$$-a_i^3 b_i \boldsymbol{f}_{14}(t) + a_i^2 \boldsymbol{f}_{11}(t) + a_i \boldsymbol{f}_{12}(t) + \boldsymbol{f}_{13}(t) = \boldsymbol{u}_i(r_i) \quad (i = 1, 2, 3, 4), \tag{54}$$

where

$$\boldsymbol{f}_{11}(t) = \boldsymbol{R}(\boldsymbol{q}_1(t))\frac{\mathrm{d}^2\boldsymbol{q}_1(t)}{\mathrm{d}t^2} + \boldsymbol{R}_A(\boldsymbol{q}_1(t))\frac{\mathrm{d}\boldsymbol{v}_1(t)}{\mathrm{d}t}$$
$$+\frac{\mathrm{d}\boldsymbol{q}_1^T(t)}{\mathrm{d}t}\boldsymbol{S}(\boldsymbol{q}_1(t))\frac{\mathrm{d}\boldsymbol{q}_1(t)}{\mathrm{d}t} + |\boldsymbol{v}_1(t)|^T \boldsymbol{D}_2(\boldsymbol{q}_1(t))\boldsymbol{v}_1(t), \tag{55}$$

$$\boldsymbol{f}_{12}(t) = \boldsymbol{B}_q \frac{\mathrm{d}\boldsymbol{q}_1(t)}{\mathrm{d}t} + \boldsymbol{D}_1(\boldsymbol{q}_1(t))\boldsymbol{v}_1(t), \tag{56}$$

$$\boldsymbol{f}_{13}(t) = \boldsymbol{g}(\boldsymbol{q}_1(t)) + \boldsymbol{b}(\boldsymbol{q}_1(t)), \tag{57}$$

$$\boldsymbol{f}_{14}(t) = \boldsymbol{R}(\boldsymbol{q}_1(t))\frac{\mathrm{d}\boldsymbol{q}_1(t)}{\mathrm{d}t} + \boldsymbol{R}_A(\boldsymbol{q}_1(t))\boldsymbol{v}_1(t). \tag{58}$$

We can derive the compensation vector f_{14} in the same manner as in the case of a robot in contact with a constraint surface.

Formation of the Desired Input Torque Pattern

The input torque pattern u_0 that realizes an arbitrary desired motion q_0 is formed from u_1, u_2, u_3, which realize q_1, q_2, q_3, and the compensation vector f_{14}.

In the same manner as Eq. (24), we can obtain

$$-\alpha_i^3 \beta_i f_{04}(r_0) + \alpha_i^2 f_{01}(r_0) + \alpha_i f_{02}(r_0) + f_{03}(r_0) = u_i(r_i) \quad (i=1,2,3) \tag{59}$$

where $\alpha_k = 1/\frac{dr_i}{dr_0}$, $\beta_k = d^2 r_i / dr_0^2$, and

$$f_{01}(r_0) = R(q_0(r_0)) \frac{d^2 q_0(r_0)}{dr_0^2} + R_A(q_0(r_0)) \frac{dv_0(r_0)}{dr_0}$$
$$+ \frac{dq_0^T(r_0)}{dr_0} S(q_0(r_0)) \frac{dq_0(r_0)}{dr_0} + |v_0(r_0)|^T D_2(q_0(r_0)) v_0(r_0), \tag{60}$$

$$f_{02}(r_0) = B_q \frac{dq_0(r_0)}{dr_0} + D_1(q_0(r_0)) v_0(r_0), \tag{61}$$

$$f_{03}(r_0) = g(q_0(r_0)) + b(q_0(r_0)), \tag{62}$$

$$f_{04}(r_0) = R(q_0(r_0)) \frac{dq_0(r_0)}{dr_0} + R_A(q_0(r_0)) v_0(r_0). \tag{63}$$

As a result, the input torque pattern u_0 to realize an arbitrary desired motion q_0 is calculated as the following.

$$u_0(r_0) = [I\ I\ I] A^{*-1} u^* = f_{01}(r_0) + f_{02}(r_0) + f_{03}(r_0)$$
$$= R(q_0(r_0)) \frac{d^2 q_0(r_0)}{dr_0^2} + R_A(q_0(r_0)) \frac{dv_0(r_0)}{dr_0}$$
$$+ \frac{dq_0^T(r_0)}{dr_0} S(q_0(r_0)) \frac{dq_0(r_0)}{dr_0} + D_1(q_0(r_0)) v_0(r_0)$$
$$+ |v_0(r_0)|^T D_2(q_0(r_0)) v_0(r_0) + B_q(q_0(r_0)) \frac{dq_0(r_0)}{dr_0}$$
$$+ g(q_0(r_0)) + b(q_0(r_0)). \tag{64}$$

4.3 Time-Optimal Control for Underwater Robot Manipulators

It is also possible to consider motion planning that realizes a minimum-time motion of an underwater robot under the following limitations:

$$u_{min}^j \leq u_{opt}^j(r_i) \leq u_{max}^j \quad (j=1,\cdots,n). \tag{65}$$

The algorithm of motion planning for underwater robots explained here is identical to that for motion planning for robots that was presented in the previous section. Therefore, the time-optimal motion $q_{opt}(r_i)$ bounded by Eq. (65) is also formed from the four basic torque patterns $u_i(r_i)$ ($i = 1, 2, 3, 4$) without using parameter estimation.

5 Experiments

This section presents some experimental results to confirm the effectiveness of the proposed control scheme based on iterative learning control and time-scale transformation. In the experiments, a robot with 3 D.O.F. moves in air and in water; the proposed control scheme is applied.

5.1 Basic Experiments using a Robot without Mechanical Constraints

As a basic experiment, a robot manipulator (A460; CRS Plus) like that shown in Fig. 3 was utilized. The robot manipulator moved in free space and learned four desired motion patterns with different speeds. The effectiveness of the time-scale transformation is demonstrated through realization of the minimum time control input.

In this experiment, only three D.O.F. in the manipulator are controlled; the input voltages to the amplifiers are regarded as input patterns.

First, one of the basic motions is set as

$$\boldsymbol{p}_1(r_1) = \boldsymbol{p}_1(t) = \begin{bmatrix} 350 \\ 0 \\ 330 \end{bmatrix} + \begin{bmatrix} \cos\zeta(t) & -\sin\zeta(t) & 0 \\ \sin\zeta(t) & \cos\zeta(t) & 0 \\ 0 & 0 & 1 \end{bmatrix} \begin{bmatrix} 50 \\ 0 \\ 0 \end{bmatrix}, \quad (66)$$

$$\zeta(t) = 2\pi \left\{ -2\left(\frac{t}{T_1}\right)^3 + 3\left(\frac{t}{T_1}\right)^2 \right\} [rad], \quad T_1 = 2.0[s]. \quad (67)$$

The other three basic motions $\boldsymbol{p}_i(r_i)$ ($i = 2, 3, 4$) are defined by the time-scale functions as

$$r_2(t) = 1.25t, \quad T_2 = 2.5[s], \quad (68)$$
$$r_3(t) = 1.5t, \quad T_3 = 3.0[s], \quad (69)$$
$$r_4(t) = \frac{2(\exp(t/2) - 1)}{\exp(1) - 1}, \quad T_4 = 2.0[s]. \quad (70)$$

The joint motion patterns $\boldsymbol{q}_i(r_i)$ are obtained from $\boldsymbol{p}_i(r_i)$ by solving inverse kinematics. The input patterns $\boldsymbol{u}_i(r_i)$ ($i = 1, 2, 3, 4$) that realize the basic motions $\boldsymbol{q}_i(r_i)$ are formed through iterative learning control. After several iterative operations, the trajectory tracking errors of the endpoint of the robot became smaller than 1 [mm] at any time of the trajectories.

Now, we set the limitations for the input voltages $u_{max}^j = 9.2[V]$, $u_{min}^j = -9.2[V]$ ($j = 1, 2, 3$). Under those limitations, we sought an optimal path in the $t - \alpha_1$ phase. The result is shown in Fig. 4. The time optimal motion $q_{opt}(r_{opt})$ and the control input $u_{opt}(r_{opt})$ in Fig. 5 were obtained. Because the terminal time of the obtained time minimum motion is 1.18 [s], it is confirmed that the four desired motions with the terminal times 2.0 [s], 2.5 [s], 3.0 [s], and 2.0 [s] can generate a faster motion while retaining the input limitations.

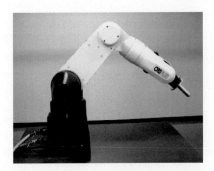

Fig. 3. 6-D.O.F. Robot Manipulator

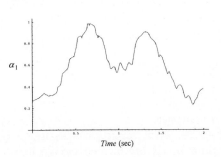

Fig. 4. $t - \alpha$ phase plane

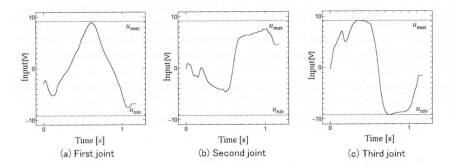

Fig. 5. Input patterns for the minimum-time motion

5.2 Experiments using an Underwater Robot Manipulator

To demonstrate the effectiveness of the proposed control scheme, the minimum time control was conducted using a 3-D.O.F. manipulator, as shown in Fig. 6. This robot was designed and developed at Ritsumeikan University. In this experiment, the input electric currents to the actuators of the manipulator are considered to be the input torques.

First, the spatial path $p_1(t)$ is chosen as

$$\boldsymbol{p}_1(r_1) = \boldsymbol{p}_1(t) = \begin{bmatrix} 350 \\ -250 + 500\sin\theta(t) \\ 750 \end{bmatrix}, \tag{71}$$

$$\theta(t) = \frac{\pi}{2}\left\{6\left(\frac{t}{T_1}\right)^5 - 15\left(\frac{t}{T_1}\right)^4 + 10\left(\frac{t}{T_1}\right)^3\right\} [rad], \quad T_1 = 1.8[s]. \tag{72}$$

For the spatial path, we consider the time-scale functions as

$$r_2(t) = 3t, \quad T_2 = 5.4[s], \tag{73}$$
$$r_3(t) = 1.5t, \quad T_3 = 2.7[s], \tag{74}$$
$$r_4(t) = \frac{t^2 + 5t}{5}, \quad T_4 = 2.45[s]. \tag{75}$$

The joint motions $\boldsymbol{q}_i(r_i)$ corresponding to $\boldsymbol{p}_i(r_i)$ are calculated using inverse kinematics.

The basic input patterns $\boldsymbol{u}_i(r_i)$ ($i = 1, 2, 3, 4$) that realize the basic motions $\boldsymbol{q}_i(r_i)$ are formed by the iterative learning control. In this case, after 15 iterative operations, the trajectory tracking errors of the endpoint of the robot became smaller than 2.5 [mm] at any time of the trajectories.

Under the following limitations of the inputs:

$$-1.0 < u^i_{opt}(r_i) < 1.0 \ [A] \ (j = 1, 2, 3), \tag{76}$$

an optimal path is sought on the $t - \alpha_1$ phase plane. The result is shown in Fig. 7. The time-optimal motion $\boldsymbol{p}_{opt}(r_{opt})$ and the control input $\boldsymbol{u}_{opt}(r_{opt})$ are obtained using α_1. Figure 8 shows that the input patterns $\boldsymbol{u}_{opt}$ are bounded by the limitations. This case also shows that the four desired motions with the terminal times of 1.8 [s], 5.4 [s], 2.7 [s] and 2.45 [s] can produce a faster motion with the terminal time 1.28 [s], while keeping the input limitations.

6 Conclusion

Through this study, we have asserted the importance of the control scheme design method based on the structure, which is physically characterized. As examples of the control scheme design method, iterative learning control and time-scale transformation were introduced in this paper. Regarding iterative learning control, the desired input pattern can be acquired without estimating the parameters of a robot and its environment. The effectiveness of iterative learning control is guaranteed by the physical characteristic of "passivity". On the time-scale transformation, a new desired input pattern for a different speed motion pattern is formed through some combination of the desired input patterns that are obtained using iterative learning control if the physical characteristic, "classification on time-scale" of the system is clear.

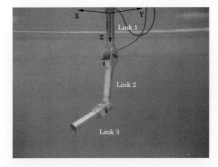

Fig. 6. 3-D.O.F. Underwater Manipulator

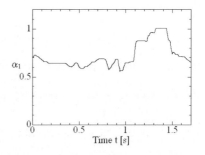

Fig. 7. $t - \alpha$ phase plane

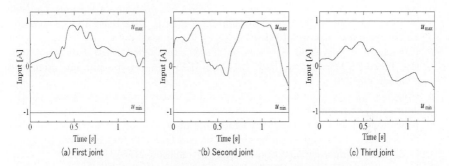

Fig. 8. Input patterns for the minimum-time motion

It is important to know whether the proposed control scheme design method based on characterization of the objective dynamics is useful for motion control of a complex system. In this paper, we set two environments for a robot as complex systems: a rigid surface and water. In both cases, the effectiveness of the time-scale transformation is theoretically proven. Moreover, it is demonstrated experimentally that the minimum control inputs for underwater robot manipulators are obtainable through iterative learning control and time-scale transformation.

7 Acknowledgments

The authors deeply appreciate Professor Suguru Arimoto for his important advice and useful suggestions and sincerely hope that he continues to encourage the young researchers of Ritsumeikan University and other universities in their research activities.

References

1. Arimoto S, Kawamura S, Miyazaki F (1984) Bettering operation of robots by learning. Journal of Robotic Systems 1(2):123–140
2. Kawamura S, Miyazaki F, Arimoto S (1985) Iterative learning control for robotic systems. In: Proc. of IECON'84 393–398
3. Kawamura S, Miyazaki F, Arimoto S (1988) Realization of robot motion based on a learning method. IEEE Transactions on Systems, Man, and Cybernetics 18(1):126–134
4. Arimoto S (1996) Control Theory of Non-linear Mechanical Systems: A Passivity-based and Circuit-theoretic Approach. Oxford Univ. Press, U.K.
5. Kawamura S, Miyazaki F, Arimoto S (1985) Hybrid position/force control of robot manipulators based on learning method. In: Proc. 2nd Int. Conference on Advanced Robotics 235–242
6. Kawamura S, Miyazaki F, Arimoto S (1985) Applications of learning method for dynamic control of robot manipulators. In: Proc. 24th IEEE Int. Conference on Decision and Control 1381–1386
7. Kawamura S, Miyazaki F, Arimoto S(1987) Intelligent control of robot motion based on learning method. In: Proc. IEEE Int. Symposium on Intelligent Control 365–370
8. Hollerbach J (1984) Dynamic scaling of manipulator trajectories. Trans. ASME, Journal of Dynamic Systems, Measurement, and Control 106:102–106
9. Kawamura S, Fukao N (1995) A time-scale interpolation for input torque patterns obtained through learning control on constrained robot motions. In: Proc. IEEE Int. Conference on Robotics and Automation 2156–2161
10. Kawamura S, Fukao N (1997) Use of feedforward input patterns obtained through learning control. In: Proc. 2nd Asian Control Conference 255–258
11. Kawamura S, Fukao N, Ichii H (1999) Planning and control of robot motion based on time-scale transformation and iterative learning control. In: Proc. 9th Int. Symposium of Robotics Research 213–220
12. Sakagami N, Inoue M, Kawamura S (2003) Theoretical and experimental studies on iterative learning control for underwater robots. International Journal of Offshore and Polar Engineering 13(2):120–127
13. Sakagami N, Kawamura S (2002) Analysis on dynamics of 3-dof underwater robot manipulators based on time-scale transformation. In: Proc. Japan-USA Symposium on Flexible Automation 1139–1145
14. McLain T, Rock S (1996) Experiments in the hydrodynamic modeling of an underwater manipulator. Int. Symposium on Autonomous Underwater Vehicle Technology 463–469
15. Leabourne K, Rock S (1998) Model development of an underwater manipulator for coordinated arm-vehicle control. In: Proc. OCEANS 98 Conference 941–946
16. Levesque B, Richard M (1994) Dynamic analysis of a manipulator in a fluid environment. The International Journal of Robotics Research 13(3):221–231

Part II

From Robot Control to Human-Like Movements

Modularity, Synchronization, and What Robotics May Yet Learn from the Brain

Jean-Jacques Slotine

Massachusetts Institute of Technology, Cambridge MA 01239 USA, jjs@mit.edu

Summary. Although neurons as computational elements are 7 orders of magnitude slower than their artificial counterparts, the primate brain grossly outperforms robotic algorithms in all but the most structured tasks. Parallelism alone is a poor explanation, and much recent functional modeling of the central nervous system focuses on its modular, heavily feedback-based architecture, the result of accumulation of subsystems throughout evolution. In our earlier work, we have extensively discussed this architecture from a global stability and convergence point of view. In this article, we describe recent work which extends these ideas to synchronization as a model of computations at different scales in the nervous system. We also describe a simple condition for a general dynamical system to globally converge to a polyrhythm, i.e., a regime where multiple groups of fully synchronized elements coexist. Applications to some classical questions in robotics and systems neuroscience are discussed.

1 Introduction

I am truly delighted to contribute to this book on the occasion of Professor Suguru's Arimoto's 70th birthday. Professor Arimoto's work is probably the one I most admire in the whole field of robotics. Like the founders of cybernetics [86, 87, 93, 3], Professor Arimoto was also deeply interested in neuroscience from the start, and actually he was one of the main contributors to the now very classical FitzHugh-Nagumo [11, 55] model of neural oscillators. So I believe it is fitting to write an article about what robotics may yet learn from the brain in the context of this celebration.

Although neurons as computational elements are 7 orders of magnitude slower than their artificial counterparts, the primate brain grossly outperforms robotic algorithms in all but the most structured tasks. Parallelism alone is a poor explanation, and much recent functional modeling of the central nervous system focuses on its modular, heavily feedback-based computational architecture, the result of accumulation of subsystems throughout evolution.

We have extensively discussed in some earlier work [76, 74, 44] this architecture from a global stability and convergence point of view. In this article, we describe recent work [88] on synchronization as a model of computations at different scales in the brain, such as pattern matching, temporal binding of sensory data, and mirror neuron response. Finally, we derive [63] a simple condition for a general dynamical system to globally converge to a regime where multiple groups of fully synchronized elements coexist. Applications of such "polyrhythms" to some classical questions in robotics and systems neuroscience are discussed.

The development makes extensive use of nonlinear contraction theory, a comparatively recent analysis tool whose main features will be briefly reviewed. In particular,

- Global results on synchronization can be obtained using most common models of neural oscillators, such as the FitzHugh-Nagumo model.
- Long-range synchronization between regions can be achieved without direct connections.
- Since contraction is preserved under most common system combinations (parallel, hierarchies, negative feedback), it represents a natural framework for motor primitives.
- In locomotion, the analysis exhibits none of the topological difficulties that may arise when coupling large numbers of phase oscillators, and it guarantees global exponential convergence.
- Replacing ordinary CPG connections by filters enables automatic frequency-based gate selection.
- Stable polyrhythmic aggregates of arbitrary size can be constructed recursively, motivated by evolution and development.
- Just as global synchronization occurs naturally and quantifiably in networks of locally coupled oscillators, it can be turned off by adding a single inhibitory connection.
- In vision and pattern recognition, detectors for various types of symmetries can be systematically constructed.

2 Modularity, Stability, and Evolution

Basically, a nonlinear time-varying dynamic system will be called *contracting* if initial conditions or temporary disturbances are forgotten exponentially fast, i.e., if trajectories of the perturbed system return to their nominal behavior with an exponential convergence rate. It turns out that relatively simple conditions can be given for this stability-like property to be verified, and furthermore that this property is preserved through basic system combinations (such as series, hierarchies, parallel combinations, and negative feedback), a state-space feature reminiscent in spirit of input-output passivity [65]. Furthermore, as discussed in Section 3, the concept of *partial contraction* allows to extend the applications of contraction analysis to include convergence to behaviors or to specific properties (such as equality of state components, or convergence to a manifold) rather than trajectories. This section is largely based on [43, 76, 74, 44] to which the reader is referred for details.

We consider general time-varying deterministic systems of the form

$$\dot{\mathbf{x}} = \mathbf{f}(\mathbf{x}, t) \qquad (1)$$

where $\mathbf{f}$ is an $n \times 1$ nonlinear vector function and $\mathbf{x}$ is the $n \times 1$ state vector. The above equation may also represent the closed-loop dynamics of a controlled system with state feedback $\mathbf{u}(\mathbf{x}, t)$. All quantities are assumed to be real and smooth, by which it is meant that any required derivative or partial derivative exists and is continuous. The basic result of [43] can then be stated as

Theorem 1. *Consider system (1), and assume there exists a uniformly positive definite metric*
$$\mathbf{M}(\mathbf{x}, t) = \boldsymbol{\Theta}'(\mathbf{x}, t)\, \boldsymbol{\Theta}(\mathbf{x}, t)$$
such that the associated generalized Jacobian
$$\mathbf{F} = \left(\dot{\boldsymbol{\Theta}} + \boldsymbol{\Theta}\frac{\partial \mathbf{f}}{\partial \mathbf{x}}\right) \boldsymbol{\Theta}^{-1}$$
is uniformly negative definite. Then all system trajectories converge exponentially to a single trajectory, with convergence rate $|\lambda_{max}|$, where λ_{max} is the largest eigenvalue of the symmetric part of $\mathbf{F}$. The system is said to be contracting.

By $\boldsymbol{\Theta}'$ we mean the Hermitian (conjugate transpose) of $\boldsymbol{\Theta}$, and by symmetric part of $\mathbf{F}$ we mean $\frac{1}{2}(\mathbf{F} + \mathbf{F}')$. It can be shown conversely that the existence of a uniformly positive definite metric with respect to which the system is contracting is also a necessary condition for global exponential convergence of trajectories. In the linear time-invariant case, a system is globally contracting if and only if it is strictly stable, with $\mathbf{F}$ simply being a normal Jordan form of the system and $\boldsymbol{\Theta}$ the coordinate transformation to that form. The results immediately extend to the case where the state is in $\mathbb{C}^n$.

Example [74] Consider the Lorenz system

$$\begin{aligned}
\dot{x} &= \sigma\,(y - x) \\
\dot{y} &= \rho\, x - y - x\, z \\
\dot{z} &= -\beta\, z + x\, y
\end{aligned}$$

with strictly positive constants σ, ρ, β, and, given measurements of x, the reduced-order identity observer [62, 58]

$$\begin{aligned}
\dot{\hat{y}} &= \rho\, x - \hat{y} - x\, \hat{z} \\
\dot{\hat{z}} &= -\beta\, \hat{z} + x\, \hat{y}
\end{aligned}$$

The symmetric part of the observer's Jacobian is $-\operatorname{diag}(1, \beta)$, and thus the observer is contracting with an identity metric. Since by construction $(\hat{y}, \hat{z}) =$

(y, z) is a particular solution, the estimated state converges exponentially to the actual state, with rate $\min(1, \beta)$.

An important property is that, under mild conditions, contraction is preserved through system combinations such as parallel, series or hierarchies, translation and scaling in time and state, and certain types of feedback [43, 45, 76, 74].

Example [74]   Consider the system

$$\dot{\mathbf{x}} = \mathbf{f}(\mathbf{x}, t) + \mathbf{B}(\mathbf{x}, t)\, \mathbf{u}$$

and assume that there exist *control primitives* $\mathbf{u} = \mathbf{p}_i(\mathbf{x}, t)$ which, for any i, make the closed-loop system contracting in some common metric $\mathbf{M}(\mathbf{x})$. Multiplying each equation

$$\dot{\mathbf{x}} = \mathbf{f}(\mathbf{x}, t) + \mathbf{B}(\mathbf{x}, t)\, \mathbf{p}_i(\mathbf{x}, t)$$

by a positive coefficient $\alpha_i(t)$, and summing, shows that any convex combination of the control primitives $\mathbf{p}_i(\mathbf{x}, t)$

$$\dot{\mathbf{x}} = \mathbf{f}(\mathbf{x}, t) + \mathbf{B}(\mathbf{x}, t) \sum_i \alpha_i(t)\, \mathbf{p}_i(\mathbf{x}, t) \quad , \quad \sum_i \alpha_i(t) = 1$$

also leads to a contracting dynamics in the same metric. For instance, the time-varying convex combination may correspond to smoothly blending learned primitives in a humanoid robot.

As our understanding of both brain function and robot design improves, common fundamental questions strongly suggest exploring the relations between integrative neuroscience and robotics beyond the most obvious analogies. While today the evolution and development of cognitive processes is seen as closely linked to the progressive refinement of sensorimotor functions, similarly robotics takes artificial intelligence beyond its classical conceptual domain by emphasizing the central role of physical interaction with the environment. Of course, the constraints and opportunities of robotics are very different from those of biology. While their physical hardware is far behind nature's, in principle robots can have perfect memory, near-perfect repeatability, can use mathematics explicitly, and can simulate (imagine) specific actions much faster than humans. The traveling speed of information through an nerve axon is significantly slower than the speed of sound, while that along an electrical wire is closer to the speed of light. Processing time at each and every chemical synapse is about 1 ms, probably a major incentive for developing parallel computational architectures. But similar delay problems can also be found in robotics, if one looks not at an autonomous robot, but rather, for instance, at telerobotics over large distances.

While most of robotic theory is founded on physical models and mathematical algorithms, the fundamental conceptual tool in biology is the theory of evolution. Evolution proceeds by accumulation and combination of stable intermediate states. Conceptually, such accumulations have also been a recurrent theme in cybernetics and AI history, and also form the basis of several recent theories of brain function. However, in themselves, accumulations and combinations of stable elements have no reason to be stable. Hence our hypothesis in [76, 74, 44] that evolution will favor a contraction-like form of stability, which automatically guarantees stability in combinations, since this would considerably reduce (in effect, avoid combinatorial explosion of) trial-and-error as the systems become large and complex. Thus, contraction theory may help guide functional modeling of the central nervous system, and conversely it provides a systematic method to build arbitrarily complex robots out of simpler elements.

Incidentally, the definition of contraction fits rather naturally with known data on biological motion perturbation, e.g. perturbation of arm movement [80, 94]. Furthermore, it is intrinsic, in the sense that the system's "nominal" behavior needs not be known. Finally, such a form of stability, at least in a local sense, is also a basic prerequisite for any learning, since it guarantees the consistency of the system's behavior in the presence of small disturbances or variations in initial conditions. Automatic contraction preservation is a a property reminiscent of input-output passivity [65], although it applies in state-space to more general forms of combinations. An interesting discussion of the application of passivity tools to recursive refinement of the control of movement can be found in [2].

In [97] we use contraction to derive a nonlinear observer which computes the continuous state of an inertial navigation system based on partial discrete measurements, the so-called strap-down problem. The mathematical statement of this problem is common to aircraft navigation, robot localization, and head stabilization in mammals and birds. Indeed, the human vestibular system uses otolithic organs measuring linear acceleration and semi-circular canals estimating angular velocity through heavily damped angular acceleration signals, an information then combined with visual data at much slower update rate. Furthermore, head stabilization is likely used to simplify control and overall balance [5], a feature yet to be implemented in humanoid or flying robots.

3 Synchronization

We use partial contraction analysis to study synchronization phenomena. This section is largely based on [88] to which the reader is referred for details. Its results will be further expanded and systematized in Section 4.

Theorem 2. *Consider a nonlinear system of the form*

$$\dot{\mathbf{x}} = \mathbf{f}(\mathbf{x}, \mathbf{x}, t)$$

and assume that the auxiliary system

$$\dot{\mathbf{y}} = \mathbf{f}(\mathbf{y}, \mathbf{x}, t)$$

is contracting with respect to $\mathbf{y}$. If a particular solution of the auxiliary $\mathbf{y}$-system verifies a smooth specific property, then all trajectories of the original $\mathbf{x}$-system verify this property exponentially. The original system is said to be *partially contracting*.

Proof: The virtual, observer-like $\mathbf{y}$-system has two particular solutions, namely $\mathbf{y}(t) = \mathbf{x}(t)$ for all $t \geq 0$ and the solution with the specific property. This implies that $\mathbf{x}(t)$ verifies the specific property exponentially.

Example [78] Consider a rigid robot model

$$\mathbf{H}(\mathbf{q})\ddot{\mathbf{q}} + \mathbf{C}(\mathbf{q},\dot{\mathbf{q}})\dot{\mathbf{q}} + \mathbf{g}(\mathbf{q}) = \tau$$

and the energy-based controller [75]

$$\mathbf{H}(\mathbf{q})\ddot{\mathbf{q}}_{\mathbf{r}} + \mathbf{C}(\mathbf{q},\dot{\mathbf{q}})\dot{\mathbf{q}}_{\mathbf{r}} + \mathbf{g}(\mathbf{q}) - \mathbf{K}(\dot{\mathbf{q}} - \dot{\mathbf{q}}_{\mathbf{r}}) = \tau$$

with $\mathbf{K}$ a constant s.p.d. matrix. The virtual $\mathbf{y}$-system

$$\mathbf{H}(\mathbf{q})\dot{\mathbf{y}} + \mathbf{C}(\mathbf{q},\dot{\mathbf{q}})\mathbf{y} + \mathbf{g}(\mathbf{q}) - \mathbf{K}(\dot{\mathbf{q}} - \mathbf{y}) = \tau$$

has $\dot{\mathbf{q}}$ and $\dot{\mathbf{q}}_{\mathbf{r}}$ as particular solutions, and furthermore is contracting, since the skew-symmetry of the matrix $\dot{\mathbf{H}} - 2\mathbf{C}$ implies

$$\frac{d}{dt}\delta\mathbf{y}^T\mathbf{H}\delta\mathbf{y} = -2\delta\mathbf{y}^T(\mathbf{C}+\mathbf{K})\delta\mathbf{y} + \delta\mathbf{y}^T\dot{\mathbf{H}}\delta\mathbf{y} = -2\delta\mathbf{y}^T\mathbf{K}\delta\mathbf{y}$$

Thus $\dot{\mathbf{q}}$ tends to $\dot{\mathbf{q}}_{\mathbf{r}}$ exponentially. Making then the usual choice $\dot{\mathbf{q}}_{\mathbf{r}} = \dot{\mathbf{q}}_{\mathbf{d}} - \lambda(\mathbf{q} - \mathbf{q}_{\mathbf{d}})$ creates a hierarchy and implies in turn that $\mathbf{q}$ tends to $\mathbf{q}_{\mathbf{d}}$ exponentially.

Example Consider a convex combination or interpolation between contracting dynamics

$$\dot{\mathbf{x}} = \sum_i \alpha_i(\mathbf{x},t)\,\mathbf{f}_i(\mathbf{x},t)$$

where the individual systems $\dot{\mathbf{x}} = \mathbf{f}_i(\mathbf{x},t)$ are contracting in a common metric $\mathbf{M}(\mathbf{x})$ and have a common trajectory $\mathbf{x}_o(t)$ (for instance an equilibrium), with all $\alpha_i(\mathbf{x},t) \geq 0$ and $\sum_i \alpha_i(\mathbf{x},t) = 1$. Then all trajectories of the system globally exponentially converge to the trajectory $\mathbf{x}_o(t)$. Indeed, the auxiliary system

$$\dot{\mathbf{y}} = \sum_i \alpha_i(\mathbf{x},t)\,\mathbf{f}_i(\mathbf{y},t)$$

is contracting (with metric $\mathbf{M}(\mathbf{y})$) and has $\mathbf{x}(t)$ and $\mathbf{x}_o(t)$ as particular solutions.

Example The main idea of the virtual system in partial contraction can be applied in a variety of ways. Consider two coupled FitzHugh-Nagumo [11, 55] neural oscillators

$$\dot{v}_1 = c(v_1 + w_1 - \tfrac{1}{3}v_1^3 + I(t)) + k(v_2 - v_1) \qquad \dot{w}_1 = -\tfrac{1}{c}(v_1 - a + bw_1)$$

$$\dot{v}_2 = c(v_2 + w_2 - \tfrac{1}{3}v_2^3 + I(t)) + k(v_1 - v_2) \qquad \dot{w}_2 = -\tfrac{1}{c}(v_2 - a + bw_2)$$

where a, b, c are strictly positive constants and $I(t)$ an external input. The above imply

$$\dot{v}_1 + (2k - c)v_1 + \tfrac{1}{3}cv_1^3 - cw_1 = \dot{v}_2 + (2k - c)v_2 + \tfrac{1}{3}cv_2^3 - cw_2$$

$$\dot{w}_1 + \tfrac{1}{c}(v_1 + bw_1) = \dot{w}_2 + \tfrac{1}{c}(v_2 + bw_2)$$

Let $g_v(t)$ be the common value of the terms in the first line, and $g_w(t)$ the common value of the terms in the second line. Both $g_v(t)$ and $g_w(t)$ depend on initial conditions. Define the virtual system

$$\dot{y}_v + (2k - c)y_v + \tfrac{1}{3}cy_v^3 - cy_w = g_v(t)$$

$$\dot{y}_w + \tfrac{1}{c}(y_v + by_w) = g_w(t)$$

By construction, both neural oscillators follow particular trajectories of the virtual system. Furthermore, using a metric transformation $\Theta = \mathrm{diag}(1, c)$ yields the generalized Jacobian

$$\mathbf{F} = \begin{bmatrix} -2k - c(y_v^2 - 1) & 1 \\ -1 & -\tfrac{b}{c} \end{bmatrix}$$

so that the virtual system is contracting if $2k > c$. Thus, if $2k > c$, the two neural oscillators synchronize with an exponential rate $\min(2k - c, \, b/c)$.

The results extend to global synchronization conditions for networks of locally coupled FitzHugh-Nagumo neural oscillators [88]. It can also be shown that in such networks, a *single* inhibitory link of the same gain between two arbitrary nodes can destroy synchronization. This can provide a simple mechanism to avoid synchronization when it is undesirable. Such inhibition properties may be useful in pattern recognition to achieve rapid desynchronization between different objects. They may also be used as simplified models of minimal mechanisms for turning off unwanted synchronization, as e.g. in epileptic seizures or oscillations in internet traffic. In such applications, small and localized inhibition may also allow one to destroy unwanted synchronization while only introducing a small disturbance to the nominal behavior of the system.

Cascades of inhibition are common in the brain, in a way perhaps reminiscent of NAND-based logic.

Distributed synchronization phenomena are the subject of intense research. In the brain such phenomena are known to occur at different scales, and are heavily studied at both the anatomical and computational levels. In particular, synchronization has been proposed as a general principle for temporal

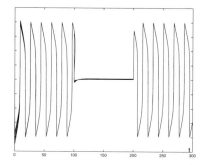

Fig. 1. An example of fast inhibition with a single inhibitory link. The plot shows the states of ten FitzHugh-Nagumo neurons as functions of time. The inhibitory link is actived at $t = 100$ and removed at $t = 200$. Thus, synchronization can be readily achieved and just as readily turned off.

binding of multisensory data [73, 20, 41, 53, 84, 37, 57], and as a mechanism for perceptual grouping [95], neural computation [6, 7, 92] and neural communication [34, 27, 69, 70]. The recently recognized pervasiveness [13] of diffusion-like, bilateral electrical synapses (gap junctions) makes these synchronization models particularly intriguing. Not entirely coincidentally, the same basic mathematics describing the collective behavior of neurons also describe the collective behavior of fish schools or bird flocks, as well as certain types of phase-transition in physics [82].

4 Polyrhythms

This section, which extends and systematizes the results of the Section 3, is based on [63], to which the reader is referred for details.

4.1 Concurrent Synchronization

In an ensemble of dynamical elements, concurrent synchronization is defined as a regime where the whole system is divided into multiple groups of fully synchronized elements[1], but elements from different groups are not necessarily synchronized [4, 96, 64] and can be of entirely different dynamics [16]. It can be easily shown that such a regime corresponds to a flow-invariant linear subspace of the global state space. Concurrent synchronization phenomena are likely pervasive in the brain, where multiple "rhythms" are known to coexist [34, 69], neurons can exhibit many qualitatively different types of oscillations [34, 26], and functional models often combine multiple oscillatory dynamics.

---

[1] In the literature, this phenomenon is often called *poly-* or *partial* synchronization. However, the latter term can also designate a regime where the elements are not fully synchronized but behave coherently [82].

A simple sufficient condition for a general dynamical system to converge to a flow-invariant subspace is introduced in [63]. The analysis is built upon nonlinear contraction theory [43, 88], and thus it inherits many of the theory's features :

- global exponential convergence and stability are guaranteed, as opposed to [4, 96, 17] where only stability in the neighborhood of the invariant manifold is discussed,
- convergence rates can be explicitly computed as eigenvalues of well-defined symmetric matrices,
- under simple conditions, convergence to a concurrently synchronized state can be preserved through system combinations,
- generalized symmetries in the sense of [16] can be systematically exploited, so that long-range synchronization between regions can be achieved without direct connections,
- robustness to variations in dynamics can be easily quantified.

Consider, in $\mathbb{R}^n$, the deterministic system

$$\dot{\mathbf{x}} = \mathbf{f}(\mathbf{x}, t) \qquad (2)$$

where $\mathbf{f}$ is a smooth nonlinear function. Assume that there exists a *flow-invariant linear subspace* $\mathcal{M}$ (i.e. a linear subspace $\mathcal{M}$ such that $\forall t : \mathbf{f}(\mathcal{M}, t) \subset \mathcal{M}$), which implies that any trajectory starting in $\mathcal{M}$ remains in $\mathcal{M}$. Let $p = \dim(\mathcal{M})$, and consider an orthonormal basis $(\mathbf{e}_1, \ldots, \mathbf{e}_n)$ where the first p vectors form a basis of $\mathcal{M}$ and the last $n - p$ a basis of $\mathcal{M}^\perp$. Define an $(n - p) \times n$ matrix $\mathbf{V}$ whose rows are $\mathbf{e}_{p+1}^\top, \ldots, \mathbf{e}_n^\top$. $\mathbf{V}$ may be regarded as a projection[2] on $\mathcal{M}^\perp$, and it verifies [24, 31] :

$$\mathbf{V}^\top \mathbf{V} + \mathbf{U}^\top \mathbf{U} = \mathbf{I}_n \qquad \mathbf{V}\mathbf{V}^\top = \mathbf{I}_{n-p} \qquad \mathbf{x} \in \mathcal{M} \iff \mathbf{V}\mathbf{x} = \mathbf{0}$$

where $\mathbf{U}$ is the matrix formed by the first p vectors.

Now let $\mathbf{z} = \mathbf{V}\mathbf{x}$. By construction, $\mathbf{x}$ converges to the subspace $\mathcal{M}$ if and only if $\mathbf{z}$ converges to $\mathbf{0}$. Multiplying (2) by $\mathbf{V}$ on the left, we get

$$\dot{\mathbf{z}} = \mathbf{V}\mathbf{f}(\mathbf{V}^\top \mathbf{z} + \mathbf{U}^\top \mathbf{U}\mathbf{x}, t)$$

Construct the auxiliary system

$$\dot{\mathbf{y}} = \mathbf{V}\mathbf{f}(\mathbf{V}^\top \mathbf{y} + \mathbf{U}^\top \mathbf{U}\mathbf{x}, t) \qquad (3)$$

By construction, a particular solution of system (3) is $\mathbf{y}(t) = \mathbf{z}(t)$. In addition, since $\mathbf{U}^\top \mathbf{U}\mathbf{x} \in \mathcal{M}$ and $\mathcal{M}$ is flow-invariant, $\mathbf{f}(\mathbf{U}^\top \mathbf{U}\mathbf{x}) \in \mathcal{M} = \mathrm{Null}(\mathbf{V})$. Thus $\mathbf{y}(t) = \mathbf{0}$ is another particular solution of system (3). If furthermore system (3) is contracting with respect to $\mathbf{y}$, then $\mathbf{z}(t)$ will converge exponentially to $\mathbf{0}$. This leads to [63]

---
[2] For simplicity we shall call $\mathbf{V}$ a "projection", although the actual projection matrix is in fact $\mathbf{V}^\top \mathbf{V}$.

Theorem 3. *If a linear subspace $\mathcal{M}$ is flow-invariant and if system (3) is contracting, then all solutions of system (2) converge exponentially to $\mathcal{M}$.*

In practice, the subspace $\mathcal{M}$ is often defined by the conjunction of $(n-p)$ linear constraints. In a synchronization context, each of the constraints may be, e.g., of the form $\mathbf{x}_i = \mathbf{x}_j$ where $\mathbf{x}_i$ and $\mathbf{x}_j$ are subvectors of the state $\mathbf{x}$. This provides directly a (generally not orthonormal) basis $(\mathbf{e}'_{p+1}, \ldots, \mathbf{e}'_n)$ of $\mathcal{M}^\perp$, and thus a matrix $\mathbf{V}'$ whose rows are ${\mathbf{e}'_{p+1}}^\top, \ldots, {\mathbf{e}'_n}^\top$, and which verifies

$$\mathbf{V}' = \mathbf{T}\mathbf{V}$$

with $\mathbf{T}$ an invertible $(n-p) \times (n-p)$ matrix, and $\mathbf{x} \in \mathcal{M} \iff \mathbf{V}'\mathbf{x} = \mathbf{0}$. For instance, for three systems, each of dimension m and state $\mathbf{x}_i$, one has

$$\mathbf{V}' = \begin{pmatrix} \mathbf{I}_m & -\mathbf{I}_m & \mathbf{0} \\ \mathbf{0} & \mathbf{I}_m & -\mathbf{I}_m \end{pmatrix}$$

Thus, using an identity metric in (3), the sufficient condition for convergence to $\mathcal{M}$

$$\forall \mathbf{x}, \qquad \mathbf{V}\left(\frac{\partial \mathbf{f}}{\partial \mathbf{x}}\right)\mathbf{V}^\top < \mathbf{0} \qquad (4)$$

can be written equivalently in the more simply evaluated form

$$\forall \mathbf{x}, \qquad \mathbf{V}'\left(\frac{\partial \mathbf{f}}{\partial \mathbf{x}}\right){\mathbf{V}'}^\top < \mathbf{0} \qquad (5)$$

Note however that to evaluate explicitly the convergence rate to $\mathcal{M}$, one has to use the orthonormal version, e.g. through a Gram-Schmidt procedure [24] on the rows of $\mathbf{V}'$.

The definition of the linear subspace may also comprise composite variables, such as sliding variables, thus reducing the dimensionality of the orthogonal subspace.

Different "rhythms" $(\alpha, \beta, \gamma, \delta)$ are known to coexist in the brain, which, in the light of the previous analysis, may be interpreted and modeled as concurrently synchronized regimes. Since contracting systems driven by periodic inputs will have states of the same period [43], different but synchronized computations could be robustly carried out by specialized areas in the brain using synchronized elements as their inputs. Such a temporal "binding" [73, 20, 41, 84, 37, 57, 95, 8, 13] mechanism would also complement the general argument in [76] that multisensory integration may occur through the interaction of contracting computational systems connected through an extensive network of feedback loops. Furthermore, because of the preservation of contraction through system combinations, Theorem 3 suggests a mechanism for stable accumulation and interaction of concurrently synchronized groups, showing [63] that the simple conditions for contraction to a linear subspace, combined with the high fan-out of typical neurons, increase the plausibility of large concurrently synchronized structures being created in the central nervous system in the course of evolution and development. Making these observations precise is the subject of current research.

4.2 Percolation

When a sync subspace is a strict superset of another, one should expect that the convergence to the former state is "easier" than the convergence to the latter [96, 4, 64]. This progressive synchronization or "percolation" effect can be quantified easily from Theorem 3, by noticing that

$$\mathcal{M}_A \supset \mathcal{M}_B \quad \Rightarrow \quad \mathcal{M}_A^\perp \subset \mathcal{M}_B^\perp \quad \Rightarrow \quad \lambda_{\min}(\mathbf{V}_A \mathbf{L}_s \mathbf{V}_A^\top) \geq \lambda_{\min}(\mathbf{V}_B \mathbf{L}_s \mathbf{V}_B^\top)$$

While in the case of identical systems and relatively uniform topologies, this effect may often be too fast to observe, the above applies to the general concurrent synchronization case and quantifies the associated and possibly very distinct time-scales.

4.3 Coincidence and Symmetry Detection

Coincidence detection is a classic mechanism proposed for segmentation and classification. In an image for instance, elements moving at a common velocity are typically interpreted as being part of a single object, and this even when the image is only composed of random dots [41, 37].

The possibility of decentralized synchronization via central diffusive couplings can be used in building a coincidence detector. In [92], inspired in part by [7], the authors consider a leader-followers network of FitzHugh-Nagumo oscillators, where each follower oscillator i receives an external input I_i as well as a diffusive coupling from the leader oscillator (the element e_1 of G_1). Oscillators i and j receiving the same input ($I_i = I_j$) synchronize, so that choosing the system output as $\sum_{1 \leq i \leq n} [\dot{v}_i]^+$ captures the moment when a large number of oscillators receive the same input.

However, the previous development also implies that this very network can detect the moments when *several* groups of identical inputs exist. Furthermore, it is possible to identify the number of such groups and their relative size. Indeed, assume that the inputs are divided into k groups, such that for each group G_m, one has $\forall i, j \in G_m, I_i = I_j$. Since the oscillators in G_m only receive as input (a) the output of the leader, which is the same for everybody and (b) the external input I_i, which is the same for every oscillator in group G_m, they are input-symmetric and should synchronize with each other.

Symmetry, in particular bilateral symmetry, has also been shown to play a key role in human perception [6]. Consider a group of oscillators having the same individual dynamics and connected together in a symmetric manner. If we present to the network an input having the same symmetry, some of the oscillators will synchronize as predicted by the earlier theoretical results. One application of this idea is to build fast symmetry detectors, extending the oscillator-based coincidence detectors of the previous section [63].

4.4 Robustness

Robustness results for contracting systems [43] can be exploited to guarantee approximate synchronization even when the elements are not exactly identical, with important practical implications for the synchronization of spiking neurons of different dynamics.

Consider again a network of n dynamical elements

$$\dot{\mathbf{x}}_i = \mathbf{f}_i(\mathbf{x}_i, t) + \sum_{j \neq i} \mathbf{K}_{ij}(\mathbf{x}_j - \mathbf{x}_i) \qquad i = 1, \ldots, n \qquad (6)$$

with now possibly $\mathbf{f}_i \neq \mathbf{f}_j$ for $i \neq j$, and let us apply the robustness result for contracting systems of [43] to the projected system of Theorem 3. For strong enough coupling strength, all trajectories of the system will exponentially converge to a boundary layer of thickness D/λ around the sync subspace $\mathcal{M}$, where λ is the contraction rate of the auxiliary system and D is a measure of the dissimilarity of the elements [63].

Consider for instance, a system of the form (similar to the model used for coincidence detection in [88] and the previous section)

$$\dot{x}_i = f(x_i) + I_i + k(x_0 - x_i)$$

In this case, $D = \frac{I_{\max} - I_{\min}}{2}$.

Assume now that two spiking neurons are approximately synchronized, as just discussed. Then, since spiking induces large abrupt variations, the neurons must spike approximately at the same time. More specifically, if the bound on their trajectory discrepancy guaranteed by the above robustness result is significantly smaller than spike size, then this bound will automatically imply that the two neurons spike approximately at the same time. This inherent robustness of spike timing may be another reason why nature chose action potential mechanisms to represent information.

4.5 Locomotion

In an animal/robotics locomotion context, central pattern generators are often modeled as coupled nonlinear oscillators delivering phase-locked signals. Consider for instance [63] a system of three coupled 2-dimensional Andronov-Hopf oscillators,

$$\begin{cases} \dot{\mathbf{x}}_1 = \mathbf{f}(\mathbf{x}_1) + k(\mathbf{R}_{\frac{2\pi}{3}}\mathbf{x}_2 - \mathbf{x}_1) \\ \dot{\mathbf{x}}_2 = \mathbf{f}(\mathbf{x}_2) + k(\mathbf{R}_{\frac{2\pi}{3}}\mathbf{x}_3 - \mathbf{x}_2) \\ \dot{\mathbf{x}}_3 = \mathbf{f}(\mathbf{x}_3) + k(\mathbf{R}_{\frac{2\pi}{3}}\mathbf{x}_1 - \mathbf{x}_3) \end{cases}$$

where $\mathbf{f}$ is the dynamics of an Andronov-Hopf oscillator and the matrix $\mathbf{R}_{\frac{2\pi}{3}}$ describes a $\frac{2\pi}{3}$ planar rotation :

$$\mathbf{f}\begin{pmatrix} u \\ v \end{pmatrix} = \begin{pmatrix} u - v - u^3 - uv^2 \\ u + v - v^3 - vu^2 \end{pmatrix} \qquad \mathbf{R}_{\frac{2\pi}{3}} = \begin{pmatrix} -\frac{1}{2} & -\frac{\sqrt{3}}{2} \\ \frac{\sqrt{3}}{2} & -\frac{1}{2} \end{pmatrix}$$

We can rewrite the dynamics as $\dot{\mathbf{x}}_{\{\}} = \mathbf{f}_{\{\}}(\mathbf{x}_{\{\}}) - k\mathbf{L}\mathbf{x}_{\{\}}$, where

$$\mathbf{L} = \begin{pmatrix} \mathbf{I}_2 & -\mathbf{R}_{\frac{2\pi}{3}} & 0 \\ 0 & \mathbf{I}_2 & -\mathbf{R}_{\frac{2\pi}{3}} \\ -\mathbf{R}_{\frac{2\pi}{3}} & 0 & \mathbf{I}_2 \end{pmatrix}$$

First, observe that the *linear* subspace

$$\mathcal{M} = \left\{ \left(\mathbf{R}_{\frac{2\pi}{3}}^2(\mathbf{x}), \mathbf{R}_{\frac{2\pi}{3}}(\mathbf{x}), \mathbf{x} \right) : \mathbf{x} \in \mathbb{R}^2 \right\}$$

is flow-invariant, and that $\mathcal{M}$ is also a subset of Null($\mathbf{L}_s$). Next, remark that the characteristic polynomial of $\mathbf{L}_s$ is $X^2(X - 3/2)^4$ so that the eigenvalues of $\mathbf{L}_s$ are 0, with multiplicity 2, and 3/2, with multiplicity 4. Now since $\mathcal{M}$ is 2-dimensional, it is exactly the nullspace of $\mathbf{L}_s$, which implies in turn that $\mathcal{M}^\perp$ is the eigenspace corresponding to the eigenvalue 3/2.

Moreover, the eigenvalues of $\mathbf{J}_s(u, v)$ are $1 - (u^2 + v^2)$ and $1 - 3(u^2 + v^2)$, which are upper-bounded by 1. Thus, using the previous development, for $k > 2/3$ the three systems will *globally exponentially converge to a $\frac{2\pi}{3}$-phase-locked state* (i. e. a state in which the difference of the phases of two consecutive elements is constant and equals $\frac{2\pi}{3}$).

4.6 Coupled CPGs

Let us further illustrate combinations of such systems on the following example, based on [71], which studies models of central pattern generators in fish and salamanders [25].

Consider again an Andronov-Hopf oscillator, now with a limit cycle of constant radius $\rho > 0$,

$$\dot{\mathbf{x}} = \mathbf{f}_\rho(\mathbf{x}) = \begin{pmatrix} -v - \left(\frac{u^2+v^2}{\rho^2} - 1 \right) u \\ u - \left(\frac{u^2+v^2}{\rho^2} - 1 \right) v \end{pmatrix},$$

where $\mathbf{x}^T = [u, v]$. Note that

- $\mathbf{f}_\rho(\mathbf{R}\mathbf{x}) = \mathbf{R}\mathbf{f}_\rho(\mathbf{x})$ for an arbitrary rotation $\mathbf{R}$
- $\mathbf{f}_\rho(k\mathbf{x}) = k\mathbf{f}_{\rho/k}(\mathbf{x})$ for $k > 0$,

Consider now a two-way chain of n such oscillators with *phase-shift diffusive couplings*,

$$\dot{\mathbf{x}}_1 = \mathbf{f}_1(\mathbf{x}_1) - \gamma b_1(\mathbf{x}_1 - \mathbf{T}_1^{-1}\mathbf{x}_2)$$
$$\vdots$$
$$\dot{\mathbf{x}}_i = \mathbf{f}_i(\mathbf{x}_i) - \gamma a_{i-1}(\mathbf{x}_i - \mathbf{T}_{i-1}\mathbf{x}_{i-1}) - \gamma b_i(\mathbf{x}_i - \mathbf{T}_i^{-1}\mathbf{x}_{i+1}) \quad (7)$$
$$\vdots$$
$$\dot{\mathbf{x}}_n = \mathbf{f}_n(\mathbf{x}_n) - \gamma a_{n-1}(\mathbf{x}_n - \mathbf{T}_{n-1}\mathbf{x}_{n-1})$$

where each $\rho_i > 0$ is the radius of the corresponding limit cycle (the $\rho_i > 0$ may be distinct) and $\mathbf{T}_i$ is defined as

$$\mathbf{T}_i = \frac{\rho_{i+1}}{\rho_i}\mathbf{R}_i,$$

with $\mathbf{R}_i$ a planar rotation of ϕ_i,

$$\mathbf{R}_i = \begin{pmatrix} \cos(\phi_i) & -\sin(\phi_i) \\ \sin(\phi_i) & \cos(\phi_i) \end{pmatrix}.$$

Hence, a pair of diffusive couplings $(\mathbf{x}_{i+1} - \mathbf{T}_i\mathbf{x}_i)$ and $(\mathbf{x}_i - \mathbf{T}_i^{-1}\mathbf{x}_{i+1})$ matches different limit cycle radii ρ_{i+1} and ρ_i, and shifts the phase as much as ϕ_i. The coupling is two-way in that the *downward* coupling $(\mathbf{x}_{i+1} - \mathbf{T}_i\mathbf{x}_i)$ that appears in the equation of $\mathbf{x}_{i+1}$ oscillator pushes $\mathbf{x}_{i+1}$ to follow $\mathbf{x}_i$ adjusted through $\mathbf{T}_i$; and the *upward* coupling $(\mathbf{x}_i - \mathbf{T}_i^{-1}\mathbf{x}_{i+1})$ pushes $\mathbf{x}_i$ to follow $\mathbf{x}_{i+1}$ through $\mathbf{T}_i^{-1}$. γa_i is the downward coupling strength and γb_i the upward. By combining the coupling coefficients multiplied by γ, we can adjust the overall strength without affecting the upward to downward ratio.

Defining collective quantities

$$\mathbf{x}_{\{\}} \triangleq \begin{pmatrix} \mathbf{x}_1 \\ \vdots \\ \mathbf{x}_n \end{pmatrix} \qquad \mathbf{f}_{\{\}}(\mathbf{x}_{\{\}}) = \begin{pmatrix} \mathbf{f}_1(\mathbf{x}_1) \\ \vdots \\ \mathbf{f}_n(\mathbf{x}_n) \end{pmatrix}$$

we can rewrite the system (7) as

$$\dot{\mathbf{x}}_{\{\}} \triangleq \mathbf{f}_{\{\}}(\mathbf{x}_{\{\}}) - \gamma \mathbf{L}\mathbf{x}_{\{\}}, \tag{8}$$

where the coupling matrix $\mathbf{L}$ is then defined as

$$\mathbf{L} = \begin{pmatrix} b_1\mathbf{I} & -b_1\mathbf{T}_1^{-1} & 0 \\ -a_1\mathbf{T}_1 & a_1\mathbf{I} & 0 \\ 0 & 0 & 0 \end{pmatrix} + \ldots + \begin{pmatrix} 0 & 0 & 0 \\ 0 & b_{n-1}\mathbf{I} & -b_{n-1}\mathbf{T}_{n-1}^{-1} \\ 0 & -a_{n-1}\mathbf{T}_{n-1} & a_{n-1}\mathbf{I} \end{pmatrix}.$$

In order to make the coupling matrix $\mathbf{L}$ symmetric, we define a coordinate transformation $\mathbf{y}_{\{\}} = \mathbf{\Phi}\mathbf{x}_{\{\}}$ where

$$\mathbf{\Phi} = \begin{pmatrix} \mathbf{I} & 0 & 0 & \cdots \\ 0 & \frac{\rho_1}{\rho_2}\sqrt{\frac{b_1}{a_1}}\mathbf{I} & 0 & \cdots \\ 0 & 0 & \frac{\rho_1}{\rho_3}\sqrt{\frac{b_1 b_2}{a_1 a_2}}\mathbf{I} & \ddots \\ \vdots & \vdots & \ddots & \ddots \end{pmatrix}$$

Then (8) becomes

$$\dot{\mathbf{y}}_{\{\}} = \mathbf{\Phi}\mathbf{f}_{\{\}}(\mathbf{\Phi}^{-1}\mathbf{y}_{\{\}}) - \gamma\mathbf{\Phi}\mathbf{L}\mathbf{\Phi}^{-1}\mathbf{y}_{\{\}}$$

where the Jacobian of the couplings is symmetric positive semidefinite,

$$\mathbf{\Phi L \Phi}^{-1} = \begin{pmatrix} b_1 \mathbf{I} & -\sqrt{a_1 b_1} \mathbf{R}_1^T & \mathbf{0} & \cdots \\ -\sqrt{a_1 b_1} \mathbf{R}_1 & a_1 \mathbf{I} & \mathbf{0} & \cdots \\ \mathbf{0} & \mathbf{0} & \mathbf{0} & \cdots \\ \vdots & \vdots & \vdots & \ddots \end{pmatrix} +$$

$$+ \begin{pmatrix} \mathbf{0} & \mathbf{0} & \mathbf{0} & \mathbf{0} & \cdots \\ \mathbf{0} & b_2 \mathbf{I} & -\sqrt{a_2 b_2} \mathbf{R}_2^T & \mathbf{0} & \cdots \\ \mathbf{0} & -\sqrt{a_2 b_2} \mathbf{R}_2 & a_2 \mathbf{I} & \mathbf{0} & \cdots \\ \mathbf{0} & \mathbf{0} & \mathbf{0} & \mathbf{0} & \cdots \\ \vdots & \vdots & \vdots & \vdots & \ddots \end{pmatrix} + \cdots$$

The corresponding linear invariant subspace in **y**-system is

$$\mathcal{M}_y = \{\sqrt{b_i} \mathbf{R}_i \mathbf{y}_i - \sqrt{a_i} \mathbf{y}_{i+1} = \mathbf{0}, \ i = 1, \ldots n-1\},$$

or equivalently, in the original **x**-system,

$$\mathcal{M}_x = \{\mathbf{x}_{i+1} = \mathbf{T}_i \mathbf{x}_i, \ i = 1, \ldots n-1\}$$

Notice that $\mathcal{M}_y$ is the eigenspace corresponding to zero eigenvalues of $\mathbf{\Phi L \Phi}^{-1}$.

We can now construct a non-orthonormal basis for $\mathcal{M}_y^\perp$ directly from $\mathcal{M}_y$ as

$$\tilde{\mathbf{V}} = \begin{pmatrix} \sqrt{b_1} \mathbf{R}_1 & -\sqrt{a_1} \mathbf{I} & 0 & \cdots & 0 \\ 0 & \sqrt{b_2} \mathbf{R}_2 & -\sqrt{a_2} \mathbf{I} & \ddots & \vdots \\ \vdots & \ddots & \ddots & \ddots & 0 \\ 0 & \cdots & 0 & \sqrt{b_{n-1}} \mathbf{R}_{n-1} & -\sqrt{a_{n-1}} \mathbf{I} \end{pmatrix},$$

whose rows form a basis of $\mathcal{M}_y^\perp$. Through orthogonalization of $\tilde{\mathbf{V}}$ we obtain the orthogonal matrix $\mathbf{V}$, and the resulting $\mathbf{V \Phi L \Phi}^{-1} \mathbf{V}^T$ is positive definite. The Jacobian of $\dot{\mathbf{x}} = \mathbf{f}_\rho(\mathbf{x})$ is

$$\frac{\partial \mathbf{f}_\rho}{\partial \mathbf{x}} = \begin{pmatrix} -2\frac{u^2}{\rho^2} - \frac{u^2+v^2}{\rho^2} + 1 & -1 - 2\frac{uv}{\rho^2} \\ 1 - 2\frac{uv}{\rho^2} & -2\frac{v^2}{\rho^2} - \frac{u^2+v^2}{\rho^2} + 1 \end{pmatrix}$$

and the eigenvalues of its symmetric part are

$$1 - \frac{u^2+v^2}{\rho^2}; \ 1 - \frac{3(u^2+v^2)}{\rho^2}$$

Hence, by choosing γ large enough so that

$$\gamma \ \lambda_{\min}(\mathbf{V \Phi L \Phi}^{-1} \mathbf{V}^T) > 1$$

the generalized projected Jacobian verifies

$$\mathbf{V}\left[\varPhi\left(\frac{\partial \mathbf{f}}{\partial \mathbf{x}} - \gamma \mathbf{L}\right)\varPhi^{-1}\right]_s \mathbf{V}^T < 0.$$

Thus, a two-way chain of oscillators with arbitrary positive couplings gains can be synchronized to the desired invariant set $\mathcal{M}_x$ for sufficiently large overall coupling strength γ. The analysis extends easily [71] to double antisynchronized chains and to the other types of gaits studied e.g. in [25].

4.7 Frequency-Based Gait Selection

To study concurrent synchronization, extensive use is made in [63] of generalized symmetries [16]. Actually, replacing ordinary connections in the CPGs by filters enables *frequency-based* symmetry selection. This idea may have powerful applications, one of which could be automatic gait selection in locomotion. An simplified analogy with horse gaits can be made, for instance, by associating the low-frequency regime with the walk (left fore, right hind, right fore, left hind), and the high-frequency regime with the trot (left fore and right hind simultaneously, then right fore and left hind simultaneously). Transitions between the two regimes can occur automatically according to the speed of the horse (the frequency of its gait).

Note that standard techniques allow sharp causal filters with frequency-independent delays to be easily constructed [61].

References

1. Amari S, Arbib M (1977) Competition and Cooperation in Neural Nets. In: Metzler J (ed) Systems Neuroscience. Academic Press, San Diego, 119–165
2. Arimoto S, Naniwa T (2002) Learnability and Adaptability from the Viewpoint of Passivity Analysis, *Intelligent Automation and Soft Computing* 8(2)
3. Ashby WR (1966) In: Design for a Brain: the Origin of Adaptive Behavior. 2nd edition, Science Paperbacks
4. Belykh V, Belykh I, Mosekilde E (2001) Cluster Synchronization Modes in an Ensemble of Coupled Chaotic Oscillators. *Phys. Rev. E*
5. Berthoz A (1999) The Brain's Sense of Movement. Harvard University Press, Cambridge Massachusetts
6. Braitenberg V (1984) *Vehicles: Experiments in Synthetic Psychology*, chap. 9. The MIT Press
7. Brody C, Hopfield J (2003) Simple Networks for Spike-Timing-Based Computation with Application to Olfactor Processing. *Neuron*
8. Crick F, Koch C (2005) What is the Function of the Claustrum? *Phil. Trans. Roy. Soc. Lond. B, 360*
9. Dayan P, Hinton G, Neal R, Zemel R (1995) The Helmholtz Machine, *Neural Computation, 7*
10. Fiedler M(1976) Algebraic Connectivity of Graphs, *Czechoslovak Mathematical Journal*

11. FitzHugh R (1961) Impulses and Physiological States in Theoretical Models of Nerve Membrane. Biophys. Journal 1:445–466
12. Fujisawa S, Matsuki N, Ikegaya Y (2006) Single Neurons Can Induce Phase Transitions of Cortical Recurrent Networks with Multiple Internal St ates. *Cerebral Cortex* 16
13. Fukuda T, Kosaka T, Singer W, Galuske RAW (2006) Gap Junctions among Dendrites of Cortical GABAergic Neurons Establish a Dense and Widespread Intercolumnar Network. *The Journal of Neuroscience* 26(13)
14. George D, Hawkins, J (2005) Invariant Pattern Recognition using Bayesian Inference on Hierarchical Sequences
15. Gerstner W (2001) A Framework for Spiking Neuron Models: The Spike Response Model. In: The Handbook of Biological Physics. 4:469–516
16. Golubitsky M, Stewart I (2003) Synchrony versus Symmetry in Coupled Cells. *Equadiff 2003: Proceedings of the International Conference on Differential Equations*
17. Golubitsky M, Stewart I, Török A (2005) Patterns of Symmetry in Coupled Cell Networks with Multiple Arrows. *SIAM J. Appl. Dynam. Sys.*
18. Gray CM (1999) The Temporal Correlation Hypothesis of Visual Feature Integration: Still Alive and Well. Neuron 24:31–47
19. Grossberg S (1982) *Studies of Mind and Brain*, Kluwer
20. Grossberg S (2000) The Complementary Brain: A Unifying View of Brain Specialization and Modularity. *Trends in Cognitive Sciences*
21. Guckenheimer J, Holmes P (1983) Nonlinear Oscillations, Dynamical Systems, and Bifurcations of Vector Fields, *Springer Verlag*
22. Hodgkin AL, Huxley AF (1952) A Quantitative Description of Membrane Current and its Application to Conduction and Excitation in Nerve, *J. Physiol.* 117:500
23. Hopfield JJ, Brody CD (2001) What Is A Moment? Transient Synchrony as a Collective Mechanism for Spatiotemporal Integration. Proc. Natl. Acad. Sci. USA 98:1282–1287
24. Horn R, Johnson C (1985) *Matrix Analysis*, Cambridge University Press
25. Ijspeert A, Crespi A, Cabelguen J-M (2005) Simulation and Robotic Studies of Salamander Locomotion: Applying Neurobiological Principles to the Control of Locomotion in Robots
26. Izhikevich E (2003) Simple Model of Spiking Neuron, *IEEE Trans. on Neural Networks*
27. Izhikevich E, Desai N, Walcott E, Hoppensteadt F (2003) Bursts as a Unit of Neural Information: Selective Communication via Resonance, *Trends in Neuroscience* 26(3)
28. Izhikevich E (2004) Which Model to Use for Cortical Spiking Neurons? *IEEE Transactions on Neural Networks*
29. Izhikevich E, Kuramoto Y (2006) Weakly Coupled Oscillators. *Encyclopedia of Mathematical Physics*, Elsevier
30. Jadbabaie A, Lin J, Morse A (2003) Coordination of Groups of Mobile Autonomous Agents using Nearest Neighbor Rules. IEEE Transactions on Automatic Control 48:988-1001
31. Jadbabaie A, Motee N, Barahona M (2004) On the Stability of the Kuramoto Model of Coupled Nonlinear Oscillators. *Proceedings of the American Control Conf.*

32. Jin DZ, Seung HS (2002) Fast Computation With Spikes in a Recurrent Neural Network. *Physical Review E* 65:051922
33. Jouffroy J, Slotine J-JE (2004) Methodological Remarks on Contraction Theory, *IEEE CDC*
34. Kandel E, Schwartz J, Jessel T (2006) *Principles of Neural Science*, 5th ed., McGraw-Hill
35. Khalil H (1995) Nonlinear Systems, 2nd Ed.,*Prentice-Hall*
36. Knuth D (1997) *The Art of Computer Programming*, 3rd Ed. Addison-Wesley
37. Koch C (2004) *The Quest for Consciousness*, Roberts and Company Publishers
38. Korner E, Gewaltig M-O, Korner U, Richter A, Rodemann T. (1999) A model of computation in neocortical architecture, *Neural Networks* 12(7-8)
39. Kumar V, Leonard N, Morse S, (eds) (2005) Cooperative Control, *Springer Lecture Notes in Control and Information Science* 309
40. Leonard NE, Fiorelli E (2001) Virtual Leaders, Artificial Potentials and Coordinated Control of Groups, *40th IEEE Conference on Decision and Control*
41. Llinas R, Leznik E, Urbano F (2002) Temporal Binding via Cortical Coincidence Detection. *PNAS 99,1*
42. Lin Z, Broucke M, Francis B (2004) Local Control Strategies for Groups of Mobile Autonomous Agents. *IEEE Trans. on Automatic Control*
43. Lohmiller W, Slotine J-JE (1998) On Contraction Analysis for Nonlinear Systems. *Automatica* 34(6)
44. Lohmiller W, Slotine J-JE (2000) Control System Design for Mechanical Systems Using Contraction Theory, *IEEE Trans. Aut. Control* 45(5)
45. Lohmiller W, Slotine J-JE (2000) Nonlinear Process Control Using Contraction Theory. *A.I.Ch.E. Journal*
46. Luettgen M, Willsky A (1995) Likelihood Calculation for a Class of Multiscale Stochastic Models, with Application to Texture Discrimination. *IEEE Transactions on Image Processing* 4(2)
47. Maass W (2000) On the Computational Power of Winner-Take-All. Neural Comput. 12(11):2519–2535
48. von der Malsburg C (1995) Binding in Models of Perception and Brain Function. Current Opinion in Neurobiology 5:520–526
49. Mataric M (1998) Coordination and Learning in Multi-Robot Systems, *IEEE Intelligent Systems* 6-8
50. May RM, Gupta S, McLean AR (2001) Infectious Disease Dynamics: What Characterizes a Successful Invader? *Phil. Trans. R. Soc. Lond. B* 356:901–910
51. Milo R, Shen-Orr S, Itzkovitz S, Kashtan N, Chklovskii D, Alon U (2002) Network Motifs: Simple Building Blocks of Complex Networks. *Science* 298
52. Mohar B (1991) Eigenvalues, Diameter, and Mean Distance in Graphs. Graphs and Combinatorics 7:53–64
53. Mountcastle V (1998) The Cerebral Cortex, *Harvard University Press*
54. Murray JD (1993) Mathematical Biology, *Berlin;New York: Springer-Verlag*
55. Nagumo J, Arimoto S, Yoshizawa S (1962) An Active Pulse Transmission Line Simulating Nerve Axon. *Proc. Inst. Radio Engineers*
56. Nayfeh AH, Blachandran B (1995) Applied Nonlinear Dynamics, *Wiley*
57. Niessing J, Ebisch B, Schmidt KE, Niessing M, Singer W, Galuske RA (2005) Hemodynamic Signals Correlate Tightly with Synchronized Gamma Oscillations. *Science, 309*
58. Nijmeijer H (2001) A dynamical control view on synchronisation. Physica D 154:219–228

59. Nowak MA, Sigmund K (2004) Evolutionary Dynamics of Biological Games. Science 303:793–799
60. Olfati-Saber R, Murray R (2004) Consensus Problems in Networks of Agents With Switching Topology and Time-Delays. *IEEE Transactions on Automatic Control*
61. Oppenheim A, Schafer R, Buck J (1999) *Discrete-Time Signal Processing, 2nd Edition* , Prentice-Hall
62. Pecora LM, Carroll TL, Johnson GA, Mar DJ (1997) Fundamentals of synchronisation in chaotic systems. Chaos 7(4)
63. Pham Q-C, Slotine J-JE (2005) Stable Concurrent Synchronization in Dynamic System Networks, *http://arxiv.org/abs/q-bio.NC/0510051*
64. Pogromsky A, Santoboni G, Nijmeijer H (2002) Partial Synchronization: from Symmetry towards Stability. *Physica D*
65. Popov VM (1973) Hyperstability of Control Systems, *Springer-Verlag*
66. Rao R, Ballard D (1999) Predictive Coding in the Visual Cortex. *Nature Neuroscience* 2(1)
67. Rao R (2005) Bayesian inference and attention in the visual cortex, *Neuroreport* 16(16)
68. Schaal S (1999) Is imitation learning the route to humanoid robots? *Trends in Cognitive Sciences* 3(6)
69. Schnitzler A, Gross J (2005) Normal and Pathological Oscillatory Communication in the Brain. *Nat. Rev. Neurosci.*
70. Schoffelen J-M, Oostenveld R, Fries P (2005) Neuronal Coherence as a Mechanism of Effective Corticospinal Interaction. *Science*
71. Seo K, Slotine J-JE (2006) Models of Global Synchronization in Fish and Salamanders, *MIT NSL Report 050106*
72. Sepulchre R, Paley D, Leonard N (2003) Collective Motion and Oscillator Synchronization. *Proceedings of the 2003 Block Island Workshop on Cooperative Control.*
73. Singer W, Gray CM (1995) Visual Feature Integration and The Temporal Correlation Hypothesis, Annu. Rev. Neurosci. 18:555-586
74. Slotine J-JE (2003) Modular Stability Tools for Distributed Computation and Control. *Int. J. Adaptive Control and Signal Processing*, 17(6)
75. Slotine J-JE, Li W (1991) Applied Nonlinear Control, *Prentice-Hall*
76. Slotine J-JE, Lohmiller W (2001) Modularity, Evolution,and the Binding Problem: A View from Stability Theory. *Neural Networks* 14
77. Slotine J-JE, Wang W (2003) A Study of Synchronization and Group Cooperation Using Partial Contraction Theory, *Block Island Workshop on Cooperative Control*, Morse S., Editor, Springer-Verlag
78. Slotine J-JE, Wang W, Elrifai K (2004) Contraction Analysis of Synchronisation and Desynchronisation in Networks of Nonlinearly Coupled Oscillators. *Proceedings of MTNS 2004*
79. Smale S (1976) A Mathematical Model of Two Cells via Turing's Equation. In: The Hopf Bifurcation and Its Applications. Spinger-Verlag 354–367
80. Soechting J, Lacquaniti F (1988) Quantitative evaluation of the electromyographic responses to multidirectional load perturbations of the human arm. *J Neurophysiol.* 59(4)
81. Strogatz SH (1994) Nonlinear Dynamics and Chaos: with applications to physics, biology, chemistry, and engineering. Addison-Wesley Pub., Reading, MA

82. Strogatz SH (2000) From Kuramoto to Crawford: Exploring the Onset of Synchronization in Populations of Coupled Oscillators. *Physica D*
83. Thorpe S, Delorme A, Rullen RV (2001) Spike-Based Strategies for Rapid Processing. Neural Networks 14:715–725
84. Tononi G, et al. (1998) Proc. Natl. Acad. Sci. USA 95:3198–3203
85. Turing A (1952) The Chemical Basis of Morphogenesis. Philos. Trans. Roy. Soc. B 237:37–72
86. Walter WG (1950) An Imitation of Life, *Scientific American*
87. Walter WG (1951) A Machine that Learns, *Scientific American*
88. Wang W, Slotine J-JE (2005) On Partial Contraction Analysis for Coupled Nonlinear Oscillators. *Biological Cybernetics*, 92(1), 2005
89. Wang W, Slotine J-JE (2004) Adaptive Synchronization in Coupled Dynamic Networks, *NSL Report 040102, MIT, Cambridge, MA*, submitted
90. Wang W, Slotine J-JE (2004) Where To Go and How To Go: a Theoretical Study of Different Leader Roles, *NSL Report 040202, MIT, Cambridge, MA*, submitted
91. Wang W, Slotine J-JE (2004) Contraction Analysis of Time-Delayed Communications Using Simplified Wave Variables
92. Wang W, Slotine J-JE (2005) Fast Computation with Neural Oscillators. *Neurocomputing* 63(5)
93. Wiener N (1961) Cybernetics, *M.I.T. Press*
94. Won J, Hogan N (1995) Stability properties of human reaching movements, *Exp Brain Res. 107(1)*
95. Yazdanbakhsh A, Grossberg S (2004) Fast Synchronization of Perceptual Grouping in Laminar Visual Cortical Circuits. *Neural Networks* 17
96. Zhang Y, et al. (2001) Partial Synchronization and Spontaneous Spatial Ordering in Coupled Chaotic Systems. *Phys. Rev. E*
97. Zhao Y, Slotine J-JE (2005) Discrete Nonlinear Observers for Inertial Navigation, *Systems and Control Letters*

Force Control with A Muscle-Activated Endoskeleton

Neville Hogan

Department of Mechanical Engineering & Department of Brain and Cognitive Sciences, Massachusetts Institute of Technology, 77 Massachusetts Avenue, Room 3-146, Cambridge, Massachusetts 02139, USA
neville@mit.edu

Summary. The advantages and challenges of producing and controlling force with a mechanism like the human skeleton driven by actuators like mammalian muscles are considered. Some counter-intuitive subtleties of musculo-skeletal biomechanics are discovered: despite the energetic cost of isometric muscle activation, exerting forces that do no work may reduce metabolic energy consumption; in some circumstances, anatomical antagonist muscles may become functional synergists; and muscle tension acts to make skeletal posture statically unstable. The latter effect can be counteracted by muscle mechanical impedance, which emerges as an essential adjunct to muscle force production.

1 Introduction

Physical interaction with the world is a commonplace of life; we touch things, squeeze them, push them. Successful interaction with objects in the world depends in part on the ability to control the force we exert on them. The human body is richly endowed with muscles; they comprise about half of our body weight and they appear to be uniquely specialized for producing force. However, because of the unique properties of muscles and the way they are connected to the skeleton, force production presents some interesting challenges and opportunities. Some of them are reviewed below.

This chapter is dedicated to Professor Suguru Arimoto on the occasion of his 70th birthday. Throughout his influential career, Professor Arimoto has advocated and articulated the value of a physics-based approach to control: understanding the physics of a manipulator and the tasks it performs leads to improved designs and more effective methods for its control. In the following, that physics-based approach is applied to "reverse-engineer" force control in the mammalian biomechanical system. Some counter-intuitive observations are discovered and the possibility is raised that some of muscle's unique properties may have evolved in response to the challenges and opportunities of force production.

Because of the molecular underpinnings of muscle contraction, exerting muscle force always consumes metabolic energy. This is true even when no mechanical work is done because muscle length remains constant or when muscle absorbs work by lengthening under load. It may therefore seem that to economize metabolic energy consumption we should avoid exerting forces that do no useful work. However, an analysis of force exertion against a kinematic constraint will show that workless forces may, in fact, be used to reduce effort and energy consumption.

Another basic fact of muscle physiology is that skeletal muscles pull but don't push; to achieve both flexion and extension of the joints, they are deployed in opposing or antagonist groups. However, an analysis of musculo-skeletal geometry will lead to the counter-intuitive observation that for certain force-production tasks, anatomically antagonist muscles may actually cooperate as agonists or synergists.

A further subtlety of musculo-skeletal biomechanics is that force production challenges skeletal stability. Examining the kinematic details of musculo-skeletal attachments will show that because muscles are deployed to surround the bones (a basic fact of mammalian anatomy) the static stability of skeletal posture is reduced in proportion as muscle tension increases.

Of course, because it is manifestly evident that muscle contraction does not, in fact, cause the skeleton to collapse, we are presented with a paradox. One solution to this puzzle lies in the properties of muscle mechanical impedance. A stabilizing muscle stiffness will be shown to be an essential requirement for controlling force with a muscle-activated endoskeleton.

All of these considerations may be derived from a straightforward mechanical analysis. To begin, the standard robotic analysis of mechanism kinematics is reviewed.

1.1 Torque Space

If the human skeleton is regarded as a mechanism[1] then a starting point for its description is a set of variables that uniquely define its configuration. These generalized coordinates, usually identified as angular degrees of freedom relating adjacent rigid limb segments, define a *configuration space* which is fundamental to the analysis of skeletal mechanics. Knowing the configuration variables and the geometry of the limb segments, the location of all points on the skeleton may be determined. Though not unique, these generalized coordinates are fundamental; for example, the inertial and gravitational dynamic equations of the skeleton are properly defined in configuration space.

To analyze force production, a starting point is to identify the corresponding generalized forces. By definition these are such that the scalar product of generalized forces with incremental displacements in configuration space defines mechanical work done on the mechanism. Just as joint angles are usually

---

[1] The term "mechanism" is used loosely herein to refer to a collection of kinematically-constrained rigid bodies.

an appropriate choice for generalized coordinates, joint torques are usually an appropriate choice for generalized forces. They define a *torque space* which is fundamental to a description of how forces are exerted on and transmitted through the skeleton.

1.2 Mapping Torque Space to Contact Space

The position and orientation of any point of contact between the skeleton and the world (i.e., a hand, a foot, etc.) defines a *contact space* which may always be expressed as a function of the generalized coordinates or configuration variables

$$\mathbf{X} = \mathbf{L}(\boldsymbol{\theta}) \qquad (1)$$

where $\mathbf{X}$ is an array containing the positions and orientations of the contact point in some appropriate external reference frame (e.g. Cartesian coordinates), $\boldsymbol{\theta}$ is an array containing the generalized coordinates and $\mathbf{L}$ is an array of algebraic functions.

Conversely, any force or torque exerted at that point of contact may be mapped into torque space, for example by considering the incremental mechanical work it does. Denoting the exerted forces and torques by $\mathbf{F}$, the incremental work is

$$dW = \mathbf{F}^t d\mathbf{X} \qquad (2)$$

The incremental displacement of the point of contact is determined by

$$d\mathbf{X} = \mathbf{J}(\boldsymbol{\theta}) d\boldsymbol{\theta} \qquad (3)$$

where $\mathbf{J}$ is the Jacobian of the function relating the two sets of coordinates. By definition, the incremental mechanical work done on the skeleton is

$$dW = \boldsymbol{\tau}^t d\boldsymbol{\theta} \qquad (4)$$

where $\boldsymbol{\tau}$ is an array containing the generalized forces (joint torques). Substituting and rearranging

$$\boldsymbol{\tau} = \mathbf{J}^t(\boldsymbol{\theta}) \mathbf{F} \qquad (5)$$

The Jacobian matrix characterizes the mechanical transmission of force and torque between the world and the skeleton. Its columns define the moment arms relating contact forces to joint torques or the "gear ratios" relating contact torques to joint torques. It is always well-defined, even when the dimension of the configuration space exceeds that of the point of contact (i.e., the skeleton is redundant with respect to this contact point).

2 Workless Forces

When we exert ourselves to push on the world, the details of the mechanical transmission affect our action. Efficiency would seem to imply that forces should be exerted only to do useful work. However, one counter-intuitive aspect of force production with a muscle-activated skeleton is that it may be advantageous and energetically efficient to exert forces that generate no mechanical work [1].

Consider the class of constrained-motion tasks represented by turning a crank. Common examples include opening a door or pedaling a bicycle. The crank defines a holonomic constraint on the motion of the limb, allowing displacement only in certain directions (e.g., tangent to the circle described by the door handle or the pedals). Displacement normal to the constraint is nominally zero; elongation or compression of the pedals is negligible as is the change of the radius of the door handle about its hinges. Any force exerted in the normal direction does negligible mechanical work and no useful work. Only forces tangent to the constraint (e.g., in line with the path of the pedals or the door handle) perform useful work. One basic fact about mammalian muscle is that generating force consumes metabolic energy even when the muscle does no work. One might therefore expect that the most energetically efficient way to turn a crank with the least muscular effort would be to exert exclusively tangential forces. Surprisingly, that turns out not to be true.

It is informative to represent the task in torque space. For simplicity, consider a two-segment model of the skeleton (e.g., describing planar motion of the arm and forearm). The elbow angle (forearm relative to arm) and shoulder angle (arm relative to thorax) may serve as configuration variables. The corresponding generalized forces are the elbow and shoulder torques and the torque space is depicted in Fig. 1.

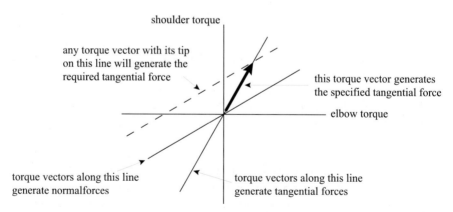

Fig. 1. Torque-space diagram illustrating how forces exerted normal to a constraint may reduce muscular effort. The sub-spaces (directions in this example) that generate normal and tangential forces at a particular limb configuration are shown by the light lines; in general they are not orthogonal. Any torque vector with its tip on the dashed line generates the specified tangential force. The torque vector (shown bold) that generates no normal force is not the shortest torque vector that meets the task requirements.

To exert a tangential force requires a particular torque vector (see Fig. 1) determined by equation (5). Consequently, at any given configuration, tangential forces exerted on the crank define a sub-space of torque space, in this two-dimensional example a line or direction in torque space. Normal forces (that stretch or compress the crank) correspond to torque vectors in a different sub-space (a different direction in this example) but, due to the limb geometry, it is rarely orthogonal to the torque vector for tangential force. If workless normal forces are allowed, the specified tangential force can be achieved by any torque vector with its tip on the (dashed) line in Fig. 1.

To assess exertion or efficiency we need some measure of effort, and one convenient candidate is the length of a vector in torque space. By that measure, effort is minimized by the shortest torque vector that meets task requirements. Referring to the figure, it becomes evident that the least muscular effort (by this measure) is generally *not* achieved by generating a purely tangential force. Except in the (unusual) case that the torque-space directions corresponding to normal and tangential forces are at right angles, minimum effort (the shortest torque vector) will be achieved by exerting a combination of normal and tangential forces.

Many alternative measures of effort (including total metabolic energy consumption, total root-mean-squared stress on the joints, etc.) have been suggested and explored. However, insofar as these measures are equivalent to defining a norm on the torque space, the above argument applies. For example, if metabolic energy consumption is described by a monotonic function of torque, with minimal consumption at zero torque, that function may be used to re-scale the axes of torque space and the argument proceeds as before. Except in the unusual case that the torque sub-spaces corresponding to normal and tangential forces are orthogonal with respect to the particular norm, the minimizing torque vector will generate both normal and tangential forces. In short, attempting to stretch or compress the crank can (almost) always be used to economize the effort need to push tangential to it.

As the dimensions of the torque space and the contact space are the same in this example, it may be analyzed equivalently in terms of the forces exerted at the contact point. However, in general the number of relevant joints exceeds the number of relevant task dimensions, often by a large margin (there are estimated to be about 200 distinct limb segments in the human skeleton). In that case it is not generally possible to identify the contact forces that result from individual joint torques. To do so would require inverting equation (5) but the Jacobian is not square and not invertible. Nevertheless, though it cannot be inverted, the Jacobian is well-defined; it is always possible to project contact forces into torque space. Torque space is fundamental for analysis of how forces are exerted on and transmitted through the skeleton.

3 Coordinating Muscle Forces

A more detailed analysis of force production should consider how muscle forces are coordinated to generate joint torques. Upon doing so, another counter-intuitive subtlety of musculo-skeletal biomechanics emerges: anatomical antagonists may be functional synergists. The best way to produce a specified force may require simultaneous contraction of opposing muscles [1].

3.1 Mapping Muscle Space to Torque Space

This arises because of the geometry of the skeleton and the way muscles are attached. For simplicity, assume that the lengths of muscles and their associated tendons are uniquely defined by the configuration of the skeleton[2]. In that case, the complete set of muscle lengths (which may be taken as the coordinates of a *muscle space*) may be expressed as a function of skeletal configuration variables (generalized coordinates)

$$\mathbf{q} = \mathbf{q}(\boldsymbol{\theta}) \tag{6}$$

where $\mathbf{q}$ is an array containing muscle lengths and $\mathbf{g}$ is an array of algebraic functions. With this information, muscle forces may be mapped to torque space. As above, one way to do this is to consider the incremental mechanical work done. Denoting the complete array of muscle forces by $\mathbf{f}$, the incremental work is

$$dW = \mathbf{f}^t d\mathbf{q} \tag{7}$$

The relation between incremental displacements in muscle length space and configuration space are determined by

$$d\mathbf{q} = \mathbf{j}(\boldsymbol{\theta}) d\boldsymbol{\theta} \tag{8}$$

where $\mathbf{j}$ is the Jacobian of the function relating the two sets of coordinates. The incremental mechanical work done on the skeleton is again given by equation (4), hence (substituting and rearranging)

$$\boldsymbol{\tau} = \mathbf{j}^t(\boldsymbol{\theta}) \mathbf{f} \tag{9}$$

The Jacobian matrix $\mathbf{j}$ characterizes the mechanical transmission of force from the muscles to the skeleton. Its columns define the moment arms of the muscles

---

[2] In fact, the relation between musculo-tendon length and skeletal configuration may depend on the force exerted by a muscle and perhaps by its neighbors. If its tendon passes through an aponeurosis or wraps around a joint, increasing tension may increase or decrease the moment arm of a muscle about the joint even when that joint does not move. Similarly, the line of action of a muscle's force may change due to the contraction (and hence shape change) of its neighbors, again without any change in skeletal configuration. However, these complications do not change the main result, that anatomical antagonists may be functional synergists.

about the joints. It is always well-defined, even though the dimension of muscle space exceeds that of configuration space, i.e., the number of muscles exceeds the number of skeletal degrees of freedom.

For any given skeletal configuration, each individual muscle force defines a sub-space of torque space, the combination of joint torques generated by that muscle. Assuming that muscles can't push[3], this sub-space is a "half-space", the set of vectors pointing in a direction in torque space, but not vectors of the opposite sign. Muscles that actuate a single degree of freedom define a sub-space co-aligned with one of the torque space axes. Consider a two-segment model of planar motion of the arm and forearm as described above. Elbow and shoulder torques define the axes of torque space as depicted in Fig. 2.

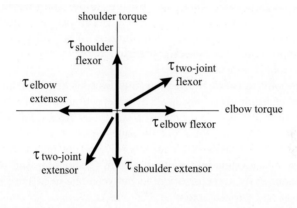

Fig. 2. Torque-space diagram illustrating the joint torque contributions due to single-joint elbow flexor ($\tau_{elbow flexor}$) and extensor ($\tau_{elbow extensor}$) muscles; single-joint shoulder flexor ($\tau_{shoulder flexor}$) and extensor ($\tau_{shoulder extensor}$) muscles; and two-joint flexor ($\tau_{twojoint\ flexor}$) and extensor ($t_{two-joint extensor}$) muscles. Torque vectors of single-joint antagonist muscles are anti-aligned but those of two-joint antagonist muscles need not be.

The action of individual muscles may be represented as vectors in this space. A muscle such as brachialis spans only the elbow joint and generates torque to flex it. In Fig. 2, single-joint elbow flexors such as brachialis define a sub-space which is the positive horizontal axis; i.e., they generate torque vectors oriented positively (but not negatively) along the horizontal axis.

Single-joint muscles have unambiguous antagonists. A muscle such as the deep head of triceps spans only the elbow joint and generates torque to extend it. In Fig. 2, single-joint elbow extensors define a sub-space which is the negative horizontal axis; i.e., they generate torque vectors oriented negatively (but not positively) along the horizontal axis.

---

[3] This is generally true of skeletal muscles but not necessarily for muscular hydrostats such as the tongue.

Similarly, single-joint shoulder flexor muscles generate torque vectors oriented along the positive vertical axis in Fig. 2 and single-joint shoulder extensor muscles generate torque vectors oriented along the negative vertical axis. The unambiguous mechanical antagonism of these muscle groups is reflected in the architecture of the nervous system that drives them. In general flexors and extensors are endowed with a different density and combination of sensors and may be activated by different neural pathways.

Other muscles span multiple degrees of freedom or multiple joints and generate torques about each of them. The torque sub-space defined by a polyarticular muscle corresponds to positive displacement in a direction that does not co-align with the torque space axes. For example, biceps brachii generates torque to flex the elbow and also to flex the shoulder. The moment arms that map the single muscle force onto torques about the two different joints may vary with limb configuration and, in general, are not equal. In Fig. 2, two-joint flexor muscles generate a combination of shoulder and elbow flexion torques represented by a vector oriented between the positive (but not negative) horizontal and vertical axes.

Like single-joint muscles, multi-joint muscles typically have anatomical antagonists. For example, the long head of triceps brachii generates torques to extend the elbow and also to extend the shoulder. Again, the moment arms that map the single muscle force onto torques about the different joints may vary with limb configuration and, in general, are not equal. In Fig. 2, two-joint extensor muscles generate a combination of shoulder and elbow flexion torques represented by a vector oriented between the negative (but not positive) horizontal and vertical axes.

For the most part, multi-joint flexor and extensor muscle groups act antagonistically; i.e., the long head of triceps generally opposes biceps brachii. However, the muscle moment arms about the two joints are not generally in the same ratio. At any given limb configuration, force in biceps brachii may generate flexion torques about elbow and shoulder in a different ratio than triceps long head generates extension torques about the same joints. Whereas the torque vectors generated by single-joint flexor and extensor muscles are always *exactly* opposed, the torque vectors generated by two-joint flexor and extensor muscles are *not* always exactly in opposition.

3.2 Antagonists or Synergists?

One important consequence is that anatomically antagonist muscles may contribute synergistically to force production. Because the two-joint flexor and extensor torque vectors are not anti-aligned (i.e., they do not exactly oppose one another), there exists a region of torque space (a sector in Fig. 2) onto which both flexor and extensor poly-articular muscles have positive projections. Any force to be generated where the limb contacts the world, (e.g., the hand, the foot, etc.) requires a combination of joint torques that may always be represented by a vector in torque space (as described above). If the required joint torque lies in the region in which flexor and extensor poly-articular muscles have positive projections, then both can make a positive contribution to producing the required force. In that case, these *anatomical antagonists become functional synergists.*

To what extent are these theoretical considerations biologically meaningful? After all, if these antagonist muscles project weakly onto the required torque, then their contribution to force production will come at a heavy cost, e.g., in the consumption of metabolic energy. To the author's knowledge, this question has not been fully explored. However, if a situation requires maximal force production, then every contribution matters, however costly. Reasoning along similar lines, if a force is to be produced with minimal effort, where the measure of effort is equivalent to defining a norm on the torque space, then for certain forces the minimum-effort solution will include contributions, albeit small, from both of these anatomical antagonists.

Given that muscles spanning multiple degrees of freedom are more the rule than the exception in the mammalian musculo-skeletal system, the identification of antagonists requires care. Though the distinction between flexors and extensors may have a sound neural basis (possibly a legacy of phylogeny) their functional definition as antagonists is ambiguous and depends sensitively on the task. The same pair of muscles may act as antagonists in one case and synergists in another—even if the configuration of the limb remains fixed and the only change is the direction of the force to be produced.

4 Kinematic Instability

In the analysis above, the configuration of the skeleton serves to parameterize the mappings between contact forces, joint torques and muscle forces. In effect, the analysis is equivalent to assuming an instantaneous static equilibrium, with the skeleton remaining at a fixed configuration as force is produced. However, a further subtlety of musculo-skeletal mechanics is that co-contraction of antagonist muscles may cause the skeletal configuration to become statically unstable.

Naturally, an equilibrium posture cannot be maintained if a joint torque is generated without any opposing equilibrating load. However, even when net joint torque is zero (a necessary condition for static equilibrium) muscle contraction may still destabilize the limbs. To understand how, consider the relation between muscle forces and joint torques described by equation (9). The columns of the Jacobian, **j**, define the moment arms of the muscles about the joints. However, those moment arms depend on limb configuration so that incremental changes of the joint angles may increase or decrease the moment arms. Any non-zero muscle force will result in a joint torque that depends on joint angle. That produces a behavior analogous to that of a spring, but with a stiffness that may be positive or negative, depending on the details of musculo-skeletal attachment.

The details may be quantified simply by differentiating equation (9), the relation between muscle force and joint torque (which has been transposed for clarity).

$$d\boldsymbol{\tau}^t = \mathbf{f}^t \left(\partial \mathbf{j}\left(\boldsymbol{\theta}\right)/\partial\boldsymbol{\theta}\right) d\boldsymbol{\theta} = \mathbf{f}^t \mathbf{h}\left(\boldsymbol{\theta}\right) d\boldsymbol{\theta} \qquad (10)$$

where the partial derivative of the Jacobian matrix $\mathbf{j}$ with respect to the array $\boldsymbol{\theta}$ denotes the Hessian $\mathbf{h}$ of the function relating configuration variables to muscle lengths. It is a three-index array of partial derivatives of the elements of the Jacobian matrix with respect to each of the configuration variables. The product of the Hessian and the muscle force vector yields a two-index matrix that defines what may be termed a configuration-dependent "kinematic stiffness", $\boldsymbol{\Gamma}$.

$$\boldsymbol{\Gamma}^t\left(\boldsymbol{\theta}\right) = \mathbf{f}^t \mathbf{h}\left(\boldsymbol{\theta}\right) \qquad (11)$$

Though it is a second-order effect of the relation between joint angles and muscle lengths, kinematic stiffness is important nonetheless. Most mammalian muscles span joints in such a manner that their torque acts to move the limbs in a direction that increases the muscle's moment arm. For example, brachialis is connected between the arm and forearm such that its origin, insertion, and the axis of the elbow form an approximate triangle. Brachialis acts to flex the elbow and as it does, the perpendicular distance between its line of action (joining its origin and insertion and forming the base of the triangle) and the elbow axis (the apex of the triangle) increases. Consequently, if brachialis exerts a constant force, the corresponding elbow torque increases as the elbow flexes and decreases as it extends. This resembles a negative spring and has a destabilizing effect on the joint. For example, if brachialis were opposed by a constant torque so that static equilibrium was achieved at a particular elbow angle, small displacements towards flexion would increase the net torque acting to flex the elbow; similarly, small displacements towards extension would increase the net torque acting to extend the elbow.

Though the details vary, this example is typical of the muscles of the mammalian skeleton: because muscles are deployed outside the bones, their moment arms tend to increase as they move the limbs. Consequently, muscles with this destabilizing kinematic connection are typically opposed by muscles which also exhibit the same behavior, and that compounds the destabilizing effect. In essence, the skeleton may be regarded as a set of columns of rigid links connected by joints of negligible torsional stiffness. The links are surrounded by tensile elements, the muscles, which act to load the columns in compression. As the column has negligible torsional stiffness, it is vulnerable to buckling. Force production in a muscle-activated endoskeleton[4] may destabilize its posture and cause it to collapse.

---

[4] It may be otherwise for an exoskeleton.

4.1 Muscle Mechanical Impedance

Despite this observation, a moment of personal experimentation will confirm that simultaneous contraction of opposing muscles (that's what we do when we tense our muscles or clench a fist) does *not* destabilize the limbs. How can this be? An answer lies in the special properties of muscle.

Muscle powers movement but is remarkably different from engineering power sources. For example, electrical voltage sources are painstakingly designed so that (as nearly as resources and engineering expertise can achieve) their output voltage is independent of current delivered. Electrical current sources (e.g., commonly used with permanent-magnet electric motors) deliver current largely independent of the required voltage. These properties are quantified by electrical *impedance*, where zero is the ideal impedance for voltage sources and infinity (equivalent to zero admittance) is the ideal for current sources. Mechanical power sources are similar; for many applications the ideal is an actuator that produces force (or torque) independent of translational (or angular) speed and displacement. This property is quantified by *mechanical impedance* (the ratio of force/torque change to motion change) with zero being the ideal for a torque or force source. Conversely, an ideal motion source has infinite mechanical impedance (zero mechanical admittance).

In striking contrast, mammalian muscle is neither an ideal force source nor an ideal motion source. Quite aside from variations due to fatigue or pathological conditions, for a constant activation of the alpha-motoneuron that drives it, the force developed by a skeletal muscle is a strong function of muscle length and its rate of change. Its mechanical impedance is certainly not infinite (the speed at which a muscle shortens depends strongly on the load it moves) but neither is it zero; force developed may change by several hundreds of percent as muscle length and shortening velocity vary within their physiological range. With mechanical impedance far from either ideal extreme, muscle appears to be a poor power source (by engineering standards).

This might reflect nothing more than a biological imperfection or an incomplete evolutionary adaptation. Skeletal muscles generate force by deforming myosin molecules after they have been attached to binding sites on actin filaments [2]. Dependence of contractile force on muscle length might be an epi-phenomenon arising in part from incomplete overlap of myosin heads and actin binding sites; or from mechanically parallel passive tissue that encapsulates a sarcomere; or from some other mechanisms. However, available evidence favors the view that finite muscle mechanical impedance is *not* an imperfection but is highly adaptive, solving some of the problems inherent in a muscle-activated endoskeleton and supporting the remarkable motor abilities of biological systems.

4.2 Mammalian Actuators

From neurophysiology we learn that mammalian muscle is richly endowed with muscle spindles [2]. These sensory organs are the origin of afferent nerve fibers carrying signals related to (at least) muscle length and its rate of change. They make monosynaptic excitatory connections in the spinal cord to at least the homologous alpha-motoneurons innervating the muscle containing the spindle. Both these afferent fibers and the efferent alpha fibers are myelinated, with nerve conduction velocities among the highest in the nervous system. The result is one of the body's fastest feedback connections, giving rise to the well-known stretch reflex: abrupt stretch of a muscle evokes a brisk, involuntary contraction a short time later. Other prominent muscle sensors include Golgi tendon organs, which respond almost exclusively to *muscle*-generated force in the tendon. Their afferents are also myelinated (hence fast-conducting) and act to inhibit the alpha-motoneurons of homonymous muscles. One plausible role of these nested feedback loops is to enhance (and perhaps regulate) the apparent mechanical impedance of the actuator that drives the skeleton [3, 4]. That is, the apparent stiffness of the actuator is increased; more precisely, because of reflex loop dynamics, apparent impedance[5] is changed.

4.3 Neural Feedback Complements Intrinsic Mechanical Impedance

Although muscle responds rapidly to activity of its embedded sensors, the delay in neural transmission is nevertheless substantial: about 30 milliseconds or more for muscles of the human upper limbs, 50 milliseconds or more for muscles of the human lower limbs. To avoid instability due to this delay, feedback gain must be limited, which in turn limits impedance bandwidth: an effective opposing force is generated in response to stretch only at lower frequencies. Feedback-generated impedance declines at higher frequencies.

It is a remarkable fact that intrinsic muscle impedance due to actin-myosin interactions exhibits a complementary variation with frequency. Broadly speaking, as muscle is stretched, deformation of each myosin molecule bound to an actin filament contributes to apparent muscle stiffness. At the same time, the rate at which myosin molecules detach from their actin binding sites increases with deformation but is limited by the dynamics of the ATP-driven reactions that provide the energy to separate myosin heads from actin filaments. The result is that the steady force generated in response to rapid stretch is much smaller than the transient force. Intrinsic muscle impedance is high at higher frequencies and declines at lower frequencies.

It therefore appears that feedback-generated impedance properties of the peripheral neuromuscular system may complement those due to intrinsic muscle contractile mechanics [5]. Rather than think of the skeleton being driven by muscles, it is probably more useful to consider the skeletal actuator to be a neuro-muscular system comprised of the muscle, its sensors and the associated reflex loops, acting in concert to manage mechanical impedance over a wide frequency range.

---

[5] Mechanical impedance may be considered a dynamic generalization of stiffness.

4.4 The Stabilizing Effect of Muscle Impedance

The analysis presented above demonstrated that the static equilibrium of the skeleton may be compromised by force production; muscle tension may cause the skeleton to collapse upon itself. Muscle stiffness (the static component of muscle mechanical impedance) counteracts this effect. A sufficient condition to ensure static stability is readily obtained by extending the differentiation of equation (9) and assuming that muscle forces **f** depend on muscle lengths **q** (as well as other variables such as neural drive)

$$\mathbf{f} = \mathbf{f}(\mathbf{q}, \mathbf{u}) \tag{12}$$

where **u** is an array containing at least the neuro-muscular control inputs[6]. Using equations (8) and (11) yields

$$\frac{\partial \boldsymbol{\tau}}{\partial \boldsymbol{\theta}} = \boldsymbol{\Gamma}(\boldsymbol{\theta}) + \mathbf{j}^t(\boldsymbol{\theta}) \left(\frac{\partial \mathbf{f}}{\partial \mathbf{q}}\right) \mathbf{j}(\boldsymbol{\theta}) \tag{13}$$

In words, the net joint stiffness is the sum of a kinematic stiffness and a neuro-muscular stiffness. Positive-definite net joint stiffness is sufficient to ensure static stability. This may be achieved if neuro-muscular stiffness is positive-definite (i.e., stabilizing) and larger than kinematic stiffness. Furthermore, note from equation (11) that the destabilizing kinematic stiffness is proportional to muscle force. To stabilize skeletal posture, neuro-muscular stiffness must increase with muscle force at least as rapidly.

One robust observation of mammalian muscle is that its neuro-muscular impedance is positive and increases with force exerted. The a-reflexic (intrinsic) contribution to muscle stiffness is positive at zero operating force and increases in proportion to force over the full range of contraction [4]. With reflexes intact, neuro-muscular stiffness is substantially larger. It is also positive at non-zero operating force and increases in approximate proportion to force (though about three times more rapidly) up to about 50% of maximum voluntary contraction [4]. Furthermore, while the torque contributions of antagonist muscles usually oppose each other, the impedance contributions of all muscles always add to net joint impedance [6].

Taken together, these properties of muscle-generated impedance offset the destabilizing effects of configuration-dependent muscle moment arms. In fact, the increase of neuro-muscular impedance with force exerted more than compensates for the static instability due to musculo-skeletal kinematics. It is easily verified that voluntary co-contraction of antagonist muscles increases the externally-observable net joint stiffness [7, 8].

---

[6] For a-reflexic muscle, the neural input is alpha-motoneuron activity. However, if we consider the basic neuro-muscular actuator to comprise the muscle and its associated reflex feedback loops, the neural control input to that actuator has not yet been identified unambiguously.

5 Concluding Remarks

One important conclusion to be drawn from this analysis is that it is inadequate or at best incomplete to consider muscle as just a force generator. While engineering actuators are often painstakingly designed to minimize mechanical impedance, that would be quite unsuitable for a machine with the kinematic structure of the mammalian musculo-skeletal system. The minimum competent description of muscle should include its mechanical output impedance, and almost certainly the fact that muscle stiffness increases with force.

Once the importance of mechanical impedance is recognized, it affords a wealth of alternative approaches to problems in both robotics and biology. For example, the stabilizing properties of neuro-muscular mechanical impedance may simplify motion control. They endow the skeleton with dynamic attractor properties (e.g., a tendency to converge to a certain pose or trajectory) so that trajectory details could then emerge from dynamic interaction between skeleton, muscle and peripheral neural circuits with minimal intervention from higher levels of the central nervous system. That is one key element of the so-called "equilibrium-point" theories of neural control, which remain appealing (though controversial!) after four decades of research [9, 10].

Modulating net joint stiffness may be used to regularize ill-posed problems such as the "inverse kinematics" of a redundant multi-joint system. The challenge is to determine trajectories of the joints that will achieve a specified trajectory of, say, the hand. Approaches equivalent to computing a pseudo-inverse of the Jacobian have been proposed, but the resulting map is not integrable; that is, closed paths of the hand do not yield closed paths of the joints. However, taking advantage of the natural stabilizing properties of joint stiffness yields an inverse map that is fully integrable [11]. In fact, the full theoretical implications of muscle mechanical impedance for motion control of redundant mechanical systems remain to be articulated; see especially the recent work by Arimoto et al. [12].

The ability to modulate externally-observable mechanical impedance also provides a robust way to control physical interaction with the world. Impedance control of robots has been investigated and applied extensively; see [13] for a review. Observations of human subjects interacting with unstable objects have verified that they can skillfully tune the stiffness of the hand to maintain stability [8]. This may be an essential requirement for effective use of tools which destabilize limb posture [14].

The unique challenges of force production in a muscle-activated endoskeleton raise the intriguing possibility that muscle mechanical impedance may be an evolutionary adaptation of the biological actuator. To prevent muscle activation from collapsing the skeleton, muscle stiffness must increase with muscle force at least as rapidly as the destabilizing effects of musculo-skeletal kinematics. If that is achieved, muscle stiffness that increases even more rapidly with muscle force enables antagonist co-contraction strategies to modulate mechanical impedance at the hand, which then enables sophisticated behavior such as the use of tools. To the author's knowledge, the biological validity of these speculations remains untested.

Acknowledgements

This work was supported in part by the New York State Spinal Cord Injury Board and by the Eric P. and Evelyn E. Newman Laboratory for Biomechanics and Human Rehabilitation. Figures 1 and 2 and portions of Sections 4.1 to 4.3 are reproduced from [15] with permission from World Scientific Publishing Co. Pte. Ltd., Singapore.

References

1. Russell D (1990) An Analysis of Constrained Motions in Manipulation. PhD thesis, Mechanical Engineering Dept., M.I.T.
2. Kandel E, Schwartz J, Jessell T (1991) Principles of Neural Science. Appleton & Lange, Norwalk, Connecticut
3. Nichols T, Houk J (1976) Improvement in linearity and regulation of stiffness that results from actions of stretch reflex. Journal of Neurophysiology 39:119–142
4. Hoffer J, Andreassen S (1981) Regulation of soleus muscle stiffness in pre-mammillary cats: Intrinsic and reflex components. Journal of Neurophysiology 45(2):267–285
5. Grillner S (1972) The role of muscle stiffness in meeting the changing postural and locomotor requirements for force development by the ankle extensors. Acta Physiol. Scand. 86:92–108
6. Hogan N (1990) Mechanical impedance of single and multi-articular systems. In Winters J, Woo S, eds.: Multiple Muscle Systems: Biomechanics and Movement Organization. Springer-Verlag, New York 149–164
7. Mussa-Ivaldi F, Hogan N, Bizzi E (1985) Neural, mechanical and geometric factors subserving arm posture in humans. Journal of Neuroscience 5(10):2731–2743
8. Burdet E, Osu R, Franklin D, Milner T, Kawato M (2001) The central nervous system stabilizes unstable dynamics by learning optimal impedance. Nature 414:446–449
9. Feldman A (1966) Functional tuning of the nervous system during control of movement or maintenance of a steady posture ii. controllable parameters of the muscle. Biofizika 11:498–508
10. Feldman A (1966) Functional tuning of the nervous system during control of movement or maintenance of a steady posture iii. mechanographic analysis of the execution by man of the simplest motor task. Biofizika 11:766–775
11. Mussa-Ivaldi F, Hogan N (1991) Integrable solutions of kinematic redundancy via impedance control. The International Journal of Robotics Research 10(5):481–491

12. Arimoto S, Hashiguchi H, Sekimoto M, Ozawa R (2005) Generation of natural motions for redundant multi-joint systems: A differential-geometric approach based upon the principle of least actions. Journal of Robotic Systems 22(11):583–605
13. Hogan N, Buerger SP (2004) Chapter 19: Impedance and interaction control. In Kurfess T, ed.: Robotics and Automation Handbook. CRC Press 149–164
14. Rancourt D, Hogan N (2001) Stability in force-production tasks. Journal of Motor Behavior 33(2):193–204
15. Hogan N (2002) Skeletal muscle impedance in the control of motor actions. Journal of Mechanics in Medicine and Biology 2(3&4):359–373

On Dynamic Control Mechanisms of Redundant Human Musculo-Skeletal System

Kenji Tahara[1] and Zhi-Wei Luo[2,1]

[1] Bio-Mimetic Control Research Center, RIKEN, 2271-130, Shimoshidami, Moriyama-Ku, Nagoya, Aichi 463-0003 Japan   tahara@bmc.riken.jp
[2] Dept. of Computer and Systems Eng., Kobe Univ., 1-1 Rokkodai, Nada-ku, Kobe, Hyogo, 657-8501 Japan   luo@gold.kobe-u.ac.jp

Summary. This chapter deals with modeling of human-like reaching and pinching movements. For the reaching movements, we construct a two-link planar arm model with six redundant muscles. A simple task-space feedback control scheme, taking into account internal forces induced by the redundant and nonlinear muscles, is proposed for this model. Numerical simulations show that our sensory-motor control can realize human-like reaching movements. The effect of gravity is also studied here and a method for the gravity compensation on the muscle input signal level is introduced. The stability of this method is proved and its effectiveness is shown through numerical simulations. For the pinching movements, realized by the index finger and the thumb, the co-contraction between the flexor and extensor digitorum muscles is analyzed. It is shown that an internal force term can be generated by the redundant muscles to modulate a damping factor in the joint space. Numerical simulations show that the co-contraction of each digitorums makes it possible to realize human-like pinching movements. Our results suggest that the central nervous system (CNS) does not need to calculate complex mathematical models based on the inverse dynamics or on the planning of optimal trajectories. Conversely, the human motor functions can be realized through the sensory-motor control by exploiting the passivity, nonlinearity and the redundancy of the musculo-skeletal systems.

1 Introduction

This paper is dedicated to Professor Suguru Arimoto on the occasion of his 70th birthday. From the early information science to signal processing and system control theory, from robotics to biological motor control studies, he has contributed a lot of original results that are recognized as the most fundamental in these fields. Specifically, he has emphasized that nonlinearity and passivity characteristics in robotics are very important in the robot control design. Recently, this has led to an interesting theory of "the stability theory on a manifold" [1, 2]. This theory may give us a hint on how animals realize their redundant dynamic motor control functions. In this chapter, we study human motions following his philosophy. By modeling the nonlinear dynamics of the human arms, fingers and redundant muscles, we show in numerical simulations that even a simple sensor-motor control can generate human-like skillful motions.

Natural human movements are smoother, more dexterous, and more sophisticated than the movements of present-day robots. The skill and the beauty of the human motor functions have attracted the attention of not only physiologists and kinesiologists, but also the robotics researchers. Studying human-like natural movements has become an important research area in modern robotics. A variety of human-like movements have been mimicked in robotic systems, and many computational models have been proposed. However, the most fundamental problem of how the CNS generates a strategy to perform these movements is still unsolved. One of the reasons is related to the ill-posed problems induced by the joint and muscle redundancies. These problems were pointed out by N.A. Bernstein [3, 4] more than half a century ago. Many physiologists and robotics researchers have challenged to solve these ill-posed problems by introducing optimization criteria. In the robotics field, there have been attempts to determine the solution of the inverse kinematics uniquely by introducing artificial performance indices and optimizing them [5, 6]. Recently, the joint redundancy problem has been approached more naturally by introducing novel concepts called "the stability on a manifold" and "the transferability to a submanifold" proposed by Arimoto [1, 2].

In physiology, there are several hypotheses to determine movements of surplus limbs or muscles. Fel'dman [7] firstly presented an equilibrium point(EP) hypothesis and later Bizzi et al. [8] observed that the joint angle can be determined by the equilibrium point when the output forces of agonist and antagonist muscles are balanced. After that, several researches treated multi-joint reaching movements based on the EP hypothesis [9, 10, 11, 12]. The attractive point of this hypothesis is that there is no need to solve the inverse dynamics. Hogan [13] extended the EP hypothesis to a control scheme called "a virtual trajectory control hypothesis". This hypothesis defines the desired trajectory by using a minimum-jerk criterion, and the virtual trajectory is derived corresponding to the equilibrium point. Morasso [14] also observed reaching movements of the human arm and concluded that the CNS central commands may be coordinated not in the joint space but in the task space. This was called "a spatial control hypothesis". This hypothesis is closely related to the Bernstein's problem introduced in 1935 [4], which states that the human motions are coordinated by the CNS in the task space, not in the joint and muscles spaces. In contrast, Kawato [15] presented a feedback-error learning scheme that is based on the learning of internal dynamical models. This scheme emphasizes that a human being controls many redundant limbs and muscles by using an internal dynamical model, and the inverse model can be obtained by learning the feedback error. It was also pointed by Katayama and Kawato [16] that the virtual trajectory hypothesis approach [13] is too complex to calculate the equilibrium point trajectory in the joint space in order to realize human-like reaching movements. Note that Kawato's approach also needs to derive the desired trajectory in advance by introducing optimization criteria [17]. However, calculating these criteria as well as computing the inverse dynamics can be too complicated.

In this chapter, in order to understand the brain-motor control strategies, we pay a special attention to the sensory-motor control mechanisms. We take the viewpoint that the human can hardly calculate in real time any complex mathematical model based on the inverse dynamics or on the planning of optimal trajectories. We address two types of human movements. One is a reaching movement produced by the upper and forearm, and the other is a pinching movement produced by the index finger and the thumb. These dexterous movements require an intelligent control.

To model the reaching movement, we construct a two-link planar arm model with six redundant muscles and introduce a simple task-space feedback control scheme. In addition, we take into account the effect of internal forces induced by the redundant muscles. The internal forces are generated by the co-contraction of agonist and antagonist muscles. Hogan [18] remarked and Gribble [19] observed that the co-contraction of agonist and antagonist muscles plays an important role in the regulation of the impedance characteristics of each joint. Arimoto et al. [2] pointed out that the damping in the joints plays a crucial role for the convergence of the end-point trajectory.

In our study we use a simple model of nonlinear muscle dynamics based on the physiological study [20]. We introduce an internal force term in addition to the simple task-space feedback control term in order to modulate the damping factors in the joint space. We also formulate the kinematics and the dynamics of the arm model, and give a proof of the convergence of the closed-loop dynamics. Numerical simulations show that our simple sensory-motor control can result in the human-like reaching movements [21, 22]. In everyday life, humans move naturally even though the gravity force affects their bodies. This is because the antigravity muscles work to compensate the gravity force without the control loop from the high level of the brain motor cortex. In physiology, it is considered as a stretch reflex. To take this effect into consideration, we propose a method for the gravity compensation at the muscle input signal level [22]. The stability of this method is proved and its effectiveness is shown through numerical simulations.

Next, we proceed to the cooperative control of a pair of fingers. In physiology, a tip-to-tip pinch produced by the index finger and the thumb is called a precision grip [23]. It is distinct from the power grip (whole-finger grasping) in both anatomical and functional sense. In fact, the muscle actuation pattern in the precision grip is quite different from that in the power grip [24]. This pinching movement can be realized by synergistic actuation [25] of the flexor and extensor digitorum muscles. The co-contraction (or co-activation) of the flexor and extensor digitorum muscles in performing the pinching motion has been observed in physiological studies [26]. In our study we focus on the co-contraction between the flexor and extensor digitorum muscles in performing the stable pinching by the index finger and the thumb. It is known that the muscle configuration of the human hand is quite complex. In particular, the index finger and the thumb consist of not only several intrinsic digitorum muscles, but also of multiarticular extrinsic digitorum muscles. They are attached to the fore-arm to perform flexion and extension movements.

In the analysis of pinching movements, we employ a model of dual 2 D.O.F. fingers. To mimic the structure of the human fingers [27], each finger in our model is actuated by one monoarticular digitorum muscle and two biarticular digitorum muscles. We formulate the kinematics and the dynamics of the finger-object system under the conditions that the overall motion of this system is in the horizontal plane, and the object has two parallel flat surfaces. Then, we derive a control law realizing the stable pinching simultaneously with the posture regulation [1, 29]. We show that the internal forces generated by the redundant muscles can modulate the damping effect in the joint space [21, 22]. Finally, we present numerical simulations and suggest that the co-contraction of each digitorums makes it possible to realize human-like pinching movements.

2 Human-Like Reaching Movements

2.1 A Two-link Arm Model with Six Redundant Muscles

In this section, we consider a two-link arm model with six muscles as shown in Fig. 1. It is assumed that the movements of the overall system are in the horizontal plane. The effects of gravity and friction are ignored. The masses of the muscles are neglected.

Kinematics of Muscle-Joint Space

The lengths of the muscles can be defined as follows:

$$l(\boldsymbol{\theta}) = \begin{pmatrix} (a_1^2 + b_1^2 + 2a_1 b_1 \cos\theta_1)^{\frac{1}{2}} \\ (a_2^2 + b_2^2 - 2a_2 b_2 \cos\theta_1)^{\frac{1}{2}} \\ (a_3^2 + b_3^2 + 2a_3 b_3 \cos\theta_2)^{\frac{1}{2}} \\ (a_4^2 + b_4^2 - 2a_4 b_4 \cos\theta_2)^{\frac{1}{2}} \\ (a_{51}^2 + a_{52}^2 + L_1^2 + 2a_{51} L_1 \cos\theta_1 \\ \quad + 2a_{52} L_1 \cos\theta_2 + 2a_{51} a_{52} \cos(\theta_1 + \theta_2))^{\frac{1}{2}} \\ (a_{61}^2 + a_{62}^2 + L_1^2 - 2a_{61} L_1 \cos\theta_1 \\ \quad - 2a_{62} L_1 \cos\theta_2 + 2a_{61} a_{62} \cos(\theta_1 + \theta_2))^{\frac{1}{2}} \end{pmatrix} \qquad (1)$$

where $l(\boldsymbol{\theta}) \in \mathbb{R}^6$ is the vector of the muscle lengths, θ_1, θ_2 are the joint angles. $a_{1\sim4}$, $b_{1\sim4}$ and $a_{51}, a_{52}, a_{61}, a_{62}$ are the positions of the muscle insertions as shown in Fig. 1. The time derivative of eq. (1) can be expressed as follows:

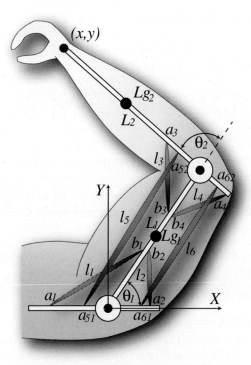

Fig. 1. A two-link arm model with six muscles

$$\dot{l} = Q(\theta)\dot{\theta} \qquad (2)$$

where $\dot{l} \in \mathbb{R}^6$ is the vector of the contractile velocity and $\dot{\theta} \in \mathbb{R}^2$ is the vector of the angular velocity. $Q(\theta) \in \mathbb{R}^{6\times 2}$ is the Jacobian matrix from the joint space to the muscle space. The relation between the muscle forces and the joint torques can be expressed, by using the principle of virtual work, as follows:

$$\boldsymbol{\tau} = \boldsymbol{W}\boldsymbol{F} \qquad (3)$$

where $\boldsymbol{F} \in \mathbb{R}^6$ is the vector of the muscle forces, $\boldsymbol{\tau} \in \mathbb{R}^2$ is the vector of the joint torques and $\boldsymbol{W} = \boldsymbol{Q}(\boldsymbol{\theta})^{\mathrm{T}} \in \mathbb{R}^{2\times 6}$ from eq. (2). The inverse relation between the joint torques and the muscle forces can be expressed as follows:

$$\boldsymbol{F} = \boldsymbol{W}^{+}\boldsymbol{\tau} + \left(\boldsymbol{I}_6 - \boldsymbol{W}^{+}\boldsymbol{W}\right)\boldsymbol{k_e} \qquad (4)$$

where $\boldsymbol{W}^{+} = \boldsymbol{W}^{\mathrm{T}}(\boldsymbol{W}\boldsymbol{W}^{\mathrm{T}})^{-1} \in \mathbb{R}^{6\times 2}$ is the pseudo-inverse matrix of $\boldsymbol{W}$, $\boldsymbol{k_e} \in \mathbb{R}^6$ is an arbitrary vector, and $\boldsymbol{I}_6 \in \mathbb{R}^{6\times 6}$ indicates an identity matrix. The physical meaning of the second term of the right-hand side in eq. (4) is the internal forces generated by the redundant muscles.

Kinematics of Joint-Task Space

The end-point position of the two-link arm, which is described by the Cartesian coordinates (x, y), is defined as:

$$\begin{pmatrix} x \\ y \end{pmatrix} = \begin{pmatrix} L_1 \cos\theta_1 + L_2 \cos(\theta_1 + \theta_2) \\ L_1 \sin\theta_1 + L_2 \sin(\theta_1 + \theta_2) \end{pmatrix} \tag{5}$$

By taking differentiation of eq. (5), we obtain:

$$\dot{\boldsymbol{x}} = \boldsymbol{J}(\boldsymbol{\theta})\dot{\boldsymbol{\theta}} \tag{6}$$

where $\dot{\boldsymbol{x}} \in \mathbb{R}^2$ is the vector of the end-point velocity and $\boldsymbol{J}(\boldsymbol{\theta}) \in \mathbb{R}^{2\times 2}$ is the Jacobian matrix from the task space to the joint space. It is defined as:

$$\boldsymbol{J}(\boldsymbol{\theta}) = \begin{pmatrix} -L_1 \sin\theta_1 - L_2 \sin(\theta_1 + \theta_2) & -L_2 \sin(\theta_1 + \theta_2) \\ L_1 \cos\theta_1 + L_2 \cos(\theta_1 + \theta_2) & L_2 \cos(\theta_1 + \theta_2) \end{pmatrix} \tag{7}$$

Modeling of Muscle Dynamics

Muscles can change their viscoelasticity depending on the muscle activation level. Here in this section, we introduce a nonlinear muscle model that is a simplified version of Hill's model [20]. We define it as:

$$f = \bar{\alpha} - (\bar{\alpha}b + b_0)\dot{l} \tag{8}$$

where f is the output force. The control input to the muscles is defined as $\bar{\alpha} = \alpha f_0 \geq 0$ (f_0 is the maximum output force of the isometric contraction model, α is muscle activation level). The control input is saturated as shown in Fig. 2.

It is noted that the approximated model (8) has two types of damping coefficients. One is b and the other is b_0. $b > 0$ is the damping coefficient which depends on the control input $\bar{\alpha}$, and $b_0 > 0$ stands for the intrinsic damping coefficient. It is known from the physiological studies [7, 8, 9, 10, 11,

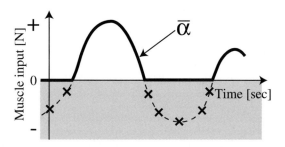

Fig. 2. The saturation of muscle input $\bar{\alpha}$

that the muscle output force behaves like a spring-like force. To take this feature into account, we introduce a simple task space position feedback control with the control input

$$\bar{\alpha} = -W^+ J^T K_p \Delta x + \left(I_6 - W^+ W\right) k_e \qquad (9)$$

where the diagonal matrix of $K_p \in \mathbb{R}^{2\times 2}$ is the position feedback gain, $x \in \mathbb{R}^2$ is the position of the end-point, $\Delta x = x - x_d \in \mathbb{R}^2$ is the position error from the desired position of the end-point. The first term of the right-hand side in eq.(9) generates a spring-like force at the end-point. The overall muscle dynamics of this system can be represented as follows:

$$F = \bar{\alpha} - (AB + B_0)\,l \qquad (10)$$

$$\begin{bmatrix} F = (f_1,\ f_2,\ \cdots,\ f_6)^T \in \mathbb{R}^6 \\ \bar{\alpha} = (\bar{\alpha}_1,\ \bar{\alpha}_2,\ \cdots,\ \bar{\alpha}_6)^T \in \mathbb{R}^6 \\ A = \mathrm{diag}(\bar{\alpha}_1,\ \bar{\alpha}_2,\ \cdots,\ \bar{\alpha}_6) \in \mathbb{R}^{6\times 6} \\ B = \mathrm{diag}(b_1,\ b_2,\ \cdots,\ b_6) \in \mathbb{R}^{6\times 6} \\ B_0 = \mathrm{diag}(b_{01},\ b_{02},\ \cdots,\ b_{06}) \in \mathbb{R}^{6\times 6} \end{bmatrix}$$

Dynamics of The Two-Link Arm Model

The dynamics of a two-link planar arm can be described by Lagrange's equation [15].

$$H(\theta)\ddot{\theta} + \left\{\frac{1}{2}\dot{H}(\theta) + S(\dot{\theta},\theta)\right\}\dot{\theta} = WF \qquad (11)$$

where $H(\theta) \in \mathbb{R}^{2\times 2}$ is the inertial matrix of the arm, $\ddot{\theta},\ \dot{\theta},\ \theta \in \mathbb{R}^2$ are the angular acceleration, velocity and position, respectively. The inside of the bracket $\{\}$ denotes the nonlinear term which include the Coriolis and centrifugal force, $S(\dot{\theta},\theta) \in \mathbb{R}^{2\times 2}$ is a skew-symmetric matrix, and WF is the vector of the input torque generated by the muscle forces. Substituting eq. (10) into the two-link arm dynamics of eq. (11) yields the overall system dynamics

$$H(\theta)\ddot{\theta} + \left\{\frac{1}{2}\dot{H}(\theta) + S(\dot{\theta},\theta)\right\}\dot{\theta} = W\left\{\bar{\alpha} - (AB+B_0)\,l\right\} \qquad (12)$$

2.2 Stability Analysis

The closed-loop dynamics, in which the input is defined as $\Delta u = 0$, can be derived from eq.(12). It is given as:

$$\begin{aligned}\Delta u = H(\theta)\ddot{\theta} + \left\{\frac{1}{2}\dot{H}(\theta) + S(\dot{\theta},\theta)\right\}\dot{\theta} \\ + J^T K_p \Delta x + W(AB+B_0)W^T \dot{\theta} = 0\end{aligned} \qquad (13)$$

Taking the inner product of the input $\boldsymbol{\Delta u}$ in eq.(13) with the output $\dot{\boldsymbol{\theta}}$ yields:

$$\dot{\boldsymbol{\theta}}^\mathrm{T}\boldsymbol{\Delta u} = \frac{\mathrm{d}}{\mathrm{d}t}\left\{\frac{1}{2}\dot{\boldsymbol{\theta}}^\mathrm{T}\boldsymbol{H}\dot{\boldsymbol{\theta}} + \frac{1}{2}\boldsymbol{\Delta x}^\mathrm{T}\boldsymbol{K_p}\boldsymbol{\Delta x}\right\} + \dot{\boldsymbol{\theta}}^\mathrm{T}\boldsymbol{W}\left(\boldsymbol{AB}+\boldsymbol{B_0}\right)\boldsymbol{W}^\mathrm{T}\dot{\boldsymbol{\theta}} \quad (14)$$

By integrating eq.(14) over time interval $[0, t]$, it becomes:

$$\int_0^t \dot{\boldsymbol{\theta}}^\mathrm{T}\boldsymbol{\Delta u}\,\mathrm{d}\tau = E(t) - E(0) + \int_0^t \dot{\boldsymbol{\theta}}^\mathrm{T}\boldsymbol{W}\left(\boldsymbol{AB}+\boldsymbol{B_0}\right)\boldsymbol{W}^\mathrm{T}\dot{\boldsymbol{\theta}}\,\mathrm{d}\tau \quad (15)$$

where

$$E(t) = \frac{1}{2}\dot{\boldsymbol{\theta}}^\mathrm{T}\boldsymbol{H}\dot{\boldsymbol{\theta}} + \frac{1}{2}\boldsymbol{\Delta x}^\mathrm{T}\boldsymbol{K_p}\boldsymbol{\Delta x} \geq 0 \quad (16)$$

Since the muscles can output only a tensile force, each diagonal element of $\boldsymbol{A} = \mathrm{diag}(\bar{\alpha}_1, \bar{\alpha}_2, ..., \bar{\alpha}_6)$ is non-negative. Therefore, the matrix $(\boldsymbol{AB}+\boldsymbol{B_0})$ is positive definite since $\boldsymbol{A} \geq 0$, $\boldsymbol{B} > 0$ and $\boldsymbol{B_0} > 0$. The control input $\bar{\boldsymbol{\alpha}} = -\boldsymbol{W}^+\boldsymbol{J}^\mathrm{T}\boldsymbol{K_p}\boldsymbol{\Delta x} + \left(\boldsymbol{I_6} - \boldsymbol{W}^+\boldsymbol{W}\right)\boldsymbol{k_e}$ becomes zero if and only if $(\boldsymbol{x} = \boldsymbol{x}_d,\ \dot{\boldsymbol{x}} = 0)$ when $\boldsymbol{k_e} = 0$. Thus, the joint damping matrix $\boldsymbol{W}\left(\boldsymbol{AB}+\boldsymbol{B_0}\right)\boldsymbol{W}^\mathrm{T}$ is positive definite as far as $\boldsymbol{W}$ is not degenerated. Therefore, from

$$\frac{\mathrm{d}}{\mathrm{d}t}E(t) = -\dot{\boldsymbol{\theta}}^\mathrm{T}\boldsymbol{W}\left(\boldsymbol{AB}+\boldsymbol{B_0}\right)\boldsymbol{W}^\mathrm{T}\dot{\boldsymbol{\theta}} \leq 0 \quad (17)$$

one concludes that the closed-loop dynamics is passive. The scalar function $E(t)$ plays the role of Lyapunov's function, and LaSalle's invariance theorem can be applied to it. Then, $\boldsymbol{x} \to \boldsymbol{x}_d$, $\dot{\boldsymbol{x}} \to 0$ when $t \to \infty$ [30].

2.3 Numerical Simulation

In this section, we present numerical simulations. The physical parameters of the model are shown in Table 1, 2 and the parameters of the control law are shown in Table 3. The internal force vector is defined as $\boldsymbol{k_e} = k_e \boldsymbol{e}$ (k_e is a positive constant and $\boldsymbol{e} \in \mathbb{R}^6$ is an unit vector). The initial posture of each joint and the desired point are shown in Table 4.

Figure 3 shows the end-point paths of the reaching movements. It can be seen from this figure that the end-point converges to the desired point. Moreover, when k_e is set to 20.0 the end-point path is slightly curved, which is similar to what is observed in human movements. However, when k_e is

Table 1. Physical parameters of the two-link arm model

	L [m]	M [kg]	I [kg·m^2]	L_g [m]
1st Link	0.31	1.93	0.0141	0.165
2nd Link	0.34	1.52	0.0188	0.19

Table 2. The muscle insertion point

Muscles	Values [m]	
Shoulder flexor (l_1)	$a_1 = 0.055$	$b_1 = 0.080$
Shoulder extensor (l_2)	$a_2 = 0.055$	$b_2 = 0.080$
Elbow flexor (l_3)	$a_3 = 0.030$	$b_3 = 0.120$
Elbow extensor (l_4)	$a_4 = 0.030$	$b_4 = 0.120$
Double-joint flexor (l_5)	$a_{51} = 0.040$	$a_{52} = 0.045$
Double-joint extensor (l_6)	$a_{61} = 0.040$	$a_{62} = 0.045$

Table 3. Gains and coefficient

$$K_p = \begin{pmatrix} 10.0 & 0 \\ 0 & 10.0 \end{pmatrix}, \; B = \begin{pmatrix} 10.0 & 0 \\ 0 & 10.0 \end{pmatrix}, \; B_0 = \begin{pmatrix} 10.0 & 0 \\ 0 & 10.0 \end{pmatrix}, \; k_e = (20.0 \; or \; 0)$$

Table 4. Initial angles and desired point

Initial posture [deg]	$(\theta_1, \; \theta_2) = (60°, \; 30°)$
Desired point [m]	$(x, \; y) = (-0.40, \; 0.40)$

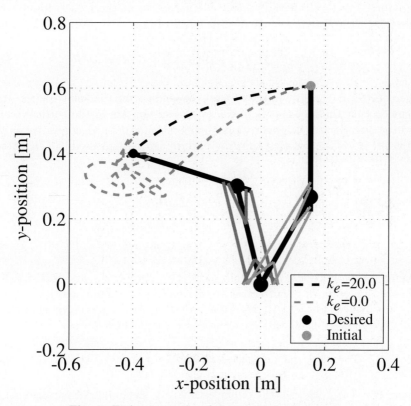

Fig. 3. End-point paths of the reaching movements

set to zero the end-point path behaves strangely and cannot be represented by a straight line. This is because the elements of the joint damping matrix $\boldsymbol{W}\left(\boldsymbol{AB} + \boldsymbol{B_0}\right)\boldsymbol{W}^{\mathrm{T}}$ are increased when k_e is increased. Note that the settling time in the case of $k_e = 20.0$ is shorter than that in the case of $k_e = 0$ (see Fig. 4).

Figure 5 shows the tangential velocities of the end-point. It can be seen that the velocity profile in the case of $k_e = 20.0$ is nearly bell-shaped, similar to what is observed in human movements. However, the velocity profile in the case of $k_e = 0$ is obviously not bell-shaped. Therefore, we suggest that the human-like reaching movement can be realized by choosing a suitable internal force to regulate the damping.

2.4 Gravity Compensation

In everyday life, humans move naturally even though the gravity force affects their bodies. The reason is that antigravity muscles always compensate the effect of gravity without a high level control loop in the brain motor cortex. In physiology it is considered as a stretch reflex. To model this reflex, we introduce a method for the gravity compensation at the muscle input level. The two-link arm dynamics under the effect of gravity is given as:

$$\boldsymbol{H}(\boldsymbol{\theta})\ddot{\boldsymbol{\theta}} + \left\{\frac{1}{2}\dot{\boldsymbol{H}}(\boldsymbol{\theta}) + \boldsymbol{S}(\dot{\boldsymbol{\theta}}, \boldsymbol{\theta})\right\}\dot{\boldsymbol{\theta}} + \boldsymbol{G}(\boldsymbol{\theta}) = \boldsymbol{WF} \qquad (18)$$

where $\boldsymbol{G}(\boldsymbol{\theta})$ is the gravitational term depending on the joint angles. It is assumed that the movements of the overall system are in the vertical plane, and the direction of the gravitational acceleration is defined as positive on the y-axis. It is well-known that the gravity term can be expressed [30] as follows:

$$\boldsymbol{G}(\boldsymbol{\theta}) = \boldsymbol{Z}(\boldsymbol{\theta})\boldsymbol{\Theta} \qquad (19)$$

where $\boldsymbol{Z}(\boldsymbol{\theta})$ is called a regressor matrix. It defined as follows:

$$\boldsymbol{Z}(\boldsymbol{\theta}) = \begin{pmatrix} L_{g1}\cos\theta_1 & L_1\cos\theta_1 + L_{g2}\cos(\theta_1 + \theta_2) \\ 0 & L_{g2}\cos(\theta_1 + \theta_2) \end{pmatrix} g \qquad (20)$$

where g is the gravity acceleration, L_1 is the length of the first link, L_{g1} and L_{g2} are the center of gravity position of each link. It is assumed that L_{g1} and L_{g2} are known. $\boldsymbol{\Theta} = (M_1, M_2)^{\mathrm{T}}$ is the vector of the unknown parameters (the link masses). Now, we define the control input as follows:

$$\bar{\boldsymbol{\alpha}} = -\boldsymbol{W}^+\left(\boldsymbol{J}^{\mathrm{T}}\boldsymbol{K}_p\Delta\boldsymbol{x} - \boldsymbol{Z}(\boldsymbol{\theta})\hat{\boldsymbol{\Theta}}\right) + \left(\boldsymbol{I}_6 - \boldsymbol{W}^+\boldsymbol{W}\right)k_e \qquad (21)$$

where $\hat{\boldsymbol{\Theta}} = (\hat{M}_1, \hat{M}_2)^{\mathrm{T}}$ is the online estimate of the unknown parameters. Its update law is defined as:

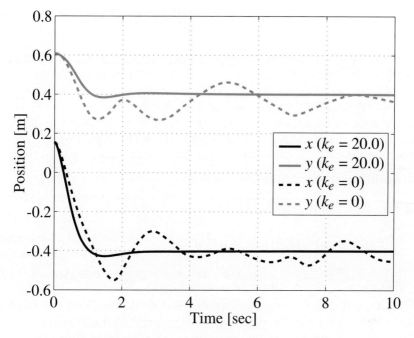

Fig. 4. Transient responses of the end-point positions

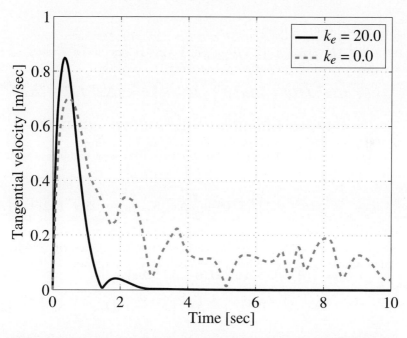

Fig. 5. Transient responses of the tangential velocities of the end-point

$$\hat{\boldsymbol{\Theta}}(t) = \hat{\boldsymbol{\Theta}}(0) - \int_0^t \boldsymbol{\Gamma}^{-1} \boldsymbol{Z}(\boldsymbol{\theta})^{\mathrm{T}} (\dot{\boldsymbol{\theta}} + \beta \boldsymbol{J}^{\mathrm{T}} \boldsymbol{\Delta x}) \mathrm{d}\tau \qquad (22)$$

where $\boldsymbol{\Gamma}^{-1}$ is a positive definite diagonal matrix and α is a positive number. Therefore, the overall dynamics under the effect of gravity can be given by:

$$\boldsymbol{H}(\boldsymbol{\theta})\ddot{\boldsymbol{\theta}} + \left\{\frac{1}{2}\dot{\boldsymbol{H}}(\boldsymbol{\theta}) + \boldsymbol{S}(\dot{\boldsymbol{\theta}}, \boldsymbol{\theta})\right\}\dot{\boldsymbol{\theta}} + \boldsymbol{G}(\boldsymbol{\theta})$$
$$= -\boldsymbol{J}^{\mathrm{T}} \boldsymbol{K}_p \boldsymbol{\Delta x} + \boldsymbol{Z}(\boldsymbol{\theta})\hat{\boldsymbol{\Theta}} - \boldsymbol{W}\left(\boldsymbol{AB} + \boldsymbol{B_0}\right)\boldsymbol{W}^{\mathrm{T}}\dot{\boldsymbol{\theta}} \qquad (23)$$

2.5 Stability under Unknown Gravity

From eq. (23), the closed-loop dynamics can be obtained as follows:

$$\boldsymbol{H}(\boldsymbol{\theta})\ddot{\boldsymbol{\theta}} + \left\{\frac{1}{2}\dot{\boldsymbol{H}}(\boldsymbol{\theta}) + \boldsymbol{S}(\dot{\boldsymbol{\theta}}, \boldsymbol{\theta})\right\}\dot{\boldsymbol{\theta}}$$
$$+ \boldsymbol{J}^{\mathrm{T}} \boldsymbol{K}_p \boldsymbol{\Delta x} + \boldsymbol{Z}(\boldsymbol{\theta})\boldsymbol{\Delta \Theta} + \boldsymbol{W}(\boldsymbol{AB} + \boldsymbol{B_0})\boldsymbol{W}^{\mathrm{T}}\dot{\boldsymbol{\theta}} = \boldsymbol{\Delta u} = 0 \qquad (24)$$

where $\boldsymbol{\Delta u} = 0$, $\boldsymbol{\Delta x} = \boldsymbol{x} - \boldsymbol{x}_d$ and $\boldsymbol{\Delta \Theta} = \boldsymbol{\Theta} - \hat{\boldsymbol{\Theta}}$. Assume that the Jacobian matrices $\boldsymbol{W}$ and $\boldsymbol{J}$ are not degenerated during the reaching movements. By taking the inner product of the input $\boldsymbol{\Delta u}$ with the output $\dot{\boldsymbol{\theta}} + \beta \boldsymbol{J}^{\mathrm{T}} \boldsymbol{\Delta x}$, we obtain:

$$(\dot{\boldsymbol{\theta}} + \beta \boldsymbol{J}^{\mathrm{T}} \boldsymbol{\Delta x})^{\mathrm{T}} \boldsymbol{\Delta u} =$$
$$\frac{\mathrm{d}}{\mathrm{d}t}\left\{\frac{1}{2}\dot{\boldsymbol{\theta}}^{\mathrm{T}} \boldsymbol{H} \dot{\boldsymbol{\theta}} + \frac{1}{2}\boldsymbol{\Delta x}^{\mathrm{T}} \boldsymbol{K}_p \boldsymbol{\Delta x} + \frac{1}{2}\boldsymbol{\Delta \Theta}^{\mathrm{T}} \boldsymbol{\Gamma} \boldsymbol{\Delta \Theta}\right\}$$
$$+ \beta \boldsymbol{\Delta x}^{\mathrm{T}} \boldsymbol{J} \boldsymbol{H} \ddot{\boldsymbol{\theta}} + \beta \boldsymbol{\Delta x}^{\mathrm{T}} \boldsymbol{J} \left\{\frac{1}{2}\dot{\boldsymbol{H}} + \boldsymbol{S}\right\}\dot{\boldsymbol{\theta}} + \beta \boldsymbol{\Delta x}^{\mathrm{T}} \boldsymbol{J} \boldsymbol{J}^{\mathrm{T}} \boldsymbol{K}_p \boldsymbol{\Delta x}$$
$$+ \dot{\boldsymbol{\theta}}^{\mathrm{T}} \boldsymbol{W}(\boldsymbol{AB} + \boldsymbol{B_0})\boldsymbol{W}^{\mathrm{T}}\dot{\boldsymbol{\theta}} + \beta \boldsymbol{\Delta x}^{\mathrm{T}} \boldsymbol{J} \boldsymbol{W}(\boldsymbol{AB} + \boldsymbol{B_0})\boldsymbol{W}^{\mathrm{T}}\dot{\boldsymbol{\theta}} = 0 \qquad (25)$$

Next, we consider the following scalar function $V(t)$ as a candidate of Lyapunov's function:

$$V(t) = \frac{1}{2}\dot{\boldsymbol{\theta}}^{\mathrm{T}} \boldsymbol{H} \dot{\boldsymbol{\theta}} + \beta \boldsymbol{\Delta x}^{\mathrm{T}} \boldsymbol{J} \boldsymbol{H} \dot{\boldsymbol{\theta}} + \frac{1}{2}\boldsymbol{\Delta x}^{\mathrm{T}} \boldsymbol{K}_p \boldsymbol{\Delta x} + \frac{1}{2}\boldsymbol{\Delta \Theta}^{\mathrm{T}} \boldsymbol{\Gamma} \boldsymbol{\Delta \Theta} \qquad (26)$$

Eq. (26) can be rewritten as follows:

$$V(t) = \frac{1}{2}(\dot{\boldsymbol{\theta}} + \beta \boldsymbol{J}^{\mathrm{T}} \boldsymbol{\Delta x})^{\mathrm{T}} \boldsymbol{H} (\dot{\boldsymbol{\theta}} + \beta \boldsymbol{J}^{\mathrm{T}} \boldsymbol{\Delta x})$$
$$+ \frac{1}{2}\boldsymbol{\Delta \Theta}^{\mathrm{T}} \boldsymbol{\Gamma} \boldsymbol{\Delta \Theta} + \frac{1}{2}\boldsymbol{\Delta x}^{\mathrm{T}} \left(\boldsymbol{K}_p - \beta^2 \boldsymbol{J} \boldsymbol{H} \boldsymbol{J}^{\mathrm{T}}\right) \boldsymbol{\Delta x} \qquad (27)$$

Therefore, if the position feedback gain $\boldsymbol{K}_p$ can be chosen large enough and the positive number β can be chosen small enough to satisfy $\boldsymbol{K}_p \geq \beta^2 \boldsymbol{J} \boldsymbol{H} \boldsymbol{J}^{\mathrm{T}}$, the scalar function $V(t)$ is semi-positive definite. Then, it plays a role of Lyapunov's function. Taking differentiation of V gives:

$$\frac{d}{dt}V(t) = \dot{\boldsymbol{\theta}}^T \boldsymbol{H}\ddot{\boldsymbol{\theta}} + \frac{1}{2}\dot{\boldsymbol{\theta}}^T \dot{\boldsymbol{H}}\dot{\boldsymbol{\theta}} + \dot{\boldsymbol{x}}^T \boldsymbol{K_p}\boldsymbol{\Delta x} + \boldsymbol{\Delta}\dot{\boldsymbol{\Theta}}^T \boldsymbol{\Gamma}\boldsymbol{\Delta\Theta}$$
$$+ \beta \dot{\boldsymbol{x}}^T \boldsymbol{J}\boldsymbol{H}\dot{\boldsymbol{\theta}} + \beta \boldsymbol{\Delta x}^T \dot{\boldsymbol{J}}\boldsymbol{H}\dot{\boldsymbol{\theta}} + \beta \boldsymbol{\Delta x}^T \boldsymbol{J}\dot{\boldsymbol{H}}\dot{\boldsymbol{\theta}} + \beta \boldsymbol{\Delta x}^T \boldsymbol{J}\boldsymbol{H}\ddot{\boldsymbol{\theta}} \quad (28)$$

Substituting eq. (28) into eq. (25) yields

$$\frac{d}{dt}V(t) - \beta \dot{\boldsymbol{\theta}}^T \boldsymbol{J}^T \boldsymbol{J}\boldsymbol{H}\dot{\boldsymbol{\theta}} - \beta \boldsymbol{\Delta x}^T \dot{\boldsymbol{J}}\boldsymbol{H}\dot{\boldsymbol{\theta}} + \beta \boldsymbol{\Delta x}^T \boldsymbol{J}\left\{-\frac{1}{2}\dot{\boldsymbol{H}} + \boldsymbol{S}\right\}\dot{\boldsymbol{\theta}}$$
$$+ \beta \boldsymbol{\Delta x}^T \boldsymbol{J}\boldsymbol{J}^T \boldsymbol{K_p}\boldsymbol{\Delta x} + \dot{\boldsymbol{\theta}}^T \boldsymbol{W}(\boldsymbol{AB} + \boldsymbol{B_0})\boldsymbol{W}^T \dot{\boldsymbol{\theta}}$$
$$+ \beta \boldsymbol{\Delta x}^T \boldsymbol{J}\boldsymbol{W}(\boldsymbol{AB} + \boldsymbol{B_0})\boldsymbol{W}^T \dot{\boldsymbol{\theta}} = 0 \quad (29)$$

Rewriting eq. (29), one can see that the scalar function $V(t)$ satisfies the following inequalities:

$$\frac{d}{dt}V(t) = -\dot{\boldsymbol{\theta}}^T \left\{\boldsymbol{W}(\boldsymbol{AB} + \boldsymbol{B_0})\boldsymbol{W}^T - \beta \boldsymbol{J}\boldsymbol{J}^T \boldsymbol{H}\right\}\dot{\boldsymbol{\theta}} + \beta h(\boldsymbol{\Delta x}, \dot{\boldsymbol{\theta}})$$
$$- \beta \boldsymbol{\Delta x}^T \boldsymbol{J}\boldsymbol{W}(\boldsymbol{AB} + \boldsymbol{B_0})\boldsymbol{W}^T \dot{\boldsymbol{\theta}} - \beta \boldsymbol{\Delta x}^T \boldsymbol{J}\boldsymbol{J}^T \boldsymbol{K_p}\boldsymbol{\Delta x}$$
$$\leq -\dot{\boldsymbol{\theta}}^T \left\{\boldsymbol{W}(\boldsymbol{AB} + \boldsymbol{B_0})\boldsymbol{W}^T - \beta \boldsymbol{J}\boldsymbol{J}^T \boldsymbol{H}\right\}\dot{\boldsymbol{\theta}} + \beta |h(\boldsymbol{\Delta x}, \dot{\boldsymbol{\theta}})|$$
$$- \beta \boldsymbol{\Delta x}^T \boldsymbol{J}\boldsymbol{W}(\boldsymbol{AB} + \boldsymbol{B_0})\boldsymbol{W}^T \dot{\boldsymbol{\theta}} - \beta \boldsymbol{\Delta x}^T \boldsymbol{J}\boldsymbol{J}^T \boldsymbol{K_p}\boldsymbol{\Delta x} \quad (30)$$

where $h(\boldsymbol{\Delta x}, \dot{\boldsymbol{\theta}})$ is defined as:

$$h(\boldsymbol{\Delta x}, \dot{\boldsymbol{\theta}}) = \boldsymbol{\Delta x}^T \dot{\boldsymbol{J}}\boldsymbol{H}\dot{\boldsymbol{\theta}} - \boldsymbol{\Delta x}^T \boldsymbol{J}\left\{-\frac{1}{2}\dot{\boldsymbol{H}} + \boldsymbol{S}\right\}\dot{\boldsymbol{\theta}} \quad (31)$$

Note that in eq. (31) $\dot{\boldsymbol{H}}$, $\dot{\boldsymbol{J}}$ and $\boldsymbol{S}$ are linear and homogeneous with respect to $\dot{\boldsymbol{\theta}}$. Therefore, the function $h(\boldsymbol{\Delta x}, \dot{\boldsymbol{\theta}})$ is quadratic in $\dot{\boldsymbol{\theta}}$, and there exist positive numbers c_0 and c_1 for h satisfying the following inequality:

$$|h| \leq c_0 ||\dot{\boldsymbol{\theta}}||^2 + c_1 ||\boldsymbol{\Delta x}||^2 \quad (32)$$

Substituting eq. (32) into eq. (30) yields

$$\frac{d}{dt}V(t) \leq -\dot{\boldsymbol{\theta}}^T \left\{\boldsymbol{W}(\boldsymbol{AB} + \boldsymbol{B_0})\boldsymbol{W}^T - \beta \boldsymbol{J}\boldsymbol{J}^T \boldsymbol{H}\right\}\dot{\boldsymbol{\theta}}$$
$$- \beta \boldsymbol{\Delta x}^T \boldsymbol{J}\boldsymbol{W}(\boldsymbol{AB} + \boldsymbol{B_0})\boldsymbol{W}^T \dot{\boldsymbol{\theta}}$$
$$+ \beta c_0 ||\dot{\boldsymbol{\theta}}||^2 + \beta c_1 ||\boldsymbol{\Delta x}||^2 - \beta \boldsymbol{\Delta x}^T \boldsymbol{J}\boldsymbol{J}^T \boldsymbol{K_p}\boldsymbol{\Delta x} \quad (33)$$

Also, since

$$-\boldsymbol{\Delta x}^T \boldsymbol{J}\boldsymbol{W}(\boldsymbol{AB} + \boldsymbol{B_0})\boldsymbol{W}^T \dot{\boldsymbol{\theta}} \leq \left\{||\boldsymbol{W}(\boldsymbol{AB} + \boldsymbol{B_0})\boldsymbol{W}^T|| ||\boldsymbol{J}||\right\} ||\dot{\boldsymbol{\theta}}|| ||\boldsymbol{\Delta x}|| \quad (34)$$

where
$$\|\dot{\boldsymbol{\theta}}\|\|\boldsymbol{\Delta x}\| \leq \frac{1}{2}\left(\|\dot{\boldsymbol{\theta}}\|^2 + \|\boldsymbol{\Delta x}\|^2\right) \tag{35}$$
then
$$-\boldsymbol{\Delta x}^{\mathrm{T}} \boldsymbol{J W}(\boldsymbol{AB} + \boldsymbol{B_0})\boldsymbol{W}^{\mathrm{T}} \dot{\boldsymbol{\theta}}$$
$$\leq \frac{1}{2}\left\{\|\boldsymbol{W}(\boldsymbol{AB} + \boldsymbol{B_0})\boldsymbol{W}^{\mathrm{T}}\|\|\boldsymbol{J}\|\right\}\left(\|\dot{\boldsymbol{\theta}}\|^2 + \|\boldsymbol{\Delta x}\|^2\right) \tag{36}$$

Therefore, eq. (33) becomes

$$\frac{\mathrm{d}}{\mathrm{d}t}V(t) \leq -\dot{\boldsymbol{\theta}}^{\mathrm{T}}\left\{\boldsymbol{W}(\boldsymbol{AB}+\boldsymbol{B_0})\boldsymbol{W}^{\mathrm{T}} - \beta \boldsymbol{J}\boldsymbol{J}^{\mathrm{T}}\boldsymbol{H}\right\}\dot{\boldsymbol{\theta}}$$
$$- \beta \boldsymbol{\Delta x}^{\mathrm{T}}\boldsymbol{J}\boldsymbol{J}^{\mathrm{T}}\boldsymbol{K_p}\boldsymbol{\Delta x} + \beta c_0\|\dot{\boldsymbol{\theta}}\|^2 + \beta c_1\|\boldsymbol{\Delta x}\|^2$$
$$+ \frac{1}{2}\beta\|\boldsymbol{W}(\boldsymbol{AB}+\boldsymbol{B_0})\boldsymbol{W}^{\mathrm{T}}\|\|\boldsymbol{J}\|\left(\|\dot{\boldsymbol{\theta}}\|^2 + \|\boldsymbol{\Delta x}\|^2\right) \tag{37}$$

Also, note that

$$-\dot{\boldsymbol{\theta}}^{\mathrm{T}}\left\{\boldsymbol{W}(\boldsymbol{AB}+\boldsymbol{B_0})\boldsymbol{W}^{\mathrm{T}}\right\}\dot{\boldsymbol{\theta}} \leq -\lambda_{\min}\left[\boldsymbol{W}(\boldsymbol{AB}+\boldsymbol{B_0})\boldsymbol{W}^{\mathrm{T}}\right]\|\dot{\boldsymbol{\theta}}\|^2 \tag{38}$$

$$\dot{\boldsymbol{\theta}}^{\mathrm{T}}\boldsymbol{J}\boldsymbol{J}^{\mathrm{T}}\boldsymbol{H}\dot{\boldsymbol{\theta}} \leq \lambda_{\max}\left[\boldsymbol{J}\boldsymbol{J}^{\mathrm{T}}\boldsymbol{H}\right]\|\dot{\boldsymbol{\theta}}\|^2 \tag{39}$$

$$-\boldsymbol{\Delta x}^{\mathrm{T}}\boldsymbol{J}\boldsymbol{J}^{\mathrm{T}}\boldsymbol{K_p}\boldsymbol{\Delta x} \leq -\lambda_{\min}\left[\boldsymbol{J}\boldsymbol{J}^{\mathrm{T}}\boldsymbol{K_p}\right]\|\boldsymbol{\Delta x}\|^2. \tag{40}$$

In the above inequalities, $\lambda_{\max}[\ ]$ and $\lambda_{\min}[\ ]$ are, respectively, the maximum and minimum eigenvalues of a matrix in the bracket $[\ ]$. The derivative of the scalar function $V(t)$ satisfies the following inequality:

$$\frac{\mathrm{d}}{\mathrm{d}t}V(t) \leq -\left\{\lambda_{\min}\left[\boldsymbol{W}(\boldsymbol{AB}+\boldsymbol{B_0})\boldsymbol{W}^{\mathrm{T}}\right]\right.$$
$$\left. - \beta\left(\lambda_{\max}\left[\boldsymbol{J}\boldsymbol{J}^{\mathrm{T}}\boldsymbol{H}\right] - \frac{1}{2}\|\boldsymbol{W}(\boldsymbol{AB}+\boldsymbol{B_0})\boldsymbol{W}^{\mathrm{T}}\|\|\boldsymbol{J}\| - c_0\right)\right\}\|\dot{\boldsymbol{\theta}}\|^2$$
$$- \beta\left\{\lambda_{\min}\left[\boldsymbol{J}\boldsymbol{J}^{\mathrm{T}}\boldsymbol{K_p}\right] - \frac{1}{2}\|\boldsymbol{W}(\boldsymbol{AB}+\boldsymbol{B_0})\boldsymbol{W}^{\mathrm{T}}\|\|\boldsymbol{J}\| - c_1\right\}\|\boldsymbol{\Delta x}\|^2 \tag{41}$$

As shown in Section 2.2, the inequality $\boldsymbol{W}(\boldsymbol{AB}+\boldsymbol{B_0})\boldsymbol{W}^{\mathrm{T}} > 0$ is satisfied if the Jacobian matrix $\boldsymbol{W}$ is not degenerated. Therefore, $V(t)$ plays the role of Lyapunov's function and LaSalle's invariance theorem can be applied to it. Then, we have $\boldsymbol{x} \to \boldsymbol{x}_d$, $\dot{\boldsymbol{x}} \to 0$ when $t \to \infty$ [30].

2.6 Simulation of the Gravity Effect

This section shows the results of numerical simulations when the gravity force affects the arm. The desired point and the parameter update gains are shown in Table 5. The physical parameters, gains, initial posture are the same as in Section 2.3. Four controllers were examined in our simulations,

1) gravity compensation & $k_e = 20.0$
2) gravity compensation & $k_e = 0$ (zero internal force)
3) no gravity compensation & $k_e = 20.0$
4) no gravity compensation & $k_e = 0$ (zero internal force)

Figure 6 shows the end-point paths of the reaching movements. The curve corresponding to the 1st controller is shown as dashed solid line, for the 2nd one as dashed gray line, for the 3rd one as dotted gray line. The simulation result for the 4th controller (not in shown in Fig. 6) is almost same as that for the 3rd one. It can be seen from Fig. 6 that if the gravity is compensated the end-point paths converge to the desired point. However, the end-point path for the controller without the gravity compensation cannot converge to the desired point, and much larger steady state error remains even if k_e is set to

Table 5. Desired point and parameter update gains

Desired point [m]	$(x, y) = (-0.40, 0.20)$
$\boldsymbol{\Gamma}^{-1} = \begin{pmatrix} 20.0 & 0 \\ 0 & 13.0 \end{pmatrix}$,	$\beta = 10.0$

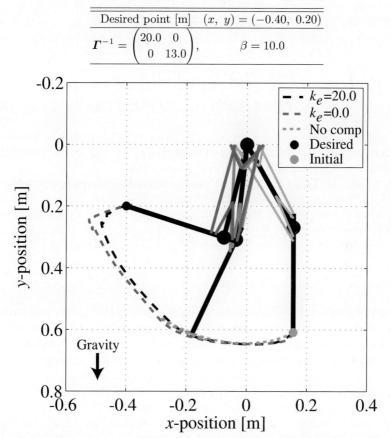

Fig. 6. End-point paths of the reaching movements under the effect of gravity

either 20.0 or zero. Of course, it is too early to say whether the human control system employs our method or not. But surely the gravity compensation is implemented in human movements. It is known that the weights of the human upper arm and forearm are about $1.5 \sim 2.5$ [kg], respectively, and the moment arms from the muscle forces to the joint torques are small (about $2 \sim 3$ [cm]). Therefore, the antigravity muscles have to generate a large force in order to support the weights. This cannot be done without the gravity compensation control.

It should be noted that the end-point path for the gravity compensation can converge to the desired point even if k_e is set to zero. It is because the control input matrix $\boldsymbol{A}$ in eq. (21) includes the gravity compensation term. Then, the damping matrix $\boldsymbol{W}(\boldsymbol{AB} + \boldsymbol{B_0})\boldsymbol{W}^\mathrm{T}$ is changed in correspondence with the estimation of the unknown parameters. Therefore, the damping force increases even if the internal force term is not introduced in the control input. It can be seen from Fig. 7 that the end-point path in the case of $k_e = 20.0$ converges to the desired point faster than that in the case of $k_e = 0$. As can be seen from Fig. 8 the tangential velocity of the end-point for $k_e = 20.0$ are smoother than that for $k_e = 0$. However, the velocity profiles are not bell-shaped. Figure 9 shows the estimated parameters. It can be seen that they are not converged to the actual values. The estimated values in the case of $k_e = 20.0$ are different from these in the case of $k_e = 0$. One of the reasons is the nonlinearity of the muscle model.

3 Human-Like Pinching Movements

3.1 A Model of Dual 2 D.O.F. Fingers with Redundant Muscles

In this section, we consider a model of dual fingers in which each finger has 2 D.O.F. and is actuated by one monoarticular digitorum muscle and two biarticular digitorum muscles. The overall system is shown in Fig. 10.

Kinematics of Muscle-Joint Space

Assume that the muscles can only be contracted straightforwardly and are not curved. In the following, the subscript $i\, (= 1, 2)$ in the variables and equations denotes the finger's number. It is known that the muscles of the index finger and the thumb are inserted in the fore-arm part. The configuration of the index finger is shown in Fig. 11. We decompose the muscle length $l_{ij}(j = 1, 2, 3)$ into a variable part $lv_{ij}(j = 1, 2, 3)$, which depends on the finger angles, and a static part $ls_{ij}(j = 1, 2, 3)$, which does not contribute to the change of the muscle moment arms. The relation between the muscle length $\boldsymbol{l}_i \in \mathbb{R}^3$ and joint angle $\boldsymbol{q}_i \in \mathbb{R}^2$ can then be expressed as follows:

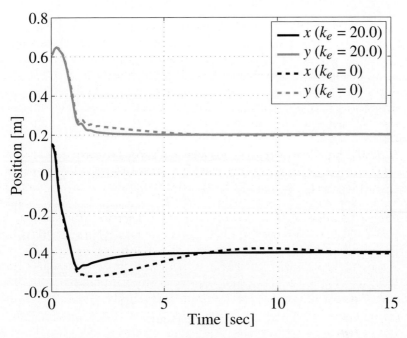

Fig. 7. Transient response of the end-point positions with the gravity compensation

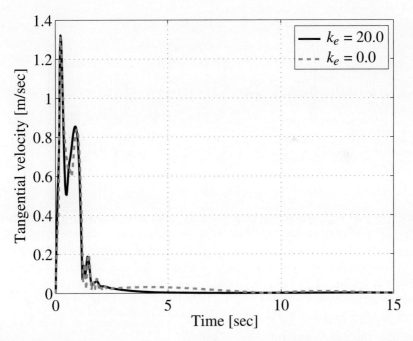

Fig. 8. Transient response of the tangential velocities of the end-point with the gravity compensation

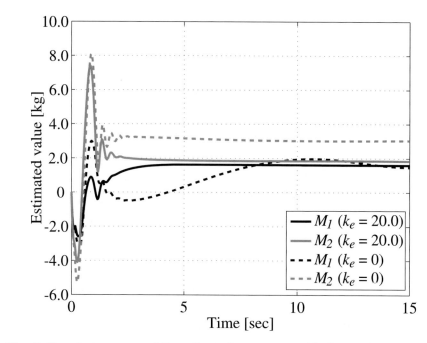

Fig. 9. Transient response of the estimated parameters with the gravity compensation

$$l_i(q_i) = (lv_{i1} + ls_{i1},\ lv_{i2} + ls_{i2},\ lv_{i3} + ls_{i3})^{\mathrm{T}} \tag{42}$$

$$= \begin{pmatrix} (a_{i1}^2 + a_{i2}^2 + 2a_{i1}a_{i2}\cos q_{i1})^{\frac{1}{2}} + ls_{i1} \\ (a_{i3}^2 + a_{i4}^2 + 2a_{i3}L_{i1}\cos q_{i2} + 2a_{i3}L_{i1}\cos q_{i1} \\ \quad + L_{i1}^2 + 2a_{i3}a_{i4}\cos(q_{i1}+q_{i2}))^{\frac{1}{2}} + ls_{i2} \\ (a_{i5}^2 + a_{i6}^2 - 2a_{i6}L_{i1}\cos q_{i2} - 2a_{i5}L_{i1}\cos q_{i1} \\ \quad + L_{i1}^2 + 2a_{i5}a_{i6}\cos(q_{i1}+q_{i2}))^{\frac{1}{2}} + ls_{i3} \end{pmatrix}$$

where $a_{im}(m = 1 \sim 6)$ are the insertion points of the muscles, and $ls_{ij}(j = 1, 2, 3)$ are constants. Now, taking derivation of eq. (42) yields:

$$\dot{l}_i = W_i^{\mathrm{T}} \dot{q}_i \tag{43}$$

where $W_i^{\mathrm{T}} \in \mathbb{R}^{3\times 2}$ is the Jacobian matrix from joint space to muscle space. The relation between the muscle output force $F_{mi} \in \mathbb{R}^3$ and the joint torque $\tau_i \in \mathbb{R}^2$ can thus be expressed from the principle of virtual work. It is defined as:

$$\tau_i = W_i F_{mi} \tag{44}$$

It is assumed that the Jacobian matrix W_i is not degenerated. Therefore, we can take the inverse relation of eq. (44):

$$F_{mi} = W_i^+ \tau_i + (I_3 - W_i^+ W_i) k_{ei} \tag{45}$$

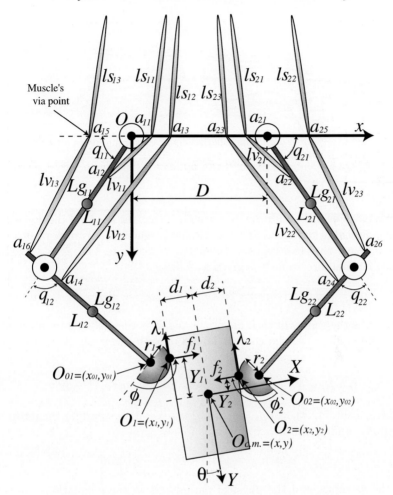

Fig. 10. Dual 2 D.O.F. fingers with redundant muscles

where $\boldsymbol{W}_i^+ = \boldsymbol{W}_i^{\mathrm{T}}(\boldsymbol{W}_i\boldsymbol{W}_i^{\mathrm{T}})^{-1} \in \mathbb{R}^{3\times 2}$ is the pseudo-inverse matrix of $\boldsymbol{W}_i$, $\boldsymbol{k}_{ei} \in \mathbb{R}^3$ is an arbitrary vector, and $\boldsymbol{I}_3 \in \mathbb{R}^{3\times 3}$ indicates an identity matrix. The physical meaning of the second term of the right-hand side in eq. (45) is the internal force generated by the redundant muscles.

Dynamics of Overall System

It is assumed that the fingers move in the horizontal plane, and the effect of gravity is ignored. It is also assumed that the shape of the object has parallel flat surfaces. Only the rolling contacts between the finger tips and the object are allowed, and each finger tip is not detached from the object surfaces

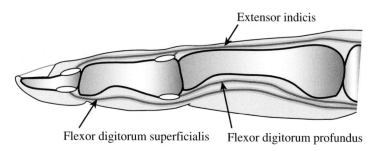

Fig. 11. Configuration of digitorum muscles in human index finger

during movement. In addition, it is assumed that the shape of the finger tips is hemispherical of radius r_i, and the friction is ignored. It is important to note that we have only two types of geometric constraints. One is the constraint on the normal direction. It is given as follows:

$$Q_i = (-1)^{i-1} \{(x - x_{0i}) \cos \theta - (y - y_{0i}) \sin \theta\} - r_i - d_i = 0 \qquad (46)$$

where r_i is the radius of each finger tip and d_i is the distance from the object mass center $O_{c.m.}$ to the contact points of each finger O_i. The other is the constraint on the tangential direction:

$$R_i = Y_i - \{n_{0i} - r_i \left(\pi + (-1)^{i-1}\theta - \boldsymbol{q}_i^{\mathrm{T}} \boldsymbol{e}_i\right)\} = 0 \qquad (47)$$

where $\boldsymbol{e}_i = (1,1)^{\mathrm{T}}$, $\boldsymbol{q}_i = (q_{i1}, q_{i2})^{\mathrm{T}}$, and n_{0i} ($i = 1, 2$) stand for the integration constants, and Y_i can be expressed as follows:

$$Y_i = (x_{0i} - x) \sin \theta + (y_{0i} - y) \cos \theta \qquad (48)$$

All the variables and the physical parameters in these equations are shown in Fig. 10. The geometric constraints (46), (47) induce four constraint forces which can be interpreted as Lagrange's multipliers f_i and λ_i:

$$Q = \sum_{i=1}^{2} Q_i f_i = 0, \quad R = \sum_{i=1}^{2} R_i \lambda_i = 0 \qquad (49)$$

Now, the kinetic energy of the overall system can be expressed as follows:

$$K = \sum_{i=1}^{2} \frac{1}{2} \dot{\boldsymbol{q}}_i^{\mathrm{T}} \boldsymbol{H}_i \dot{\boldsymbol{q}}_i + \frac{1}{2} \dot{\boldsymbol{z}}^{\mathrm{T}} \boldsymbol{M} \dot{\boldsymbol{z}} \qquad (50)$$

where $\boldsymbol{H}_i \in \mathbb{R}^{2 \times 2}$ and $\boldsymbol{M} = \mathrm{diag}(M, M, I) \in \mathbb{R}^{3 \times 3}$ are the inertia matrices for each finger and the object. They are symmetric and positive definite, and $\boldsymbol{z} = (x, y, \theta)^{\mathrm{T}}$. Therefore, the Lagrangian can be defined as follows:

$$L = K + Q + R \tag{51}$$

Applying Hamilton's principle to the Lagrangian L gives:

$$\int_{t_0}^{t_1} \left(\delta L + \sum_{i=1}^{2} \tau_i^T \delta q_i \right) dt = 0 \tag{52}$$

where $\tau_i \in \mathbb{R}^2$ is the input to each joint. Therefore, Lagrange's equations can be stated as follows:
For the dual 2 D.O.F fingers

$$\boldsymbol{H}_i \ddot{\boldsymbol{q}}_i + \left\{ \frac{1}{2} \dot{\boldsymbol{H}}_i + \boldsymbol{S}_i \right\} \dot{\boldsymbol{q}}_i - \frac{\partial Q_i^T}{\partial \boldsymbol{q}_i} f_i - \frac{\partial R_i^T}{\partial \boldsymbol{q}_i} \lambda_i = \boldsymbol{\tau}_i \tag{53}$$

For the object

$$\boldsymbol{M} \ddot{\boldsymbol{z}} - \sum_{i=1}^{2} \frac{\partial Q_i^T}{\partial \boldsymbol{z}} f_i - \sum_{i=1}^{2} \frac{\partial R_i^T}{\partial \boldsymbol{z}} \lambda_i = 0 \tag{54}$$

where $\boldsymbol{S}_i \in \mathbb{R}^{2 \times 2}$ is a skew-symmetric matrix, and the constraint Jacobian matrices in eqs. (53) and (54) are given as follows:

$$\begin{aligned}
\frac{\partial Q_i^T}{\partial \boldsymbol{q}_i} &= (-1)^i \left(\frac{\partial x_{0i}}{\partial \boldsymbol{q}_i}, \frac{\partial y_{0i}}{\partial \boldsymbol{q}_i} \right) \begin{pmatrix} \cos\theta \\ -\sin\theta \end{pmatrix} \\
\frac{\partial R_i^T}{\partial \boldsymbol{q}_i} &= -r_i \boldsymbol{e}_i + \left(\frac{\partial x_{0i}}{\partial \boldsymbol{q}_i}, \frac{\partial y_{0i}}{\partial \boldsymbol{q}_i} \right) \begin{pmatrix} \sin\theta \\ \cos\theta \end{pmatrix} \\
\frac{\partial Q_i^T}{\partial \boldsymbol{z}} &= \begin{pmatrix} (-1)^i \cos\theta \\ (-1)^{i-1} \sin\theta \\ (-1)^{i-1} Y_i \end{pmatrix}, \quad \frac{\partial R_i^T}{\partial \boldsymbol{z}} = \begin{pmatrix} -\sin\theta \\ -\cos\theta \\ (-1)^{i-1} d_i \end{pmatrix}
\end{aligned} \tag{55}$$

3.2 Nonlinear Muscle Model Inspired by Physiological Studies

In this section, we consider a muscle model that is more plausible than that analyzed in Section 2.1. We assume that the masses of all the muscles are included into the masses of each link. The well-known Hill muscle model [20] is defined as:

$$(f_{mij} + a_{ij})(\dot{l}_{ij} + b_{ij}) = b_{ij}(f_{m0ij} + a_{ij}) \tag{56}$$

where f_{mij} is the tensile force of the muscle, $\dot{l}_{ij}$ is the contractile velocity, f_{m0ij} is the maximum isometric tensile force, a_{ij} is the heat constant, b_{ij} is the rate constant of the energy liberation. The subscript j is the number of muscles. It is known [28] that the tensile force in the lengthening phase is larger than that in the shortening phase. To take into account this feature, the Hill muscle model (56) can be modified as follows:

$$f_{mij}(\alpha_{ij}, \dot{l}_{ij}) = \begin{cases} \dfrac{b_{ij}f_{m0ij} - a_{ij}\dot{l}_{ij}}{|\dot{l}_{ij}| + b_{ij}}\alpha_{ij}, & \dot{l}_{ij} \geq 0 \\ \dfrac{b_{ij}f_{m0ij} - (2f_{m0ij} + a_{ij})\dot{l}_{ij}}{|\dot{l}_{ij}| + b_{ij}}\alpha_{ij}, & \dot{l}_{ij} < 0 \end{cases} \tag{57}$$

where $0 \leq \alpha_{ij} \leq 1$ is the muscle activation level, and the constants a_{ij}, b_{ij} defined as $a_{ij} = 0.25 f_{m0ij}$, $b_{ij} = 0.9 l_{0ij}$ [28]. Here, l_{0ij} is the intrinsic rest length of the muscles. Now, we define the control input to the muscles $\bar{\alpha}_{ij} = f_{m0ij}\alpha_{ij}$ and introduce a muscle intrinsic viscosity $c_{0ij} > 0$ independently from $\bar{\alpha}_{ij}$. The muscle dynamic model can be represented as follows:

$$f_{mij}(\bar{\alpha}_{ij}, \dot{l}_{ij}) = p_{ij}\left\{\bar{\alpha}_{ij} - (\bar{\alpha}_{ij}c_{ij} + c_{0ij})\dot{l}_{ij}\right\} \tag{58}$$

where

$$p_{ij} = \frac{0.9 l_{0ij}}{0.9 l_{0ij} + |\dot{l}_{ij}|}, \quad c_{ij} = \begin{cases} \dfrac{0.25}{0.9 l_{0ij}} > 0, & \text{if } \dot{l}_{ij} \geq 0 \\ \dfrac{2.25}{0.9 l_{0ij}} > 0, & \text{if } \dot{l}_{ij} < 0 \end{cases}$$

It is known that muscle can generate only a contractile force. Therefore, the control input $\bar{\alpha}_{ij}$ can be defined as a saturated function so as to keep $\bar{\alpha}_{ij} \geq 0$. There are three muscles for the i-th finger, and the muscles dynamics for the i-th finger can be expressed as follows:

$$\boldsymbol{F}_{mi} = \boldsymbol{P}_i \bar{\boldsymbol{\alpha}}_i - \boldsymbol{P}_i \left(\boldsymbol{A}_i \boldsymbol{C}_i + \boldsymbol{C}_{0i}\right) \dot{\boldsymbol{l}}_i \tag{59}$$

$$\begin{bmatrix} \boldsymbol{F}_{mi} = (f_{mi1}, f_{mi2}, f_{mi3})^\mathrm{T} \in \mathbb{R}^3 \\ \boldsymbol{P}_i = \mathrm{diag}(p_{i1}, p_{i2}, p_{i3}) \in \mathbb{R}^{3\times 3} \\ \bar{\boldsymbol{\alpha}}_i = (\bar{\alpha}_{i1}, \bar{\alpha}_{i2}, \bar{\alpha}_{i3})^\mathrm{T} \in \mathbb{R}^3 \\ \boldsymbol{A}_i = \mathrm{diag}(\bar{\alpha}_{i1}, \bar{\alpha}_{i2}, \bar{\alpha}_{i3}) \in \mathbb{R}^{3\times 3} \\ \boldsymbol{C}_i = \mathrm{diag}(c_{i1}, c_{i2}, c_{i3}) \in \mathbb{R}^{3\times 3} \\ \boldsymbol{C}_{0i} = \mathrm{diag}(c_{0i1}, c_{0i2}, c_{0i3}) \in \mathbb{R}^{3\times 3} \end{bmatrix}$$

It is important to note that the control input matrix $\boldsymbol{A}_i$ is given not as a column vector but as a diagonal matrix [21, 22]. This is attributed by the nonlinearity of the muscle model, and by the fact that diagonal matrices $\boldsymbol{P}_i$, $\boldsymbol{C}_i$, and $\boldsymbol{C}_{0i}$ are positive definite, and $\boldsymbol{A}_i$ is semi-positive definite.

3.3 Stable Pinching and Posture Regulation

Let us design the control input $\bar{\boldsymbol{\alpha}}$ for the simultaneous realization of the stable pinching and the posture regulation. It is assumed that the observable state variables are the joint angles $\boldsymbol{q}_i$ and the orientation of the object θ. The control law τ_{p_i}, realizing the stable pinching of the object can be defined as follows [1, 29]:

$$\tau_{p_i} = -\frac{\partial Q_i^{\mathrm{T}}}{\partial q_i} f_d + (-1)^i \frac{r_i f_d}{r_1 + r_2}(Y_1 - Y_2) e_i \qquad (60)$$

where f_d stands for the desired pinching force. Now, it should be noted that in eq. (60) the distance $Y_1 - Y_2$ can be expressed through the observable variables q_i and θ:

$$Y_1 - Y_2 = (x_{01} - x_{02}) \sin \theta + (y_{01} - y_{02}) \cos \theta \qquad (61)$$

where (x_{0i}, y_{0i}) $(i = 1, 2)$ are the end-positions of the links. Next, the control law, regulating the object angle, can be defined as follows:

$$\tau_{\theta i} = (-1)^i \frac{\partial R_i^{\mathrm{T}}}{\partial q_i} k_\theta \Delta\theta \qquad (62)$$

where $\Delta\theta = \theta - \theta_d$, and k_θ is a positive constant. The diagonal matrix P_i in eq. (59) is positive definite. The bound of its diagonal elements p_i satisfies $0 < p_i \leq 1$. Therefore, we can take the pseudo-inverse matrix $\bar{W}_i^+ = (W_i P_i)^+$ as long as the Jacobian matrix W does not become singular during the pinching movement. Then, the control input $\bar{\alpha}_i$ can be given as:

$$\bar{\alpha}_i = \bar{W}_i^+ \left(\tau_{p_i} + \tau_{\theta i}\right) + \left(I_3 - \bar{W}_i^+ \bar{W}_i\right) k_{ei} \qquad (63)$$

where $k_{ei} \in \mathbb{R}^3$ is an arbitrarily vector, and $\bar{W}_i = W_i P_i$. The second term of the right-hand side in eq. (63) plays a role of the internal force generated by the redundant muscles. Therefore, the closed-loop dynamics of the finger-object system can be expressed by substituting eqs. (43), (44), (59) and (63) into eq. (53). It is defined as follows:

For the dual 2 D.O.F fingers

$$H_i \ddot{q}_i + \left\{\frac{1}{2}\dot{H}_i + S_i\right\} \dot{q}_i - \frac{\partial Q_i^{\mathrm{T}}}{\partial q_i} \Delta f_i - \frac{\partial R_i^{\mathrm{T}}}{\partial q_i} \bar{\lambda}_i$$
$$- (-1)^i \frac{r_i f_d}{r_1 + r_2}(Y_1 - Y_2) e_i + W_i P_i (A_i C_i + C_{0i}) W_i^{\mathrm{T}} \dot{q}_i = 0 \qquad (64)$$

For the object

$$M \ddot{z} - \sum_{i=1}^{2} \frac{\partial Q_i^{\mathrm{T}}}{\partial z} \Delta f_i - \sum_{i=1}^{2} \frac{\partial R_i^{\mathrm{T}}}{\partial z} \bar{\lambda}_i = \left\{ f_d (Y_1 - Y_2) - \sum_{i=1}^{2} d_i k_\theta \Delta\theta \right\} e_\theta \qquad (65)$$

where

$$\Delta f_i = f_i - f_d, \quad \bar{\lambda}_i = \left(\lambda_i + (-1)^i k_\theta \Delta\theta\right), \quad e_\theta = (0, 0, 1)^{\mathrm{T}}$$

It should be noted that in eq. (64) the last term of the left-hand side plays a role of the damping factor. It can be modulated by the arbitrary vector $\boldsymbol{k}_{ei}$ which is generated by the redundant muscles. Now, taking the inner product of the closed-loop dynamics of the overall system (64), (65) with $\dot{\boldsymbol{q}}_i$, $\dot{\boldsymbol{z}}$ yields:

$$\frac{d}{dt} E(t) = - \sum_{i=1}^{2} \dot{\boldsymbol{q}}_i^T \boldsymbol{W}_i \boldsymbol{P}_i \left(\boldsymbol{A}_i \boldsymbol{C}_i + \boldsymbol{C}_{0i} \right) \boldsymbol{W}_i^T \dot{\boldsymbol{q}}_i \leq 0 \qquad (66)$$

where

$$E(t) = K + \frac{f_d (Y_1 - Y_2)^2}{2 (r_1 + r_2)} + \frac{d_1 + d_2}{2} k_\theta \Delta \theta^2 \geq 0 \qquad (67)$$

It should be noted that in eq. (67) the scalar function $E(t)$ is semi-positive definite with respect to the state vector $(\boldsymbol{q}, \boldsymbol{z}, \dot{\boldsymbol{q}}, \dot{\boldsymbol{z}}) \in \mathbb{R}^{14}$ even though the holonomic constraints Q_i and R_i are taken into consideration. Therefore, $E(t)$ cannot be taken as Lyapunov's function. Here, the concept of "stability on a manifold [1]" can be fruitful in establishing the stability and the convergence of the closed-loop dynamics of this redundant system.

3.4 Numerical Simulation

In this section, we present the results of numerical simulations. All the physical parameters used in the simulations are shown in Table. 6. The initial condition and the desired states for the system configuration are shown in Table. 7. Figure 12 depicts the pinching movement in the case of $\theta_d = -5$ [deg]. In this figure, thin lines correspond to the initial state and thick lines correspond to the final state of the system. It can be seen from this figure that the pinching is stable when $Y_1 - Y_2 \to 0$, which cancels the rotational moment affecting the object. At the same time the object angle θ converges to the desired angle $\theta_d = -5$ [deg]. Figures 13 and 14 show the transient responses of the pinching forces f_i and the rolling constraint forces λ_i, respectively. It can be seen from these figures that the pinching forces converge to the desired value of $f_d = 0.2$ [N] quickly. The rolling constraint forces λ_i converge to zero with the settling time about 3 [sec]. Figures 15 and 16 show the transient responses for $Y_1 - Y_2$ and the object angle θ, respectively. It can be seen from these figures that the value of $Y_1 - Y_2$ converge to zero with the settling time about 3 [sec], and the object angle θ converges to the desired angle $\theta_d = -5$ [deg] with the settling time about 2 [sec]. Figure 17 depicts the pinching movement in the case of $\theta_d = 5$ [deg]. It can be seen from this figure that the movement stability is realized by $Y_1 - Y_2 \to 0$. At the same time the object angle θ converges to the desired angle $\theta_d = 5$ [deg]. Figures 18 and 19 show the transient responses of the pinching forces f_i and the rolling constraint forces λ_i in the case of $\theta_d = 5$ [deg], respectively. It can be seen from these figures that the pinching forces converge to the desired value of $f_d = 0.2$ [N] and the rolling constraint

Table 6. Parameters of the dual 2 D.O.F. fingers model

Physical parameter	Value
Length of L_{i1} [m]	0.035
Length of L_{i2} [m]	0.030
Center of mass $L_{g_{i1}}$ [m]	0.0175
Center of mass $L_{g_{i2}}$ [m]	0.015
Mass of L_{i1} [kg]	0.035
Mass of L_{i2} [kg]	0.025
Inertia of L_{i1} [kg·m^2]	1.25×10^{-7}
Inertia of L_{i2} [kg·m^2]	1.88×10^{-6}
Radius of finger tip r_i [m]	0.006
Distance of each finger D [m]	0.040
Width of the object $d_1 + d_2$ [m]	0.020
Mass of the object M [kg]	0.010
Inertia of the object I [kg·m^2]	6.93×10^{-7}
Insertion points of the muscles a_{ij} [m]	$a_{i1} = 0.005, \quad a_{i2} = 0.015$ $a_{i3} = 0.004, \quad a_{i4} = 0.003$ $a_{i5} = 0.005, \quad a_{i6} = 0.004$
Intrinsic length of the muscles l_{0ij} [m]	$l_{0i1} = 0.264$ $l_{0i2} = 0.290$ $l_{0i3} = 0.280$

Table 7. Initial condition and desired states

Initial condition		Desired states and gains	
Y_1	-0.004 [m]	f_d	0.2 [N]
Y_2	-0.001 [m]	θ_d	-5.0 or 5.0 [deg]
θ	0.0 [deg]	k_θ	0.8
		$\boldsymbol{k}_{e i}$	$(1.5,\ 1.8,\ 1.8)^{\mathrm{T}}$

forces λ_i converge to zero. Figures 20 and 21 show the transient responses for $Y_1 - Y_2$ and the object angle θ in the case of $\theta_d = 5$ [deg], respectively. It can be seen from these figures that the value of $Y_1 - Y_2$ converges to zero. Also, the object angle θ converges to the desired angle $\theta_d = 5$ [deg]. The settling times for these variables are about 3 [sec]. A slight overshoot is presented at 1.8 [sec].

Figure 22 shows the pinching movement in the case of $\theta_d = -5$ [deg] with zero internal force ($\boldsymbol{k}_{ei} = \boldsymbol{0}$). It can be observed from this figure that the final posture takes a configuration impossible for the human fingers. Note that the Jacobian matrix $\boldsymbol{W}_i$ degenerates during this pinching movement. It is because the joints cannot support the reaction force from the opposite finger if the damping factor modulated by the internal force coming from the co-contraction between agonist and antagonist digitorum muscles is small. We

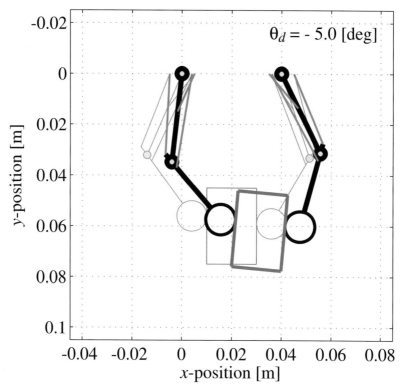

Fig. 12. Pinching movement in the case of $\theta_d = -5$ [deg]

have thus shown that both the stable pinching and the posture regulation of the object can be realized by the sensory-motor control law (63) with the adequate internal force, modulating the damping factor in the joint space.

4 Summary

In this chapter we studied human-like reaching and pinching movements from the viewpoint of robot control. First, a two-link planar arm model with six redundant muscles has been formulated, and a simple task-space feedback control scheme, taking into account the internal force induced by the redundant nonlinear muscles, has been proposed. In addition, the effect of gravity has been studied and a method for the gravity compensation on the muscle input signal level has been introduced. The stability of this method has been proved, and its effectiveness has been shown through numerical simulations. In particular, it has been demonstrated that our sensory-motor control can realize human-like reaching movements .

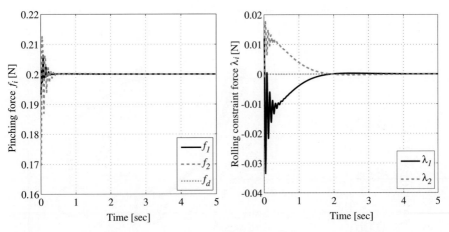

Fig. 13. The pinching forces f_i in the case of $\theta_d = -5$ [deg]

Fig. 14. The rolling constraint forces λ_i in the case of $\theta_d = -5$ [deg]

Next, an analysis of human-like pinching movements has been undertaken. Here, the co-contraction between the flexor and extensor digitorum muscles in the stable pinching and the posture regulation of the object has been analyzed. It has been shown that the internal force term generated by the redundant muscles could modulate the damping factor in the joint space. Numerical simulations has shown that the co-contraction of each digitorums makes it possible to realize both the stable pinching and the orientation regulation of of the object.

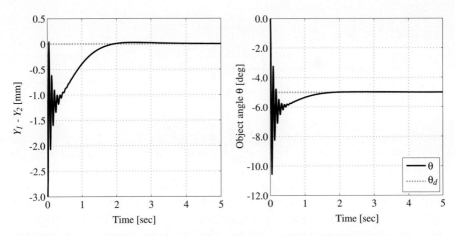

Fig. 15. The value of $Y_1 - Y_2$ in the case of $\theta_d = -5$ [deg]

Fig. 16. The object angle θ in the case of $\theta_d = -5$ [deg]

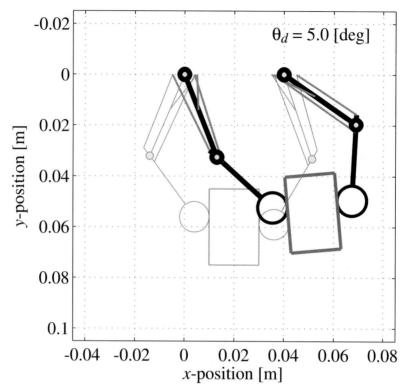

Fig. 17. Pinching movement in the case of $\theta_d = 5$ [deg]

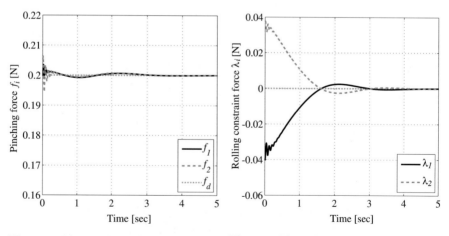

Fig. 18. The pinching forces f_i in the case of $\theta_d = 5$ [deg]

Fig. 19. The rolling constraint forces λ_i in the case of $\theta_d = 5$ [deg]

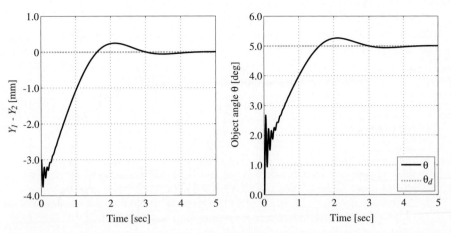

Fig. 20. The value of $Y_1 - Y_2$ in the case of $\theta_d = 5$ [deg]

Fig. 21. The object angle θ in the case of $\theta_d = 5$ [deg]

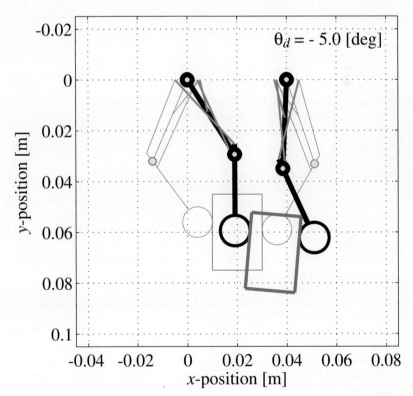

Fig. 22. Pinching movement in the case of $\theta_d = -5$ [deg] with zero internal force ($k_{e_i} = 0$)

As a conclusion, we suggest that the CNS does not need to calculate complex mathematical models based on the inverse dynamics or on the planning of optimal trajectories. Conversely, the human motor functions could be realized through a simple sensory-motor control exploiting the passivity, nonlinearity, and the redundancy of the musculo-skeletal systems.

References

1. Arimoto S, Tahara K, Bae JH, Yoshida M (2003) A stability theory on a manifold: concurrent realization of grasp and orientation control of an object by a pair of robot fingers. Robotica 21(2):163–178
2. Arimoto S, Sekimoto M, Hashiguchi H, Ozawa R (2005) Natural resolution of ill-posedness of inverse kinematics for redundant robots: a challenge to Bernstein's degrees-of-freedom problem. Adv. Robot. 19(4):401–434
3. Bernstein N (1996) On dexterity and its development. Lawrence Erlbaum Associates, New Jersey (translated from the Russian by Latash, M.L.)
4. Bernstein N (1935) The problem of the interrelation of coordination and localization. Arch. Biol. Sci. 38:15–59 (reprinted in N.A. Bernstein, "The Coordination and Regulation of Movements", Pergamon, London, 1967.)
5. Hollerbach J, Suh K (1987) Redundancy resolution of manipulators through torque optimization. IEEE Trans. Robot. Autom. RA-3(4):308–316
6. Nakamura Y (1991) Advanced Robotics: Redundancy and Optimization. Addison-Wesley, Reading, MA
7. Fel'dman A (1966) Functional tuning of the nervous system with control of movement or maintenance of steady posture. iii. mechanographic analysis of the execution by man of the simplest robot tasks. Biophysics 11:766–775
8. Bizzi E, Polit A, Morasso P (1976) Mechanisum underlying achievement of final head position. J. Neurophysiol. 3(9):435–444
9. Fel'dman A (1986) Once more on the equilibrium-point hypothesis (λ model) for motor control. J. Mot. Behav. 18:17–54
10. Flash T, Hogan N (1985) The coordination of arm movements: An experimentally confirmed mathematical model. J. Neurosci. 5(7):1688–1703
11. Flash T (1987) The control of hand equilibrium trajectories in multi-joint arm movements. Biol. Cybern. 57(4–5):257–274
12. Bizzi E, Hogan N, Mussa-Ivaldi F, Giszter S (1992) Does the nervous system use equilibrium-point control to guide single and multiple joint movements? Behav. Brain Sci. 15:603–613
13. Hogan N (1984) An organizing principle for a class of voluntary movements. J. Neurosci. 4(11):2745–2754
14. Morraso P (1981) Spatial control of arm movements. Exp. brain res. 42:223–227
15. Kawato M, Furukawa K, Suzuki R (1987) A hierarchical neural network model for control and learning of voluntary movement. Biol. Cybern. 57(3):169–185
16. Katayama M, Kawato M (1993) Virtual trajectory and stiffness ellipse during multijoint arm movement predicted by neural inverse model. Biol. Cybern. 69(5–6):353–362
17. Kawato M, Maeda Y, Uno Y, Suzuki R (1990) Trajectory formation of arm movements by cascade neural network model based on minimum torque-change criterion. Biol. Cybern. 62(4):275–288

18. Hogan N (1984) Adaptive control of mechanical impedance by coactivation of antagonist muscles. IEEE Trans. Automat. Contr. AC-29(8):681–690
19. Gribble P, Mullin L, Cothros N, Mattar A (2003) Role of cocontraction in arm movement accuracy. J. Neurophysiol. 89:2396–2405
20. Hill A (1938) The heat of shortening and the dynamic constants of muscles. In: Proc. R. Soc. Volume B. 136–195
21. Tahara K, Luo ZW, Arimoto S, Kino H (2005) Task-space feedback control for a two-link arm driven by six muscles with variable damping and elastic properties. In: Proc. IEEE Int. Conf. Robot. Autom. 224–229
22. Tahara K, Luo ZW, Arimoto S, Kino H (2005) Sensory-motor control mechanism for reaching movements of a redundant musculo-skeletal arm. J. Robot. Sys. 22(11):639–651
23. Napier J (1956) The prehensile movements of the human hand. J. Bone and Joint Surgery 38B(4):902–913
24. Close J, Kidd C (1969) The functions of the muscles of the thumb, the index and the long fingers. J. Bone and Joint Surgery 51A(8):1601–1620
25. Ferrier D (1876) The Function of the Brain. G.P. Putnams Sons, New York
26. Wing A, Haggard P, Flanagan J, eds. (1996) Hand and Brain: The Neurophysiology and Psychology of Hand Movements. Academic Press Inc., California
27. Tahara K, Luo ZW, Ozawa R, Bae JH, Arimoto S (2006) Bio-mimetic study on pinching motions of a dual-finger model with synergistic actuation of antagonist muscles. In: Proc. IEEE Int. Conf. Robot. Autom., Orlando, FL 994–999
28. Mashima H, Akazawa K, Kushima H, Fujii K (1972) The force-load-velocity relation and the viscous-like force in the frog skeletal muscle. Jap. J. Physiol. 22:103–120
29. Arimoto S, Nguyen PTA, Han HY, Doulgeri Z (2000) Dynamics and control of a set of dual fingers with soft tips. Robotica 28:71–80
30. Arimoto S (1996) Control Theory of Non-linear Mechanical Systems: A Passivity-based and Circuit-theoretic Approach. Oxford Univ. Press, U.K.

Principle of Superposition in Human Prehension

Mark L. Latash and Vladimir M. Zatsiorsky

Department of Kinesiology, Rec. Hall-268N, The Pennsylvania State University, University Park, PA 16802, USA mll11@psu.edu

Summary. The principle of superposition introduced by Prof. S. Arimoto and his colleagues for the control of robotic hand has been shown to be applicable to the control of prehensile actions by humans. In particular, experiments have shown that static human hand actions can be viewed as a superposition of two independent synergies controlling the grasping force and the orientation of the object. Studies of elderly persons have shown that they are impaired in both synergies and show worse stabilization of the grasping force and of the total moment of forces applied by the digits to a hand-held object. Recent studies have also shown that the principle of superposition holds with respect to reactions to expected and unexpected mechanical perturbations applied to a hand-held object. Indices of the two synergies have shown different changes following a perturbation. Generalization of the principle of superposition to human prehension is an important step towards understanding the principles of control of the human hand.

1 History of the Principle of Superposition

The idea that biological processes obey some kind of a superposition principle has been debated for many years. This principle implies that output of a set of elements with several inputs equals the sum of the outputs produced by each of the inputs applied separately. Violations of this principle in neurophysiology are many and varied. They do not come as a surprise because of the well known highly nonlinear properties of neurons, in particular their threshold properties and the all-or-none generation of units of information, action potentials (reviewed in [1]). For example, the phenomena of spatial and temporal summation of synaptic inputs illustrate situations when individual stimuli cannot bring a target neuron to its threshold for action potential generation while several stimuli coming simultaneously or at a high frequency can. A number of recent studies questioned applicability of the principle of superposition in studies of such diverse objects as motor unit firing patterns and maintenance of vertical posture [2, 3].

Despite the obvious highly nonlinear features of neurons, muscles, and reflex loops (reviewed in [4]), several research teams tried to find regularities of behaviors of large populations of such elements that would behave in a nearly linear fashion and obey the principle of superposition. Ruegg and Bongioanni [5] reported nearly linear input-output properties of motoneuronal pools in tasks that involved a superposition of steady-state and ballistic contractions. Studies of neuronal populations in different areas of the brain have also supported applicability of the principle of superposition [6, 7]. In a recent study, Fingelkurts and Fingelkurts [8] have come up with a conclusion that integrative brain functions can be manifested in the superposition of distributed multiple oscillations according to the principle of superposition (see also [9]).

In the area of motor control, the idea of the principle of superposition has been developed by proponents of the equilibrium-point (EP) hypothesis [10, 11]. According the EP-hypothesis, voluntary actions are controlled with variables that define spatial coordinates of muscle activation thresholds (reviewed in [12]). Actual levels of muscle activation, as well as externally manifested mechanical variables (such as endpoint force and/or speed), are defined by control signals, properties of the tonic stretch reflex loop, and external force field (load). Feldman suggested that control of a joint could be described with two variables, a coactivation command (c) and a reciprocal command (r). Effects of these two control variables affect activation of individual muscles according to the principle of superposition. Experimental support for this principle has been obtained in studies of fast arm movements [13, 14] and postural control [15].

2 Arimoto's Principle of Superposition

To our knowledge, prior to the third millennium, only one study applied the principle of superposition at the level of mechanical variables; this was a study of kinematic synergies during stepping in rats [16]. Seminal papers by Arimoto and colleagues [17, 18] have shown that the principle of superposition may be applied at the level of mechanical variables for the control of robotic hand action. The main idea of this approach is to separate complex motor tasks into sub-tasks that are controlled by independent controllers. The output signals of the controllers converge onto the same set of actuators where they are summed up. This method of control has been shown to lead to a decrease in the computation time as compared to control of the action as a whole.

This idea is far from being trivial. For example, if a robotic hand tries to manipulate a grasped object, it has to produce adequate grasping and rotational actions. However, a straightforward change in the grasping force, in general, leads to a change in the total moment produced by the digits on the object. Hence, the grasping and rotational components of the action are not independent, and if a controller responsible for the grasping changes its output, a controller responsible for the rotational action also has to change its output. Having independent controllers for the two action components becomes possible for a mechanically redundant system; the redundancy allows to decouple the two action components.

Hence, the principle of superposition in robotics becomes tightly linked to the problem of motor redundancy, arguably the most famous problem in motor control. Although approaches to the problem of redundancy have been rather different in robotics and motor control, the principle of superposition represents an example when progress in robotics proves to be fruitful for studies of the control of the human hand.

3 Major Issues in the Control of Redundant Biological Systems

Over half-a-century ago, a Russian scientist, N.A. Bernstein suggested that the essence of control of human movement was in elimination of redundant degrees-of-freedom [19, 20]. This problem, known as the Bernstein problem [21] has been in the center of attention of the motor control community for many years. Typical approaches involved application of optimization methods (reviewed in [22, 23, 24] to find unique solutions for the generally ill-posed problems such as, for example, defining joint trajectories based on a desired endpoint trajectory of a multi-joint limb (the problem of inverse kinematics [25]) or finding joint torques that would produce a desired endpoint trajectory (the problem of inverse dynamics [26]).

However, already in the nineteen-sixties, Gelfand and Tsetlin [27] suggested that degrees-of-freedom during biological motion could be not eliminated (for example, by adding a sufficient number of additional constraints) but used to form task-specific structural units (synergies). Later, this idea was developed as a principle of abundance [28], which views the availability of numerous degrees-of-freedom not as a source of computational problems but rather as a powerful apparatus that has to be properly organized. This approach has resulted in a definition of synergies as task-specific neural organizations of elemental variables with the purpose to stabilize a desired value or a time profile of a performance variable.

In different studies, depending on the selected levels of analysis, elemental variables have been associated with joint rotations, digit forces, hypothetical signals to fingers (finger modes), hypothetical signals to muscle groups (muscle modes), while performance variables could represent endpoint trajectory of a multi-joint limb, the total gripping force and the total moment of forces produced by a set of digits, shifts of the center of pressure in a standing person, etc ([29, 30, 31, 32, 33, 34, 35, 36]).

In this Chapter, we are primarily interested in the combined action of the human digits during prehensile tasks. The human hand is a complex mechanical structure that combines both serial and parallel manipulators. Individual digits can be viewed as serial mechanisms, with several joints linking the fingertip to the wrist. Such mechanisms are redundant in kinematic tasks that typically have fewer task constraints than the number of degrees of freedom in the joints. However, in force-producing isometric tasks, serial mechanisms are over-constrained because a force vector at the endpoint defines unambiguously all the joint torques [37]. In contrast, when several fingers grasp a rigid object, they represent a parallel mechanism that is over-constrained in kinematics, because movement of one finger induces movement of all other fingers. However, the parallel mechanism is redundant in statics because an infinite number of combinations of finger forces can produce a required total force [38, 39].

The design of the human hand makes it a very attractive object to study the organization of multi-element synergies. These synergies are learned by humans over the lifetime and may be expected to show reproducible behaviors across a variety of tasks. Obviously multi-digit synergies are very important for everyday activities. Besides, studies of multi-digit synergies have obvious practical applications in such areas as robotics, prosthetics, and rehabilitation.

4 Prehension Synergies: The Hierarchical Control

We are going to accept a definition of synergies as conjoint changes in mechanical variables produced by individual digits that are seen over repetitions at a multi-digit task [40]. We view such conjoint changes not as results of "noise" but as a purposeful control strategy that does not select a unique combination of elemental variables based on some optimization criterion but facilitates large families of such combinations that stabilize the explicit task variables and also provide for an ability of the system to handle other tasks simultaneously.

This definition is close to the framework of the uncontrolled manifold (UCM) hypothesis in motor control. The UCM hypothesis [29] assumes that the neural controller acts in the space of elemental variables and selects in that space a sub-space (a UCM) corresponding to a desired value of a performance variable. Further, the controller limits variability in the space of elemental variables in directions orthogonal to the UCM while it allows relatively large variability within the UCM. The ideas of UCM and its orthogonal complement are close to the notions of range and self-motion manifolds in robotics [41].

It is of no surprise that much experimental support for the UCM hypothesis has come from studies of multi-finger action [31, 42, 43, 44, 45, 46]. In those studies, to meet the requirement of orthogonality of elemental variables, analyses were performed not in the space of digit forces but in the space of hypothetical commands to digits, finger modes. Finger forces are not independent of each other because of the well documented phenomenon of enslaving, that is unintended force production by fingers when another finger of the hand produces force [47, 38, 48, 49]. Difficulties with estimating enslaving in prehensile tasks has been a major obstacle that has not allowed to use the notion of finger modes for analysis of such tasks. This resulted in using finger forces and moments of forces as elemental variables.

If an object is grasped using the so-called prismatic grasp with all five digits of the hand, there are constraints of statics that have to be met for the object to be in an equilibrium. For a relatively simple case of an external load acting vertically (along the direction of gravity) and an external moment acting in the plane of the grasp (a vertical plane that contains all the points of digit contacts with the object), three equations have to be satisfied:

$$F_{th}^n = F_i^n + F_m^n + F_r^n + F_l^n, \tag{1}$$

$$L = F_{th}^t + F_i^t + F_m^t + F_r^t + F_l^t, \tag{2}$$

$$M = \underbrace{F_{th}^n d_{th} + F_i^n d_i + F_m^n d_m + F_r^n d_r + F_l^n d_l}_{\text{Moment of the normal forces} \equiv M^n} \tag{3}$$
$$+ \underbrace{F_{th}^t r_{th} + F_i^t r_i + F_m^t r_m + F_r^t r_r + F_l^t r_l}_{\text{Moment of the tangential forces} \equiv M^t},$$

where the subscripts th, i, m, r, and l refer to the thumb, index, middle, ring, and little finger, respectively; the superscripts n and t stand for the normal and tangential force components, respectively; L is load (weight of the object), and coefficients d and r stand for the moment arms of the normal and tangential force with respect to a pre-selected center, respectively.

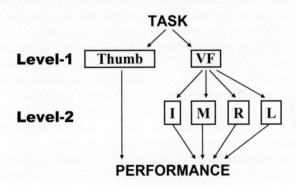

Fig. 1. A two-level hierarchical scheme of control. The first level (Level-1) distributed the task between the thumb and the virtual finger (VF). At the second level (Level-2), the command to the VF is distributed among actual fingers of the hand (I – index, M – middle, R – ring, and L – little).

To simplify analysis of multi-finger grasps, a hierarchical control scheme has been suggested that includes two levels [50, 51] (Figure 1). At the upper level, the total mechanical effect is distributed between the thumb and a virtual finger – an imagined digit whose mechanical action is equal to the combined action of all four fingers. At the lower level, the action of the virtual finger is distributed among the fingers of the hand. A number of studies have supported the idea of such two-level hierarchical control by demonstrating variable behavior at the level of individual digits that was accompanied by stable behavior at the thumb-VF level [33, 45, 46, 52].

5 Prehension Synergies: The Principle of Superposition in Static Tasks

The principle of superposition suggests that two synergies may form the foundation for a variety of prehensile actions. One of them is likely to be responsible for the stabilization of the grasping force, while the other controls the rotational action of the hand (the total moment applied by the digits on the grasped object).

Grasping itself is associated with two groups of synergies. First, there are kinematic synergies that stabilize the grasp aperture. For example, if a person tries to grip a small object between the index finger and the thumb, a mechanical perturbation applied to one of the digits produces quick adjustments in motion of both digits such that the time profile of the grip aperture remains relatively unchanged [53]. Besides, there are kinetic synergies that stabilize the mechanical action of the hand on the grasped object. If a person holds an object steadily, there are fluctuations in the forces applied by individual digits. The fluctuations of the forces produced by the thumb and the combined force produced by the opposing fingers tend to be in-phase, while fluctuations of forces of a pair of finger acting in parallel are more likely to be out-of-phase [54]. The former relation makes sure that the net force applied to the object is small, while the second relation prevents large variations in the total force applied by the set of fingers opposing the thumb.

Analysis of prehensile synergies can be performed at two levels of the hierarchical control scheme (Figure 1). For two-dimensional tasks, each digit (the thumb and the VF) can produce two components of force, normal to the surface and parallel to the surface, and also change the point of force application in the direction parallel to the surface of the object. The three static constraints introduced earlier (eqs. 1-3) are:

$$F_{th}^n = F_{VF}^n, \qquad (4)$$

$$L = F_{th}^t + F_{VF}^t, \qquad (5)$$

$$M = \underbrace{F_{th}^n d_{th} + F_{VF}^n d_{VF}}_{\text{Moment of the normal forces} \equiv M^n} \qquad (6)$$

$$+ \underbrace{F_{th}^t r_{th} + F_{VF}^t r_{VF}}_{\text{Moment of the tangential forces} \equiv M^t},$$

where the abbreviations are the same as in eqs. 1-3, VF stands for virtual finger. The thumb-VF system is apparently redundant because is has fewer constraints than elemental variables.

In one of the studies, variations of elemental variables were analyzed when subjects held statically a handle, to which different external loads and different external torques were applied in different trials [55]. Elemental variables showed strong effects of both load and torque, but they did not show any significant interaction between the two factors (Figure 2) suggesting that commands to adjust elemental variables when one of the two external parameters (load or torque) changed did not interfere with commands related to the other parameter.

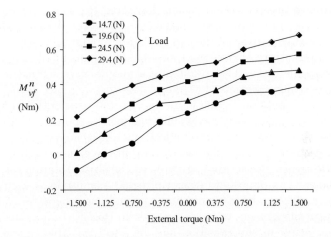

Fig. 2. An illustration of dependencies between one of the elemental variables (M_{vf}^n) and task parameters (external load and external torque). The variable showed significant changes with each of the two task parameters but no interaction between the two. Typical data for a representative subject.

A study of the patterns of variability of the elemental variables at the thumb-VF level in experiments when the subjects were required to hold the same object repeatedly many times [33] have also shown the existence of two groups of the elemental variables. Variables within a group showed strong correlations across trials, while variables that belonged to different groups did not. The two groups could be associated with two sub-components of the task, "to grasp the object stronger/weaker" and "to maintain rotational equilibrium of the object".

These observations have suggested that the principle of superposition, originally formulated by Arimoto and his colleagues [18] for robotic prehension, is also valid for human prehension. The central nervous system apparently decouples the control of two essential components of prehensile tasks and forms two synergies or two null-spaces in the space of elemental variables.

Recently, these results have been generalized to static grasping tasks in three dimensions [45]. The sets of elemental variables associated with the moment production about the vertical axis in the grasp plane and the axis orthogonal to the grasp plane consisted of two non-correlated subsets each; one subset of variables was related to the control of grasping forces (grasp control) and the other subset was associated with the control of the orientation of the hand-held object (torque control). Hence, the principle of superposition has been shown to be valid for the prehension in three dimensions as well.

The principle of superposition has also shown its applicability to applied studies of movements. In one of the studies, the ability of young and elderly persons to stabilize the total gripping force and the total moment of forces applied to a hand-held object was studied [34]. Based on the principle of superposition, separate analysis of the two synergies, force-stabilizing and moment-stabilizing, has shown that elderly persons are impaired in both components of the prehensile tasks: Their indices of synergies were significantly smaller than those in young persons.

People commonly manipulate objects with the center of mass beyond the hand grip, e.g. mugs with the handle, hammers, spoons, etc. Such objects exert torques on the hand. During object manipulation, e.g. during motion in the vertical direction, both the load force and the external torque change in parallel with object acceleration. To prevent the object rotation the performer should generate an opposing moment of equal magnitude at each instant of time. Hence, during object manipulation, simultaneous control of the grasping force and the moment of forces is required. Recent studies have suggested that the principle of superposition may be also applicable to tasks that involve moving a hand-held object [56]. In particular, the observed finger forces have been explained as the outcome of summation of two central commands whose goals are to achieve: (a) slip prevention and (b) tilt prevention. The first command is proportional to the load force, i.e. to object acceleration. It causes stronger/weaker forces by all involved fingers and hence changes in the grip force. The second command potentiates flexion commands to the fingers that produce moments of forces acting against the external torque (moment agonists) and inhibits the commands to the moment antagonist fingers. As a result, the force of the agonist fingers increases and the force of the antagonist fingers decreases. This pattern was actually observed experimentally in agreement with the principle of superposition.

6 Prehension Synergies: The Principle of Superposition in Reactions to Perturbations

In a recent series of experiments, subjects were required to hold a handle with a set of three loads that altogether produced no external torque with respect to a horizontal axis passing through the center of the handle (Figure 3). Further, one of the loads could be lifted (removed) either by the subject himself/herself or unexpectedly, by an experimenter. This action always produced the same perturbation in the weight of the handle but it could produce different torque perturbations depending on the load that was lifted, either a close to zero toque perturbation or a non-zero torque perturbation in either pronation or supination.

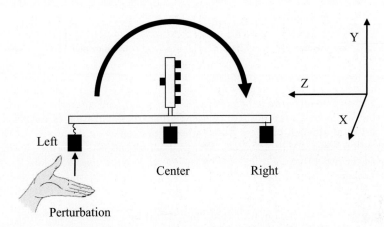

Fig. 3. An illustration of the experimental setup. The subject grasped the handle instrumented with force/torque sensors (shown as black boxes). Three loads were attached to the handle. One of them was removed (lifted) in each trial either by the subject or by the experimenter (unexpectedly).

Indices of covariation of elemental mechanical variables that contributed to the grasping force and to the total moment produced by the digits were computed separately over multiple repetitions at each task. These indices reflected the strength of two synergies, force-stabilizing and moment-stabilizing, ΔV_F and ΔV_M in Figure 4.

When the subjects held the object steadily, both indices corresponded to a strong negative covariation among elemental variables contributing to both total grasping force and total moment of forces, that is, two synergies were acting in parallel. When the subjects introduced the perturbation themselves, the indices of both synergies showed a decline about 100-150 ms prior to the actual perturbation. Such adjustments were not seen when the same perturbations were triggered unexpectedly by the experimenter. These phenomena, anticipatory synergy adjustments were described in earlier studies prior to quick voluntary changes in the total force produced by a set of fingers [57, 46]. They have been interpreted as purposeful weakening of a force-stabilizing synergy in preparation to a quick change in the force. This experiment has shown that anticipatory synergy adjustments can also be seen in preparation to a

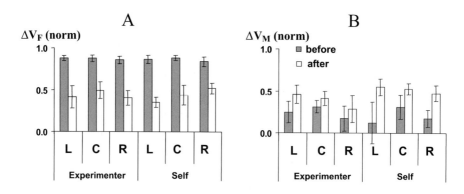

Fig. 4. Indices of two synergies, force-stabilizing (ΔV_F) and moment-stabilizing (ΔV_M) at two steady-states, before load perturbation and after the perturbation. Note that all indices were positive corresponding to stabilization of both grasping force and total moment of forces across trials. However, ΔV_F decreased following a perturbation while ΔV_M increased. Averaged across subjects data are shown with standard deviation bars. L – left load location, C – center load location, R – right load location.

self-triggered perturbation and that they could be organized with respect not only to the total force but also with respect to the total moment of forces.

Following a perturbation, there was a drop in indices of both synergies and, after about 1-2 s, the synergies recovered. It is of interest that the indices of the force-stabilizing synergies decreased in a new steady-state, while indices of the moment-stabilizing synergies increased (Figure 4). This was true for both self- and experimenter-triggered perturbations across the magnitudes of the external torque perturbation. These contrasting changes in the synergy indices provide additional support for the relative independence of the two synergies involved in the grasping and rotational components of prehensile actions and show that the principle of superposition holds for multi-digit reactions to force/torque perturbations.

7 Concluding Comments

The principle of superposition has a long history in neurophysiology and motor control. However, the recent development of this principle by Prof. Arimoto has been very fruitful not only for the area, for which it was originally meant (robotics), but also for analysis of human hand actions. This principle has received support in a variety of experimental studies involving both static hand actions and reactions to perturbations, young and elderly persons, and two- and three-dimensional tasks. Applications of this principle to prosthetic design and development of rehabilitation strategies seem very promising.

References

1. Rothwell J (1994) Control of human voluntary movement. Second edn. Chapman & Hall, London
2. Demieville H, Partridge L (1980) Probability of peripheral interaction between motor units and implications for motor control. American Journal of Physiology 238:R119–R137
3. Latt L, Sparto P, Furman J, Redfern M (2003) The steady-state postural response to continuous sinusoidal galvanic vestibular stimulation. Gait and Posture 18:64–72
4. Latash M (1998) Neurophysiological basis of movement. Human Kinetics, Urbana Champaign, IL
5. Ruegg D, Bongioanni F (1989) Superposition of ballistic on steady contractions in man. Experimental Brain Research 77:412–420
6. Kowalski N, Depireux S, Shamma S (1996) Analysis of dynamic spectra in ferret primary auditory cortex. part ii: Prediction of unit responses to arbitrary dynamic spectra. Journal of Neurophysiology 76:3524–3534
7. Glazer V, Gauzelman V (1997) Linear and nonlinear properties of simple cells of the striate cortex of the cat: two types of nonlinearity. Experimental Brain Research 117:281–291
8. Fingelkurts A, Fingelkurts A (2005) Stability, reliability and consistency of the compositions of brain oscillations. International Journal of Psychophysiology (electronic publication)
9. Karakas S, Erzengin O, Basar E (2000) A new strategy involving multiple cognitive paradigms demonstrates that erp components are determined by the superposition of oscillatory responses. Clinical Neurophysiology 111:1719–1732
10. Feldman A (1966) Functional tuning of the nervous system with control of movement or maintenance of a steady posture. part ii: Controllable parameters of the muscle. Biofizika 11:565–578
11. Feldman A (1986) Once more on the equilibrium-point hypothesis (λ-model) for motor control. Journal of Motor Behavior 18:17–54
12. Feldman A, Levin M (1995) Positional frames of reference in motor control: their origin and use. Behavioral and Brain Sciences 18:723–806
13. Adamovich S, Feldman A (1984) A model of central regulation of movement parameters. Biofizika 29:306–309
14. Pigeon A, Yahia L, Mitnitski A, Feldman A (2000) Superposition of independent units of coordination during pointing movements involving the trunk with and without visual feedback. Experimental Brain Research 131:336–349
15. Slijper H, Latash M (2000) The effects of instability and additional hand support on anticipatory postural adjustments in leg, trunk, and arm muscles during standing. Experimental Brain Research 135:81–93
16. Ganor I, Golani I (1980) Coordination and integration in the hindleg step cycle of the rat: kinematic synergies. Brain Research 195:57–67
17. Arimoto S, Nguyen P, Han H, Doulgeri Z (2000) Dynamics and control of a set of dual fingers with soft tips. Robotica 18:71–80
18. Arimoto S, Tahara K, Yamaguchi M, Nguyen P, Han H (2001) Principles of superposition for controlling pinch motions by means of robot fingers with soft tips. Robotica 19:21–28
19. Bernstein N (1947) On the construction of movements. Medgiz, Moscow (in Russian)

20. Bernstein N (1967) The co-ordination and regulation of movements. Pergamon Press, Oxford
21. Turvey M (1990) Coordination. American Journal of Psychology 45:938–953
22. Nelson W (1983) Physical principles for economies of skilled movements. Biological Cybernetics 46:135–147
23. Seif-Naraghi A, Winters J (1990) Optimized strategies for scaling goal-directed dynamic limb movements. In Winters J, Woo SY, eds.: Multiple Muscle Systems: Biomechanics and Movement Organization. Springer-Verlag, New York 312–334
24. Prilutsky B, Zatsiorsky V (2002) Optimization-based models of muscle coordination. Exercise and Sport Science Reviews 30:32–38
25. Mussa-Ivaldi F, Morasso P, Zaccaria R (1989) Kinematic networks: A distributed model for representing and regularizing motor redundancy. Biological Cybernetics 60:1–16
26. Hollerbach J, Atkeson C (1987) Deducing planning variables from experimental arm trajectories pitfalls and possibilities. Biological Cybernetics 56:279–292
27. Gelfand I, Tsetlin M (1966) On mathematical modeling of the mechanisms of the central nervous system. In Gelfand I, Gurfinkel V, Fomin S, Tsetlin M, eds.: Models of the structural-functional organization of certain biological systems. Nauka, Moscow 9–26 (in Russian, a translation is available in 1971 edition by MIT Press: Cambridge MA)
28. Gelfand I, Latash M (2002) On the problem of adequate language in biology. In Latash M, ed.: Progress in Motor Control: Structure-Function Relations in Voluntary Movement. Volume 2. Human Kinetics, Urbana Champaign, IL 209–228
29. Scholz J, Schöner G (1999) The uncontrolled manifold concept: Identifying control variables for a functional task. Experimental Brain Research 126:289–306
30. Scholz J, Schöner G, Latash M (2000) Identifying the control structure of multi-joint coordination during pistol shooting. Experimental Brain Research 135:382–404
31. Latash M, Scholz J, Danion F, Schöner G (2001) Structure of motor variability in marginally redundant multi-finger force production tasks. Experimental Brain Research 141:153–165
32. Danion F, Schöner G, Latash M, Li S, Scholz J, Zatsiorsky V (2003) A force mode hypothesis for finger interaction during multi-finger force production tasks. Biological Cybernetics 88:91–98
33. Shim J, Latash M, Zatsiorsky V (2003) Prehension synergies: Trial-to-trial variability and hierarchical organization of stable performance. Experimental Brain Research 152:173–184
34. Shim J, Lay B, Zatsiorsky V, Latash M (2004) Prehension synergies in three dimensions. Journal of Applied Physiology 97:213–224
35. Krishnamoorthy V, Latash M, Scholz J, Zatsiorsky V (2003) Muscle synergies during shifts of the center of pressure by standing persons. Experimental Brain Research 152:281–292
36. Krishnamoorthy V, Latash M, Scholz J, Zatsiorsky V (2004) Muscle modes during shifts of the center of pressure by standing persons: Effects of instability and additional support. Experimental Brain Research 157:18–31
37. Zatsiorsky V (2002) Kinetics of Human Motion. Human Kinetics, Urbana Champaign, IL
38. Li Z, Latash M, Zatsiorsky V (1998) Force sharing among fingers as a model of the redundancy problem. Experimental Brain Research 119:276–286

39. Burstedt M, Flanagan J, Johansson R (1999) Control of grasp stability in humans under different frictional conditions during multidigit manipulation. Journal of Neurophysiology 82:2393–2405
40. Zatsiorsky V, Latash M (2004) Prehension synergies. Exercise and Sport Science Reviews 32:75–80
41. Murray R, Sastry S, Li Z (1994) A mathematical introduction to robotic manipulation. CRC Press, Boca Raton
42. Latash M, Scholz J, Schöner G (2002) Motor control strategies revealed in the structure of motor variability. Exercise and Sport Science Reviews 30:26–31
43. Scholz J, Danion F, Latash M, Schöner G (2002) Understanding finger coordination through analysis of the structure of force variability. Biological Cybernetics 86:29–39
44. Scholz J, Kang N, Patterson D, Latash M (2003) Uncontrolled manifold analysis of single trials during multi-finger force production by persons with and without down syndrome. Experimental Brain Research 153:45–48
45. Shim J, Latash M, Zatsiorsky V (2005) Prehension synergies in three dimensions. Journal of Neurophysiology 93:766–776
46. Shim J, Olafsdottir H, Zatsiorsky V, Latash M (2005) The emergence and disappearance of multi-digit synergies during force production tasks. Experimental Brain Research 164:260–270
47. Kilbreath S, Gandevia S (1994) Limited independent flexion of the thumb and fingers in human subjects. Journal of Physiology 479:487–497
48. Zatsiorsky V, Li Z, Latash M (1998) Coordinated force production in multi-finger tasks: Finger interaction and neural network modeling. Biological Cybernetics 79:139–150
49. Zatsiorsky V, Li Z, Latash M (2000) Enslaving effects in multi-finger force production. Experimental Brain Research 131:187–195
50. Arbib M, Iberall T, Lyons D (1985) Coordinated control programs for movements of the hand. In Goodwin A, Darian-Smith I, eds.: Hand function and the neocortex. Springer-Verlag, Berlin 111–129
51. MacKenzie C, Iberall T (1994) The grasping hand. North-Holland, Amsterdam
52. Gao F, Latash M, Zatsiorsky V (2005) Control of finger force direction in the flexion-extension plane. Experimental Brain Research 161:307–315
53. Cole K, Abbs J (1987) Kinematic and electromyographic responses to perturbation of a rapid grasp. Journal of Neurophysiology 57:1498–1510
54. Santello M, Soechting J (2000) Force synergies for multifingered grasping. Experimental Brain Research 133:457–467
55. Zatsiorsky V, Gao F, Latash M (2003) Prehension synergies: Effects of object geometry and prescribed torques. Experimental Brain Research 148:77–87
56. Gao F, Latash M, Zatsiorsky V (2006) Maintaining rotational equilibrium during object manipulation: Linear behavior of a highly non-linear system. Experimental Brain Research 169:519–531
57. Olafsdottir H, Yoshida N, Zatsiorsky V, Latash M (2005) Anticipatory covariation of finger forces during self-paced and reaction time force production. Neuroscience Letters 381:92–96

Motion Planning of Human-Like Movements in the Manipulation of Flexible Objects

Mikhail Svinin[1], Igor Goncharenko[2] and Shigeyuki Hosoe[1]

[1] Bio-Mimetic Control Research Center, RIKEN, Anagahora, Shimoshidami, Nagoya 463-0003, Japan, `svinin@bmc.riken.jp, hosoe@bmc.riken.jp`
[2] 3D Incorporated, 1-1 Sakaecho, Kanagawa-ku, Yokohama, Kanagawa 221-0052, Japan `igor@ddd.co.jp`

Summary. The paper deals with modeling of human-like reaching movements in dynamic environments. A simple but not trivial example of reaching in a dynamic environment is the rest-to-rest manipulation of a multi-mass flexible object (underactuated system) with the elimination of residual vibrations. This a complex, sport-like movement task where the hand velocity profiles can be quite different from the classical bell shape and may feature multiple phases. First, we establish the Beta function as a model of unconstrained reaching movements and analyze it properties. Based on this analysis, we construct a model where the motion of the most distal link of the object is specified by the lowest order polynomial, which is not uncommon in the control literature. Our experimental results, however, do not support this model. To plan the motion of the system under consideration, we develop a minimum hand jerk model that takes into account the dynamics of the flexible object and show that it gives a satisfactory prediction of human movements.

1 Introduction

This chapter is dedicated to Professor Suguru Arimoto on the occasion of his 70th birthday. Recently, modeling of human-like reaching movements and resolving Bernstein's problem of redundant degrees of freedom has become an active topic of his research activity. In the past few years we have had the pleasure to discuss with him this research subject in our personal communications, and now we are delighted to write about it in this book.

When humans make point-to-point movements in free space, there is, in principle, an infinite choice of trajectories. However, many studies have shown that human subjects tend to choose unique trajectories with invariant features. First, hand paths in point-to-point movements tend to be straight and smooth. Second, for relatively fast movements the velocity profile of the hand trajectory is bell-shaped [1, 2]. These invariant features give hints about the internal representation of the movements in the central nervous system (CNS) [3].

Understanding the trajectory formation in human reaching movements is a very important research problem in computational neuroscience and modern robotics. This problem can be attacked from different directions. One of the research lines deals with the sensory-motor feedback control and looks for a natural resolution of the redundancy of human movements [4, 5]. Another, complimentary research line deals with the open-loop control and employs optimization approaches [6, 7].

These two research lines are quite different but both important because, in our opinion, it is the blending of the feedforward and feedback control that defines the appearance of reaching movements. Indeed, experiments show that for relatively slow movements the velocity profiles are left-skewed [8, 9], a feature that is well-captured by the feedback control. As the time of movement decreases, the velocity profiles tend to a symmetric bell-shaped form[3], that can be captured by the feedforward control [11]. Thus, the resulting movement can be considered as the superposition of a major ballistic (feedforward) component and a corrective (feedback) component [12, 13]. How specifically the blending of these components works is one the greatest secrets of the CNS, and recent ideas of Arimoto [14, 15] may prove fruitful in opening it.

In this chapter we will deal with the feedforward component of reaching movements and model it via optimization theory. In this approach, the trajectory of the human arm is predicted by minimizing, over the movement time T, an integral performance index $\mathcal{J}$ subject to boundary conditions imposed on start and end points. The performance index can be formulated in the joint space or in the task space normally associated with the human hand. It is well established that for the unconstrained reaching movements the trajectory of human hand can be predicted with reasonable accuracy by the minimum hand jerk criterion [11]. Another popular model is based on the minimum joint torque change criterion [16].

While the above criteria captures well the invariant features of reaching movements in the free space, it remains to be seen if they are applicable to the modeling of reaching movements in dynamic environments. A simple but not trivial example of reaching in a dynamic environment is the rest-to-rest manipulation of a linear chain of flexible objects with compensation of structural vibration. Considering this manipulation task from the engineering point of view, one can find different control strategies to generate rest-to-rest motion commands that eliminate residual vibrations [17, 18, 19, 20, 21]. Perhaps the most simple open-loop control strategy is to specify the motion of the most distal link of the flexible object by the lowest-order polynomial satisfying the system boundary conditions [22]. Recently this control strategy has been successfully applied to the modeling of reaching movements in the manipulation of one flexible object [23]. Despite seeming simplicity, this task

---

[3] Experiments also show that for very fast movements the velocity profiles can be even slightly skewed to the right [8, 10]. However, it remains to be tested how far from zero the end-point accelerations are in such movements.

requires a lot of skill that must be acquired by practicing. An interesting feature of this task, experimentally established in [23], is that a human subject controls the object in a very non-trivial way, keeping two distinct phases in the hand velocity profile.

In this paper we analyze the lowest-order polynomial approach for the general case of a multi-mass flexible object and compare it with the minimum hand jerk model. Our theoretical and experimental analysis show that the minimum hand jerk model is a more consistent candidate for the prediction of reaching movements in the manipulation of flexible objects. For simplicity, we deal with a simplified, one-dimensional model of human movements. In this model the configuration dependance of the human arm is ignored and the motion is considered at the hand level. Inertial properties of the arm are not taken into consideration. It is thus assumed that the human arm, controlled by the CNS, is a perfect actuator driving the hand in the task space.

The paper is organized as follows. First, in Section 2 we consider a generalization of the classical minimum jerk model of the free-space reaching movements, relate it to the lowest polynomial approach, and show that the solution is given by the regularized incomplete Beta function. Next, in Section 3, we introduce a mathematical model of the flexible object and derive analytical solutions for the lowest-order-polynomial model and for the minimum hand jerk model. The analytical solutions are tested against experimental data in Section 4. Finally, conclusions are summarized in Section 5.

2 Beta Function as a Model of Reaching Movements

Consider the problem of finding a function $x(t)$ minimizing the criterion

$$\mathcal{J} = \frac{1}{2} \int_{t_0}^{t_1} \left(\frac{d^n x}{dt^n} \right)^2 dt \tag{1}$$

under the following boundary conditions:

$$x(t_0) = x_0,\ \dot{x}(t_0) = 0,\ \ddot{x}(t_0) = 0, \ldots, x^{(n-1)}(t_0) = 0, \tag{2}$$
$$x(t_1) = x_1,\ \dot{x}(t_1) = 0,\ \ddot{x}(t_1) = 0, \ldots, x^{(n-1)}(t_1) = 0. \tag{3}$$

This problem is often raised in the prediction of human movements, normally for $n = 3$ (the minimum hand jerk model [11]) and occasionally for $n = 4$ (the minimum snap model [24, 25]). Solutions of the problem under considerations are also often employed in motion planning for robotic systems, again most commonly for $n = 3$. Higher derivatives are used when the motion smoothness is important. For example, the solutions for $n = 6$ and $n = 8$ were used in, respectively, [26] and [27] to generate reference signals for motion planning of robots with flexible links.

From the engineering point of view the use of higher-order feedforward trajectories, having the inherent advantage of smoothing, implies a lower energy content at higher frequencies [28]. This results in a lower high-frequency content of the error signals, which facilitates the construction of the feedback control loop [29]. This reasoning can be also applied to the human movements. It is often argued in the literature on the computational neuroscience that the brain plans and controls the reaching movements in such a way so that to produce highly smooth trajectories [30, 31]. The argument comes from the comparison of the optimal trajectories obtained by the minimum acceleration and the minimum jerk criteria. The minimum jerk criterion does not produce acceleration jumps at the start and end points, while the minimum acceleration criterion does. Based on this observation, some researchers suppose that the human brain implicitly adopts smoothness as the criterion for motion planning.

A general formula for the optimal velocity profiles minimizing the criterion (1) under the boundary conditions (2,3) has been correctly guessed in [22, 32]. Here, we present a formal derivation, establish the solution, and recognize it as a famous special mathematical function. To pose the optimization problem in non-dimensional settings, introduce the following change of variables:

$$x(t) = x_0 + (x_1 - x_0)\, y(\tau(t)) \tag{4}$$

where

$$\tau = \frac{t - t_0}{t_1 - t_0}, \quad \tau \in [0, 1]. \tag{5}$$

In the new variables y and τ we have the criterion

$$\mathcal{J} = \frac{1}{2} \int_0^1 \left(\frac{d^n y}{d\tau^n}\right)^2 d\tau, \tag{6}$$

and the boundary conditions

$$y(0) = 0,\ y'(0) = 0,\ y''(0) = 0, \ldots, y^{(n-1)}(0) = 0, \tag{7}$$

$$y(1) = 1,\ y'(1) = 0,\ y''(1) = 0, \ldots, y^{(n-1)}(1) = 0, \tag{8}$$

where

$$y^{(k)}(\tau) = \frac{(t_1 - t_0)^k}{(x_1 - x_0)} x^{(k)}(t),\ k = 1, \ldots, n-1. \tag{9}$$

As can be easily shown, the Euler-Lagrange equation for the problem under consideration is $y^{(2n)}(\tau) = 0$. The general solution for this equation is given by the polynomial $y(\tau) = \sum_{i=0}^{2n-1} c_i \tau^i$, where the coefficients c_i are established from the boundary conditions (7,8). Thus, the optimization problem under consideration is equivalent to the construction of the lowest order polynomial satisfying the boundary conditions (7,8).

To clarify the analytical structure of the solution, we recast the optimization task as the optimal control problem. Let us introduce the state vector

$\boldsymbol{y} = \{y, y', \ldots, y^{(n-1)}\}^{\mathrm{T}}$ and define the control input $u = y^{(n)}$. The state dynamics are the n-th order integrator $\dot{\boldsymbol{y}} = \boldsymbol{A}\boldsymbol{y} + \boldsymbol{b}u$, where $\boldsymbol{b} = \{0, \ldots, 0, 1\}^{\mathrm{T}}$, and the elements of the matrix $\boldsymbol{A}$ are defined as

$$[\boldsymbol{A}]_{ij} = \begin{cases} 1 & \text{if } j = i+1, \\ 0 & \text{otherwise.} \end{cases} \quad (10)$$

The analytical solution for the minimum effort control problem, seeking the control minimizing $\mathcal{J} = \frac{1}{2}\int_0^1 u^2\, d\tau$ for the dynamic system $\dot{\boldsymbol{y}} = \boldsymbol{A}\boldsymbol{y} + \boldsymbol{b}u$ and the boundary conditions $\boldsymbol{y}(0) = \boldsymbol{y}_0$, $\boldsymbol{y}(1) = \boldsymbol{y}_1$, is well-established in the control literature [33]. It can be represented as

$$\boldsymbol{y}(\tau) = e^{\boldsymbol{A}\tau}\left(\left\{\boldsymbol{I} - \boldsymbol{W}(\tau)\boldsymbol{W}^{-1}(1)\right\}\boldsymbol{y}_0 + \boldsymbol{W}(\tau)\boldsymbol{W}^{-1}(1)e^{-\boldsymbol{A}}\boldsymbol{y}_1\right), \quad (11)$$

where

$$\boldsymbol{W}(\tau) = \int_0^\tau e^{-\boldsymbol{A}s}\boldsymbol{b}\boldsymbol{b}^{\mathrm{T}}e^{-\boldsymbol{A}^{\mathrm{T}}s}\, ds. \quad (12)$$

For the n-th order integrator the matrix exponent is defined as

$$\left[e^{\boldsymbol{A}\tau}\right]_{ij} = \begin{cases} \dfrac{\tau^{j-i}}{(j-i)!} & \text{if } j \geq i, \\ 0 & \text{if } j < i, \end{cases} \quad (13)$$

and by direct calculations one finds

$$[\boldsymbol{W}(\tau)]_{ij} = \frac{(-1)^{i+j}\tau^{2n+1-i-j}}{(2n+1-i-j)(n-i)!(n-j)!}. \quad (14)$$

As shown in [34], for the n-th order integrator the matrix $\boldsymbol{W}(\tau)$ can be decomposed as $\boldsymbol{W}(\tau) = \tau\boldsymbol{P}(-\tau)\boldsymbol{H}\boldsymbol{P}(-\tau)$, where

$$[\boldsymbol{P}(\tau)]_{ij} = \begin{cases} \dfrac{\tau^{n-j}}{(n-i)!} & \text{if } i = j, \\ 0 & \text{otherwise.} \end{cases} \quad (15)$$

and $\boldsymbol{H}$ is the Hilbert matrix

$$[\boldsymbol{H}]_{ij} = \frac{1}{(2n+1-i-j)}, \quad (16)$$

the inverse of which is known to be

$$\left[\boldsymbol{H}^{-1}\right]_{ij} = \frac{(-1)^{i+j}(2n-i)!(2n-j)!}{(2n+1-i-j)\{(n-i)!(n-j)!\}^2(i-1)!(j-1)!}. \quad (17)$$

The elements of the inverse matrix $\boldsymbol{W}^{-1}(\tau) = \tau^{-1}\boldsymbol{P}^{-1}(-\tau)\boldsymbol{H}^{-1}\boldsymbol{P}^{-1}(-\tau)$ can now be easily established:

$$\left[\boldsymbol{W}^{-1}(\tau)\right]_{ij} = \frac{(2n-i)!(2n-j)!}{(2n+1-i-j)(n-i)!(n-j)!(i-1)!(j-1)!\tau^{2n+1-i-j}}. \quad (18)$$

For the rest-to-rest movements $\boldsymbol{y}_0 = \boldsymbol{0}$, $\boldsymbol{y}_1 = \{1,0\ldots,0\}^{\mathrm{T}}$, and $e^{-\boldsymbol{A}}\boldsymbol{y}_1 = \boldsymbol{y}_1$. Therefore, $\boldsymbol{y}(\tau) = e^{\boldsymbol{A}\tau}\boldsymbol{W}(\tau)\boldsymbol{W}^{-1}(1)\boldsymbol{y}_1$. The components of the vector $\boldsymbol{y}(\tau)$, which are the derivatives of the corresponding order of $y(\tau)$, are found from (13,14,18) by direct computation:

$$y^{(i-1)}(\tau) = \sum_{j=i}^{n}\sum_{s=1}^{n} \frac{(-1)^{j+s}(2n-1)!(2n-s-1)!\tau^{2n+1-i-s}}{(j-i)!(2n+1-s-j)\{(n-s)!\}^2(n-j)!(n-1)!(s-1)!}. \quad (19)$$

To find the velocity $y'(\tau)$, we set $i=2$ in (19) and take into account that

$$\sum_{j=2}^{n}\frac{(-1)^j}{(j-2)!(2n+1-s-j)(n-j)!} = \frac{1}{(s-n-1)(s-n-2)\ldots(s-n-(n-1))}$$

$$= \frac{(-1)^n\Gamma(1+n-s)}{\Gamma(2n-s)} = \frac{(-1)^n(n-s)!}{(2n-s-1)!}, \quad (20)$$

where Γ is the usual Gamma function. Then, after some simple manipulations using the binomial expansion for $(1-\tau)^{n-1}$, one obtains

$$y'(\tau) = \frac{(2n-1)!\tau^{n-1}}{(n-1)!}\sum_{s=1}^{n}\frac{(-1)^{n+s}\tau^{n-s}}{(n-s)!(s-1)!} = \frac{\tau^{n-1}(1-\tau)^{n-1}}{B(n,n)}, \quad (21)$$

where

$$B(n,n) = \frac{(n-1)!(n-1)!}{(2n-1)!} \quad (22)$$

is the symmetric Beta function. Therefore, the solution $y(\tau)$ can be formally represented as

$$y(\tau) = \frac{\int_0^\tau p^{n-1}(1-p)^{n-1}dp}{B(n,n)} \triangleq \frac{B(\tau;n,n)}{B(n,n)} \triangleq \bar{B}(\tau;n,n). \quad (23)$$

This expression is known as the regularized incomplete Beta function [35]. While it is widely used in many fields, it is best known for its applications in statistics [36]. The basic properties of the Beta distribution are well-known [35]. In particular, the velocity $y'(\tau)$ is symmetric with respect to the middle point $\tau = 1/2$, the position $y(1/2) = 1/2$, and the even derivatives $y^{(2k)}(1/2) = 0$ for $k = 1,2,3\ldots$. The first 15 solutions for $y(\tau)$ and $y'(\tau)$ are plotted in Fig. 1.

Several comments are in order.

1°. The opinion that the brain maximizes the movement smoothness is sometimes expressed in the literature [30, 31]. It can be appealing to associate the order of the optimization problem with the smoothness of the resulting solution and measure the smoothness by, say, the L_2-norm:

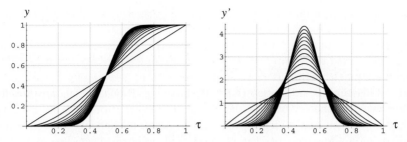

Fig. 1. Position and velocity profiles for $n = 1, 2, \ldots, 15$.

$$\|y_n(\tau)\|_{L_2}^2 = \int_0^1 y_n^2(\tau)\, d\tau. \tag{24}$$

Here, with slight abuse of notation we use $y_n(\tau)$ to denote the solution of the optimization problem (6-8) for a given n. It can be shown that indeed $\|y_n(\tau)\|_{L_2}$ is a monotonously increasing function of n, and $\lim_{n\to\infty} \|y_n(\tau)\|_{L_2} = 1/\sqrt{2}$ is well-defined. However, while the solutions $y_n(\tau)$ are continuous functions of τ for any fixed n, in the limiting case of $n \to \infty$ the solution is discontinuous. In this case the velocity is the unit impulse function $y'_\infty(\tau) = \delta(\tau - 1/2)$, $\int_0^1 y'_\infty(\tau)d\tau = 1$. The position $y_\infty(\tau)$ is the unit step function, which is unity for $\tau > 1/2$ and is 0 for $\tau < 1/2$.

$2°$. It is noticed in [11] that while, in general, the minimum jerk model ($n = 3$) is a good candidate for mimicking human-like movements, the minimum snap model ($n = 4$) sometimes also provide a reasonable fit to the experimental data. In this connection, it should be noted that the Beta function is defined not only for integer but also for real n. This suggests that the criterion (1) can be generalized to non-integer orders using, for example, Riemann-Liouville fractional integrals and derivatives[37]. It also suggests that in fitting experimental data of reaching movements we can use real numbers $n \in [3, 4]$.

$3°$. The symmetry of the Beta function with respect to its last two arguments in (21,23) is explained by the symmetric placement of the boundary conditions. It can be shown that for the asymmetric placement (n_a boundary conditions at the start point and n_b at the end point, $n_a + n_b = 2n$) the solution is expressed through the asymmetric Beta function

$$y(\tau) = \frac{\int_0^\tau p^{n_a-1}(1-p)^{n_b-1}dp}{B(n_a, n_b)} \triangleq \frac{B(\tau; n_a, n_b)}{B(n_a, n_b)}, \tag{25}$$

where

$$B(n_a, n_b) = \frac{(n_a - 1)!(n_b - 1)!}{(n_a + n_b - 1)!}. \tag{26}$$

Note that while the function (25) is the lowest order polynomial satisfying the asymmetric boundary conditions, it does not have a variational meaning.

Strictly speaking, it does not minimize (6) because in the classical variational formulation the boundary conditions (those that are imposed directly, as in (7,8), and those obtained automatically from the variational formulation and sometimes called the complementary or the natural boundary conditions [38]) are always placed symmetrically at the end-points.

4°. This remark addresses the boundary conditions, which are not less important than the criterion itself. In our optimization problem we fixed the first $n-1$ derivatives, which, as far as the human movements are concerned, might not be plausible from the physiological point of view[4]. Let us now fix (directly impose) to zero only the first k derivatives and assume that the order n of the optimization problem can be changed. Then, for any given n the natural boundary conditions that follow from the variational problem will be imposed on the derivatives from n to $2n-2-k$. It can be shown that the lowest order polynomial solutions $y_n(\tau)$ minimizing criterion (6) for different n are the same for $n=k+1$ and $n=2k+2, 2k+3, \ldots, \infty$. They are given by $\bar{B}(\tau; k+1, k+1)$, which has the minimal L_2-norm compare to the solutions obtained for $n=k+2, \ldots, 2k+1$. This gives a different interpretation for the optimal solution.

Let us illustrate this point for the case of $k=2$. The optimization problems for $n=1$ and $n=2$ are over-constrained (in terms of the boundary conditions) and do not make sense. For $n=3$ there are no natural boundary conditions. For any $n>3$ the structure of the boundary conditions has the following form

$$y(0)=0, y'(0)=0, y''(0)=0, \mid y^{(n)}(0)=0, \ldots, y^{(2n-4)}(0)=0, \quad (27)$$
$$y(1)=1, y'(1)=0, y''(1)=0, \mid y^{(n)}(1)=0, \ldots, y^{(2n-4)}(1)=0. \quad (28)$$

$\underbrace{}_{\text{direct boundary conditions}}$ $\underbrace{\phantom{y^{(n)}(1)=0, \ldots, y^{(2n-4)}(1)=0.}}_{\text{natural boundary conditions}}$

It is not difficult to show that the lowest order polynomial solutions $y_n(\tau)$, minimizing (6) under the directly imposed boundary conditions and the natural boundary conditions, are

$$y_n(\tau) = \begin{cases} 10\tau^3 - 15\tau^4 + 6\tau^5 = \bar{B}(\tau; 3,3), & \|y(\tau)\|_{L_2} \approx 0.626, & n=3 \\ 7\tau^3 - 21\tau^5 + 21\tau^6 - 6\tau^7, & \|y(\tau)\|_{L_2} \approx 0.629, & n=4 \\ 9\tau^3 - 10.5\tau^4 + 6\tau^7 - 4.5\tau^8 + \tau^9, & \|y(\tau)\|_{L_2} \approx 0.627, & n=5 \\ 10\tau^3 - 15\tau^4 + 6\tau^5 = \bar{B}(\tau; 3,3), & \|y(\tau)\|_{L_2} \approx 0.626, & n=6,7\ldots\infty \end{cases} \quad (29)$$

This gives a different interpretation for the classical minimum jerk model ($\bar{B}(\tau; 3,3)$). Namely, for the fixed end-point positions, velocities and accelerations, among the lowest order polynomials $y_n(\tau)$, minimizing criterion (6) for $n=3,4,5,\ldots\infty$, the solution given by $\bar{B}(\tau; 3,3)$ has the minimal L_2-norm.

5°. The minimum jerk model can be thought of as a phenomenological model [7]. Its biological relevance is often questioned as there are no plausible physiological mechanisms through which it can be implemented directly [39, 40]. Yet, numerous experiments reported in the literature on free-space

---

[4] Can the reader imagine the end-point control of, say, the 21st derivative?

reaching movements show that this model can be a reasonable (yet indirect) approximation to what really takes place in the human control system. Can the minimum jerk model enjoy the overall success in the prediction of reaching movements without a physiological reason? The remarks 2° and 4° may be helpful in the clarification of this issue.

If we assume that the CNS is pre-wired in such a way that, given the boundary conditions for $y, y', y'', \ldots, y^{(k)}$, among all the smooth curves $y(t)$, represented by the lowest order polynomials minimizing (6) for $n = k+1, k+2, \ldots \infty$, it selects the curve of the minimal L_2 norm, the main accent in the issue of the physiological plausibility is shifted from the optimality criterion to the boundary conditions. The question now is, what k is physiologically plausible? It is known that in reaching a human being can control the end-point position, velocity, and acceleration, and this observation can rationalize the minimum jerk model ($\bar{B}(\tau; 3, 3)$). It is not known, however, if the human can control the end-point jerk or snap, and until such evidence is presented the corresponding models ($\bar{B}(\tau; 4, 4)$ or $\bar{B}(\tau; 5, 5)$) cannot be considered as plausible.

This discussion can be generalized if we assume that k is a real number. The assumption sounds biologically plausible because the natural sensors and actuators in the human body are likely to deal with the fractional derivatives. The generalization can go even further if we observe that, in general, the human sense of position is better than that of velocity, that of velocity is better than that of acceleration, and so on. It would be then reasonable to introduce a continuous, monotonically decreasing strength function and associate it with the positional derivatives. The development of a macroscopic model of reaching movements in these settings would require an essential reformulation of the model considered in this section and remains the subject of future research.

3 Reaching Movements in Dynamic Environments

In this section we consider reaching movements in a dynamic environment that is modeled as a multi-mass flexible object [41]. First, we construct a mathematical model of the object and formulate a reaching task. Then we proceed to an analysis of two models for motion planning, the lowest order polynomial model and the minimum hand jerk model.

3.1 Multi-Mass Flexible Object

Consider a chain system of $n + 1$ masses connected by n springs as shown in Fig. 2. Let $x_h \triangleq x_0$ be the hand coordinate. Assume that the first mass in the system is that of the hand, $m_h = m_0$, and the object is composed of the remaining n masses. The hand dynamics is given as

$$m_0\ddot{x}_0 + k_1(x_0 - x_1) = F, \tag{30}$$

where F is the driving force, while the dynamic equations for the object read

$$m_i\ddot{x}_i + k_i(x_i - x_{i-1}) + k_{i+1}(x_i - x_{i+1}) = 0, \tag{31}$$
$$m_n\ddot{x}_n + k_n(x_n - x_{n-1}) = 0, \tag{32}$$

where $i = 1, \ldots, n-1$.

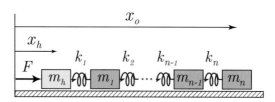

Fig. 2. Multi-mass object.

The system described by equations (30) and (31,32) can be considered as force-actuated. However, in what follows we will assume that the human arm, controlled by the CNS, is an ideal driver of the hand. Under such an assumption the inertial properties of the hand do not need be not taken into consideration. Thus, the hand dynamics (30) can be ignored, the hand position can be treated as the control input, and the object dynamics (31,32) can be considered as kinematically actuated. For the kinematically actuated system the hand dynamics represented by equation (30) does not play an independent role in finding optimal solutions. It can be used later on, for example, for the estimation of the driving force when the optimal solution is found. Note, however, that equation (30) plays a role in setting up the boundary conditions.

Let m_o be the total mass of the flexible object. Assume that the masses $m_i = m_o/n$, $i = 1, \ldots, n$. Also assume that the stiffness coefficients are equal and assign the stiffness distribution from the condition that the n-mass system is statically equivalent[5] to a virtual one mass flexible system (of mass m_o and stiffness k_o), which gives $k_i = k_o n$, $i = 1, \ldots, n$. For the symmetric mass and stiffness distribution the object dynamics (31,32) can be rearranged as

$$x_{i-1} = \frac{1}{(n\omega)^2}\ddot{x}_i + 2x_i - x_{i+1}, \quad i = 1, \ldots, n-1, \tag{33}$$

$$x_{n-1} = \frac{1}{(n\omega)^2}\ddot{x}_n + x_n, \tag{34}$$

where $\omega = \sqrt{k_o/m_o}$.

---

[5] The equivalence is defined in the following sense: pre-loading by the same static force causes equal end-point displacements for the n-mass system under consideration and for the virtual one mass flexible system.

3.2 Eigen Values

Consider the eigen-value problem for the system (33,34) with $x_h = 0$ (clamped-free vibration mode). For the n-mass object the characteristic equation is defined as $D_n = 0$, where

$$D_n = \begin{vmatrix} 2+z & -1 & 0 & \cdots & 0 & 0 & 0 \\ -1 & 2+z & -1 & \cdots & 0 & 0 & 0 \\ \vdots & \vdots & \vdots & \ddots & \vdots & \vdots & \vdots \\ 0 & 0 & 0 & \cdots & -1 & 2+z & -1 \\ 0 & 0 & 0 & \cdots & 0 & -1 & 1+z \end{vmatrix}, \tag{35}$$

and

$$z = (\lambda/n\omega)^2. \tag{36}$$

The characteristic determinant D_n can be unfolded by the following recurrence relations

$$D_k = (2+z)D_{k-1} - D_{k-2}, \quad k = 2, \ldots, n, \tag{37}$$

with the initial conditions

$$D_1 = (1+z), \quad D_0 = 1. \tag{38}$$

Using these relations and the principle of mathematical induction, it is easy to prove that

$$D_n = \sum_{s=0}^{n} C_{n+s}^{n-s} z^s, \tag{39}$$

where $C_p^q = p!/(q!(p-q)!)$ denotes the binomial coefficient.

As the system under consideration is conservative, the eigen-values are purely imaginary: $\lambda_k = \pm i p_k$, where p_k, $k = 1, \ldots, n$, are the natural frequencies of the object. To find them, replace in (36) λ^2 by $-p^2$. Then, if one defines

$$\cos\theta = 1 - \frac{1}{2}\left(\frac{p}{n\omega}\right)^2, \tag{40}$$

the characteristic determinants D_k turn into the Chebyshev polynomials of the 3rd kind for which the general expression is known [42]:

$$D_n = \frac{\cos(n+\tfrac{1}{2})\theta}{\cos\tfrac{1}{2}\theta}. \tag{41}$$

Solving $D_n = 0$, one obtains from (40) the natural frequencies

$$p_k = 2\omega n \cos\frac{(n-k+1)\pi}{2n+1}, \quad k = 1, 2, \ldots, n. \tag{42}$$

Notice that in the limiting case of $n \to \infty$, $p_k = \omega\frac{2k-1}{2}\pi$, as in the clamped-free rod with distributed stiffness.

3.3 Reaching Task and the Boundary Conditions

Assume that a human subject is requested to make a reaching movement of length L and time T and stop the hand and all the masses of the object without excitation of oscillations. In other words, the subject is requested to generate a rest-to-rest motion command that eliminates residual vibrations. Without loss of generality we assume that the subject transports the masses from the initial state

$$x_i(0) = 0, \ \dot{x}_i(0) = 0, \quad i = 1, \ldots, n, \tag{43}$$

to the final state

$$x_i(T) = L, \ \dot{x}_i(T) = 0, \quad i = 1, \ldots, n. \tag{44}$$

Then, assuming $F(0) = F(T) = 0$ in (30), one defines the following boundary conditions for the hand

$$x_h(0) = 0, \quad \dot{x}_h(0) = 0, \quad \ddot{x}_h(0) = 0, \tag{45}$$
$$x_h(T) = L, \quad \dot{x}_h(T) = 0, \quad \ddot{x}_h(T) = 0. \tag{46}$$

In what follows, we will represent the optimization problems in terms of the object end position $x_o \triangleq x_n$ that is the coordinate of the most distal mass. For this purpose we need to formulate the boundary conditions for x_o which can be obtained from the boundary conditions (43,44) and (45,45). Differentiating equations (33,34) sequentially, i-th equation $2i$ times, and considering them at $t = 0$ and $t = T$ one obtains $x_i^{(2)}(0) = x_i^{(3)}(0) = \ldots = x_i^{(2i+2)}(0) = 0$, and $x_i^{(2)}(T) = x_i^{(3)}(T) = \ldots = x_i^{(2i+2)}(T) = 0$. For $i = n$ one thus obtains the following boundary conditions for the object end position:

$$x_o(0) = 0, \ \dot{x}_o(0) = \ldots = x_o^{(2n+2)}(0) = 0, \tag{47}$$
$$x_o(T) = L, \ \dot{x}_o(T) = \ldots = x_o^{(2n+2)}(T) = 0. \tag{48}$$

3.4 Minimum Order Polynomial Model

One way to plan the reaching movements under consideration is to specify the motion of the most distal link $x_o(t)$ in the form of the lowest order polynomial satisfying the boundary conditions (47,48). This approach is not uncommon in the control literature. The examples can be found in [22, 43, 44]. Recently, it was successfully used for the modeling of reaching movements with a one-mass flexible object [23].

For the boundary conditions (47,48) the minimal order of the polynomial $x_o(t)$ is $2n + 3$. As shown in Section 2, the solution is given by the regularized incomplete Beta function. For the dimensions under consideration we have

$$x_o(t) = L\bar{B}(\tau; 2n+3, 2n+3), \tag{49}$$

$$v_o(t) = \frac{L}{T} \frac{\{\tau(1-\tau)\}^{2n+2}}{B(2n+3, 2n+3)}, \tag{50}$$

where $\tau = t/T$. Note that for a given number of masses n the normalized object velocity $\bar{v}_h = v_h/(L/T)$ does not depend on the object mass and stiffness and is defined exclusively by the non-dimensional movement time τ.

Having defined the object trajectory, one can proceed to the hand trajectory. The hand position x_h can be expressed in terms of the object coordinate x_o and its derivatives. Solving equations (33,34) recurrently for $i = n-1, n-2, \ldots, 0$, one obtains

$$x_h(t) = \sum_{l=0}^{n} \frac{C_{n+l}^{n-l}}{(n\omega)^{2l}} x_o^{(2l)}(t). \tag{51}$$

The hand velocity is therefore defined as

$$v_h(t) = \frac{L}{T} \frac{1}{B(2n+3, 2n+3)} \sum_{l=0}^{n} \frac{C_{n+l}^{n-l}}{(n\omega T)^{2l}} \frac{d^{(2l)}}{d\tau^{(2l)}} \{\tau(1-\tau)\}^{2n+2}. \tag{52}$$

As can be shown, the resulting solution $x_h(t)$ keeps the main properties of the regularized incomplete Beta function. In particular, the even derivatives are antisymmetric and the odd derivatives are symmetric with respect to the middle point $t = T/2$. Also, the position $x_h(T/2) = L/2$, and the even derivatives $x_h^{(2k)}(T/2) = 0$ for $k = 1, 2, 3\ldots$

For a given number of masses n, the normalized hand velocity $\bar{v}_h = v_h/(L/T)$ is a function of two parameters: the normalized movement time $\tau \in [0,1]$ and $\kappa = \omega T$. The latter parameter can be thought of as the non-dimensional reaching time or the non-dimensional frequency. It can be also interpreted as the non-dimensional stiffness of the reaching movement. Qualitatively, changing T has the same effect as changing ω. The hand trajectories with the same ωT have the same normalized profiles and, therefore, can be called isochronous. For the isochronous trajectories the increase of T is equivalent to the decrease of ω and vice versa.

An interesting feature of the hand velocity profiles in the minimum order polynomial model is the change of phases. Depending on the number of masses n they may have up to $n+1$ phases. This is illustrated in Figures 3-5, where we show the normalized velocities $\bar{v}_h$ and $\bar{v}_o$ as function of κ and τ. The maximal number of phases is attained for relatively small values of the reaching stiffness κ. As κ increases the number of phases declines. As follows from (52), the difference between the hand and the object velocities becomes smaller as $\kappa \to \infty$, and in the limit both these velocities are defined by (50).

It should be noted that from the engineering point of view there is nothing wrong in the minimum order polynomial model. It makes sense and is feasible. However, when considering it in the context of human movements,

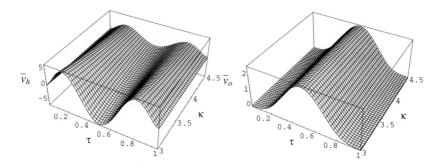

Fig. 3. Normalized hand (left) and object (right) velocity profiles for $n = 1$

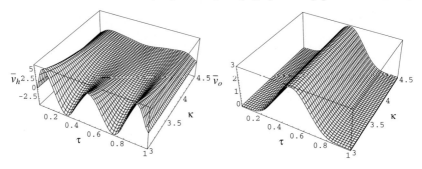

Fig. 4. Normalized hand (left) and object (right) velocity profiles for $n = 2$

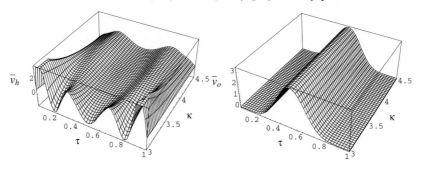

Fig. 5. Normalized hand (left) and object (right) velocity profiles for $n = 3$

one can see the following drawbacks. First, the fact that in the limiting case of the minimum order polynomial model the hand trajectory is given by the regularized incomplete Beta function $\bar{B}(\tau; 2n+3, 2n+3)$ contradicts to the minimum hand jerk hypothesis, which offers $\bar{B}(\tau; 3, 3)$ for the absolutely rigid object. Next, the criterion of optimality associated with the solution (49) is defined in the task space of object coordinates and has the following form:

$$\mathcal{J} = \frac{1}{2} \int_0^T \left(\frac{d^{(2n+3)} x_o}{dt^{(2n+3)}} \right)^2 dt. \tag{53}$$

This criterion is not invariant with respect to the number of masses n. For one-mass object it is defined through the 5-th derivative of the object position[6], for two-mass object through the 7-th, for three-mass object through the 9-th, and so on. It cannot be well-defined for the object with distributed mass and stiffness ($n \to \infty$). In the latter case the object velocity is defined by the Dirac delta-function $v_o(t) = L\delta(T/2)$, which does not seem plausible for human movements. To illustrate the general behavior of the prediction obtained by criterion (53), we plot the object and hand velocity profiles in Fig. 6 for $n = 1, 2, \ldots, 15$. Here we set $L = 0.2$ m, $T = 2.5$ s, $m_o = 3$ kg, $k_o = 10$ N/m.

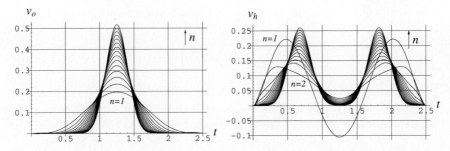

Fig. 6. Object (left) and hand (right) velocity profiles predicted by criterion (53) for $n = 1, 2, \ldots, 15$.

3.5 Minimum Hand Jerk Model

Consider now a minimum hand jerk model where the criterion of optimality is defined as

$$\mathcal{J} = \frac{1}{2} \int_0^T \left(\frac{d^3 x_h}{dt^3} \right)^2 dt. \tag{54}$$

To find the optimal solution corresponding to this criterion, one can solve the minimum norm control problem [33] for the dynamic system (31,32) with the

---

[6] It was named the minimum object crackle criterion in [23].

control $u = \ddot{x}_h$. However, due to the specific structure of the flexible object (linear chain) the use of the calculus of variations is the shortest way to solve the optimization problem.

Differentiating the hand position (51) three times and substituting the result into (54), one can represent the minimum hand jerk criterion in the following form:

$$J = \frac{1}{2} \int_0^T \left(\sum_{l=0}^n \frac{C_{n+l}^{n-l}}{(n\omega)^{2l}} x_o^{(2l+3)} \right)^2 dt. \tag{55}$$

Denote by $\mathcal{L}$ the integrand of the criterion (55). The object trajectory $x_o(t)$ minimizing criterion (55) under the boundary conditions (47,48) must satisfy the Euler-Lagrange equation

$$\sum_{k=0}^{2n+3} (-1)^k \frac{d^{(k)}}{dt^{(k)}} \left\{ \frac{\partial \mathcal{L}}{\partial x_o^{(k)}} \right\} = 0, \tag{56}$$

which for the given structure of $\mathcal{L}$ reduces to the following linear differential equation

$$\sum_{s=0}^n \sum_{l=0}^n \frac{C_{n+s}^{n-s} C_{n+l}^{n-l}}{(n\omega)^{2s} (n\omega)^{2l}} x_o^{(2\{l+s+3\})} = 0. \tag{57}$$

It is easy to show that the characteristic equation corresponding to (57) can be represented as

$$\lambda^6 \left(\sum_{s=0}^n \frac{C_{n+s}^{n-s}}{(n\omega)^{2s}} \lambda^{2s} \right)^2 = 0. \tag{58}$$

This equation has 6 zero roots and, as follows from (39), $2n$ pairs of imaginary roots $\lambda = \pm \imath p_s$, where p_s, $s = 1, 2, \ldots, n$, are the natural frequencies of the object defined by (42). The optimal object trajectory is therefore defined as

$$x_o(t) = \sum_{i=0}^5 \alpha_i t^i + \sum_{i=1}^n (\beta_{1i} + \beta_{2i} t) \sin p_i t + (\beta_{3i} + \beta_{4i} t) \cos p_i t, \tag{59}$$

where α_i and β_{ji} are the constant coefficients. For any fixed n, $T > 0$ and $\omega \neq 0$ these coefficients are defined uniquely from the boundary conditions (47,48). This follows from the fact that in the equivalent minimum-norm optimal control problem the dynamic system (31,32) is controllable unless $\omega = 0$, and this implies the uniqueness of the optimal solution [45].

To find the solution for the hand position, we first establish the following expression for the even derivatives of the object position:

$$x_o^{(2l)}(t) = \sum_{i=0}^5 \alpha_i A_i^{2l} t^{i-2l} + (-1)^l 2l \sum_{i=1}^n p_i^{2l-1} (\beta_{4i} \sin p_i t - \beta_{2i} \cos p_i t)$$

$$+ (-1)^l \sum_{i=1}^n p_i^{2l} \{(\beta_{1i} + \beta_{2i} t) \sin p_i t + (\beta_{3i} + \beta_{4i} t) \cos p_i t\}, \tag{60}$$

where $A_i^k = i!/(i-k)!$ are the permutation numbers. Next, taking into account that

$$\sum_{l=0}^{n}(-1)^l \frac{C_{n+l}^{n-l}}{(n\omega)^{2l}} p_i^{2l} = 0, \tag{61}$$

which follows from the system characteristic equation (39), one obtains

$$\begin{aligned} x_h(t) &= \sum_{i=0}^{5}\sum_{l=0}^{n} \frac{C_{n+l}^{n-l}}{(n\omega)^{2l}} \alpha_i A_i^{2l} t^{i-2l} \\ &+ \sum_{i=1}^{n}\sum_{l=0}^{n} \frac{C_{n+l}^{n-l}}{(n\omega)^{2l}} (-1)^l 2l p_i^{2l-1} (\beta_{4i} \sin p_i t - \beta_{2i} \cos p_i t) \\ &\triangleq \sum_{i=0}^{5} \tilde{\alpha}_i t^i + \sum_{i=1}^{n} \tilde{\beta}_{1i} \sin p_i t + \tilde{\beta}_{2i} \cos p_i t. \end{aligned} \tag{62}$$

Thus, the optimal hand trajectory is composed of a 5-th order polynomial and trigonometric terms. Note that in (62) there are no secular terms that are featured in the optimal object trajectory (59).

As can be shown, the solution (62) possesses the property $x_h(t) = x_h(T) - x_h(T-t)$. Therefore, one concludes that the even derivatives are antisymmetric and the odd derivatives are symmetric with respect to the middle point $t = T/2$. Also, the position $x_h(T/2) = L/2$, and the even derivatives $x_h^{(2k)}(T/2) = 0$ for $k = 1, 2, 3 \ldots$ It can also be shown that exactly the same properties hold for the object trajectory $x_o(t)$.

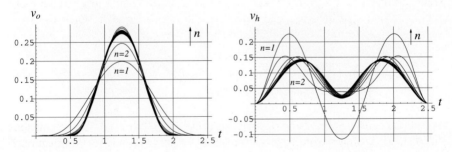

Fig. 7. Object (left) and hand (right) velocity profiles predicted by criterion (55) for $n = 1, 2, \ldots, 15$.

To illustrate the general behavior of the prediction obtained by criterion (55), we conduct a simulation, solving the boundary value problem (47,48,57) for $n = 1, 2, \ldots, 15$ and $L = 0.2$m, $T = 2.5$s, $m_o = 3$kg, $k_o = 10$N/m. The results are shown in Fig. 7 where we plot the object and hand velocity profiles As can be seen, the trajectories predicted by criterion (55) appear to converge to single profiles.

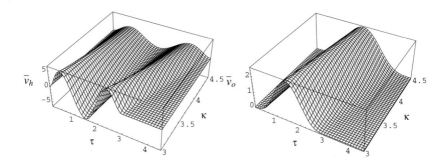

Fig. 8. Normalized hand (left) and object (right) velocity profiles for $n = 1$

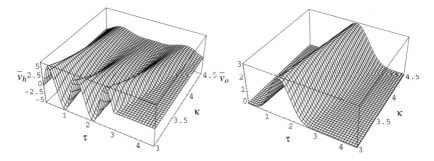

Fig. 9. Normalized hand (left) and object (right) velocity profiles for $n = 2$

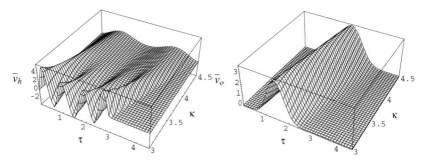

Fig. 10. Normalized hand (left) and object (right) velocity profiles for $n = 3$

For a given number of masses n, the normalized hand velocity $\bar{v}_h = v_h/(L/T)$ and the normalized object velocity $\bar{v}_h = v_h/(L/T)$ are characterized by the normalized frequencies $p_i/\omega, i = 1, \ldots, n$ (constant numbers) and two parameters: the non-dimensional stiffness $\kappa = \omega T$ and the non-dimensional movement time $\tau = \omega t$. Note that τ differs from the non-dimensional movement time in the lowest order polynomial model, while κ is defined exactly as in Section 3.4.

Similar to what has been established for the lowest order polynomial model, the hand velocity profiles in the minimum hand jerk model may have up to $n+1$ phases. The same holds true for the object velocity profiles. However, the phase transitions in the object velocity profiles show up for relatively large values of κ and are manifested weakly. In contrast, the maximal number of phases in the hand velocity profiles is attained for relatively small values of the reaching stiffness κ. This is illustrated in Figures 8-10, where we plot the normalized velocities $\bar{v}_o$ and $\bar{v}_h$ as function of κ and τ. As κ is increasing the number of phases in the hand velocity profiles is declining. Note that the jerk component in (55) becomes dominant as $\omega \to \infty$. As follows from (51), the difference between the hand and the object trajectories becomes smaller as $\kappa \to \infty$, and in the limit both these velocities are defined by the classical minimum jerk solution $\bar{B}(\tau; 3, 3)$.

Finally, we would like to note that if instead of the minimum hand jerk criterion (54) one employs a more general criterion

$$\mathcal{J} = \frac{1}{2} \int_0^T \left(\frac{d^m x_h}{dt^m} \right)^2 dt \tag{63}$$

the structure of the optimal solution for the reaching movement undergoes changes only on the polynomial part. More specifically, it can be shown that in this case

$$x_o(t) = \sum_{i=0}^{2m-1} \alpha_i t^i + \sum_{i=1}^{n} (\beta_{1i} + \beta_{2i} t) \sin p_i t + (\beta_{3i} + \beta_{4i} t) \cos p_i t, \tag{64}$$

$$x_h(t) = \sum_{i=0}^{2m-1} \tilde{\alpha}_i t^i + \sum_{i=1}^{n} \tilde{\beta}_{1i} \sin p_i t + \tilde{\beta}_{2i} \cos p_i t. \tag{65}$$

We do not conduct the formal analysis of the generalized solution because we think that setting $m = 3$ is enough for the generation of human-like reaching movements. To confirm this point, we turn to the experiments.

4 Experimental Results

To check the velocity profiles of reaching movements with multi-mass objects, we conducted an experiment. In the experimental setup, shown in Fig. 11, a haptic device (PHANToM Premium 151A/3DOF, maximum exertable force 8.5N) is connected to computer (dual core CPU, Intel Pentium 4, 3.0 GHz) through PCI interface.

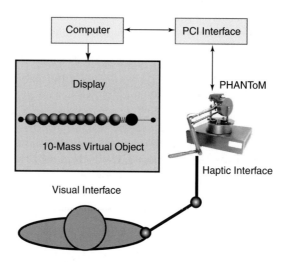

Fig. 11. Experimental setup.

Five right-handed naïve subjects (males, aged between 25 and 35 years old) participated in the experiments. The subjects were instructed to move a multi-mass virtual flexible object, with the 1st mass (shown in black color in Fig. 11) "connected" to human hand by haptic feedback generated by the PHANToM motors. The hand & object system was at rest at the start point. Before starting the movement the subject positioned the PHANToM pointer to the 1st mass and "connected" it to the hand by pressing a button on the computer keyboard. The subjects were requested to move the flexible object and stop the hand and all the masses at a target point. The subjects made these rest-to-rest movements along a line (in the direction from left to right) in the horizontal plane using the PHANToM stylus. The one-dimensional movements were implemented in software by setting a higher constraint force in the direction orthogonal to the movement line. The positions of the hand and the object were displayed on the computer monitor, providing the subject with real-time visual feedback. The object dynamics were simulated in the computer (4th-order Runge-Kutta method with constant step $h = 0.001$s) and real-time haptic feedback was supplied to the subject through the PHANToM stylus. The hand position and velocity were measured by the PHANToM hardware.

As the reaching movements under consideration are quite unusual and different from what we experience in daily life, the experiment was conducted in three days. On the preliminary day we conducted a general evaluation of the subjects' performance. The subjects familiarized with the experimental setup, comprehended the reaching task, performed movement trials and learned the unusual dynamic environment. The subjects were requested to produce reaching movements in a natural way, on their own pace, trading off the speed and the comfortability. A trial was considered to be successful if the task was completed within certain position, and velocity tolerances. The subject was given an audio feedback, generated by the computer, if a trial was successful. No data were recorded during the preliminary evaluation as the main purpose was to select such parameters of the movement task that would guarantee an acceptable rate of the successful trials and facilitate the learning process.

In the course of the preliminary evaluation, we selected the parameters of the object (the total mass $m_o = 3$kg, stiffness $k_o = 10$N/m, and the number of masses $n = 10$). For the experimental analysis we selected the reaching task with the traveling distance $L = 0.2$m, the movement time $T = 2.35$s, and the time tolerance $\Delta T = \pm 0.4$s. For each mass of the object we selected the following position, velocity, and acceleration tolerances to be satisfied at the start and end points: $\Delta x = \pm 0.012$m, $\Delta v = \pm 0.024$m/s, $\Delta a = \pm 0.16$m/s^2. For the hand, the start and end point position and velocity tolerances were set as $\Delta x = \pm 0.012$m and $\Delta v = \pm 0.05$m/s. A movement trial now was defined as successful if the subject was able to complete the reaching task within the above-specified tolerances.

Table 1. Progress in motor training (success rate)

Subject	Day 1 (600 trials)	Day 2 (600 trials)	Total (1200 trials)
Subject 1	54 (9.00%)	68 (11.33%)	122 (10.17%)
Subject 2	68 (11.33%)	94 (15.67%)	162 (13.50%)
Subject 3	84 (14.00%)	89 (14.83%)	173 (14.41%)
Subject 4	80 (13.33%)	96 (16.00%)	176 (14.67%)
Subject 5	100 (16.67%)	119 (19.83%)	219 (18.25%)

Having selected the parameters of the reaching task, we proceeded further and conducted experiments in two days. On the 1st day the subjects performed the reaching task and completed 600 trials. The subject was given an audio feedback, generated by the computer, if a trial was successful. The experiment was conducted in two blocks. Upon completing 300 trials the subjects rested for about 30 minutes. The overall success rate achieved on the 1st day of the experiment is shown in Table 1. On the 2nd day (recording phase) the experiment was repeated in the similar manner. The data regarding the position and the velocity of the hand and those of the simulated masses were collected for analysis. These data were recorded at 100 Hz.

Fig. 12. Learning history of subject 1.

Fig. 13. Learning history of subject 2.

Fig. 14. Learning history of subject 3.

Fig. 15. Learning history of subject 4.

Fig. 16. Learning history of subject 5.

As can be seen from Table 1, in the 2nd day of the experiment the success rate slightly increased for all the subjects in all the reaching tasks. This confirms the recovery and the increase of the motor memory. Partially it is reflected in the history of learning presented in Fig. 12-16. The resulting success rate is still far from perfect after two days of practicing. The relatively low success rate can be explained by the complexity of the reaching task which features a sport-like movement. The behavior of the flexible object in this movement is non-trivial, which requires from the subject a good coordination skill. It was observed that in the successful trials the subjects produced the following movement strategy. In the beginning, the positions of all the masses coincide at the start point. During the first half of the movement the first (driving) mass, shown as black sphere in Fig. 11, is ahead of the last (most distal) mass, with the distance between the first and last masses being about 5cm. During the second half of the movement the configuration is reversed symmetrically. The driving mass becomes behind the last mass, and finally all the masses reach the target point. Clearly, more training is necessary to achieve a higher success rate for such a complex movement. However, the experimental data of the successful trials can be taken for the comparison with the theoretical predictions.

Table 2. Movement times

Subject	Average	Minimum	Maximum	RMS
Subject 1	2.258s	1.97s	2.72s	0.215s
Subject 2	2.280s	2.01s	2.75s	0.206s
Subject 3	2.261s	1.96s	2.74s	0.213s
Subject 4	2.260s	1.96s	2.75s	0.225s
Subject 5	2.237s	1.96s	2.75s	0.189s

The average, minimum and maximum times shown by the subjects in the successful trials are summarized in Table 2. Also listed in Table 2 are the root-mean-squares (RMS) of the movement times, characterizing the standard deviations from the averages. Examination of the hand and object velocity profiles demonstrated that for all the subjects the experimental data were in favor of the minimum hand jerk model. Qualitatively, the resulting velocity patterns for all the subjects were similar. A quantitative measure for the comparisons was represented by the integrated RMS of the velocity errors,

$$\varepsilon = \sqrt{\frac{1}{N}\sum_{i=1}^{N}\{v_{pred}(t_i) - v_{exp}(t_i)\}^2}, \qquad (66)$$

over the trajectories between the theoretical predictions (by criteria (53) and (55)) and the experimental data. Here, N is the number of sampled data in one trial. The integrated RMS of the velocity errors, calculated for all the successful trials for all the subjects, are summarized in Table 3. It should be noted that, overall, the worst matching (maximal RMS error) by criterion (55) is better than the best matching (minimal RMS error) by (53).

Table 3. Matching of experimental data by integrated RMS

Velocity prediction	Average	Minimum	Maximum
Subject 1			
$v_h(t)$ by (55)	0.0265 m/s	0.0209 m/s	0.0293 m/s
$v_o(t)$ by (55)	0.0269 m/s	0.0161 m/s	0.0391 m/s
$v_h(t)$ by (53)	0.0508 m/s	0.0456 m/s	0.0565 m/s
$v_o(t)$ by (53)	0.0655 m/s	0.0562 m/s	0.0760 m/s
Subject 2			
$v_h(t)$ by (55)	0.0273 m/s	0.0212 m/s	0.0313 m/s
$v_o(t)$ by (55)	0.0360 m/s	0.0259 m/s	0.0442 m/s
$v_h(t)$ by (53)	0.0556 m/s	0.0506 m/s	0.0605 m/s
$v_o(t)$ by (53)	0.0764 m/s	0.0663 m/s	0.0852 m/s
Subject 3			
$v_h(t)$ by (55)	0.0251 m/s	0.0136 m/s	0.0292 m/s
$v_o(t)$ by (55)	0.0304 m/s	0.0170 m/s	0.0485 m/s
$v_h(t)$ by (53)	0.0522 m/s	0.0402 m/s	0.0586 m/s
$v_o(t)$ by (53)	0.0705 m/s	0.0579 m/s	0.0777 m/s
Subject 4			
$v_h(t)$ by (55)	0.0258 m/s	0.0140 m/s	0.0322 m/s
$v_o(t)$ by (55)	0.0293 m/s	0.0199 m/s	0.0413 m/s
$v_h(t)$ by (53)	0.0519 m/s	0.0411 m/s	0.0567 m/s
$v_o(t)$ by (53)	0.0693 m/s	0.0604 m/s	0.0808 m/s
Subject 5			
$v_h(t)$ by (55)	0.0251 m/s	0.0155 m/s	0.0289 m/s
$v_o(t)$ by (55)	0.0304 m/s	0.0168 m/s	0.0425 m/s
$v_h(t)$ by (53)	0.0511 m/s	0.0392 m/s	0.0576 m/s
$v_o(t)$ by (53)	0.0694 m/s	0.0539 m/s	0.0822 m/s

For the graphical illustration of the velocity profiles, we take the data of last 15 successful trials of each subject and compare them with the theoretical predictions. To compare motions of different durations, the velocity profiles

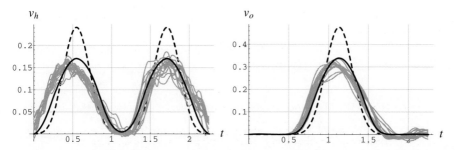

Fig. 17. Hand (left) and object (right) velocity profiles, predicted by criteria (55) (solid line) and (53) (dashed line), in comparison with experimental data (gray lines) of the subject 1.

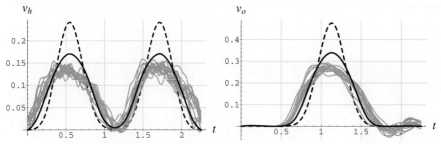

Fig. 18. Hand (left) and object (right) velocity profiles, predicted by criteria (55) (solid line) and (53) (dashed line), in comparison with experimental data (gray lines) of the subject 2.

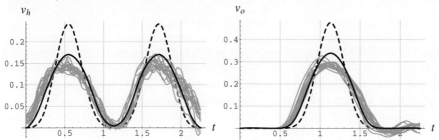

Fig. 19. Hand (left) and object (right) velocity profiles, predicted by criteria (55) (solid line) and (53) (dashed line), in comparison with experimental data (gray lines) of the subject 3.

are time scaled using the average time of successful trials, $T = 2.259$s. The experimental velocity profiles are shown in Fig. 17-21 in gray color. The hand and object velocity profiles, predicted by the minimum hand jerk criterion, are shown in solid lines in Fig. 17-21. The predictions by criterion (53), are shown there in dashed lines. As can be seen from Fig. 17-21, the collected experimental data are clearly in favor of the minimum hand jerk criterion.

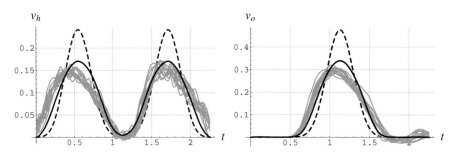

Fig. 20. Hand (left) and object (right) velocity profiles, predicted by criteria (55) (solid line) and (53) (dashed line), in comparison with experimental data (gray lines) of the subject 4.

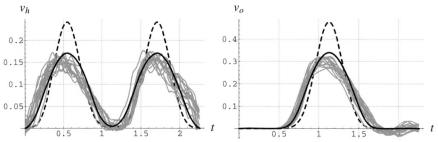

Fig. 21. Hand (left) and object (right) velocity profiles, predicted by criteria (55) (solid line) and (53) (dashed line), in comparison with experimental data (gray lines) of the subject 5.

While our initial experiments support the minimum hand jerk model, one should realize that its applicability is apparently limited to some range of the parameter T that can be associated with comfortable movements. It is known that for comfortable reaching movements in free space the movement time is in the range of $[0.25, 0.8]$s for the traveling distance varying from 0.2m to 0.5m [46]. For the movement in dynamic environments, the time range of comfortable reaching is of course environment-dependent. In this connection, it should be mentioned that the fastest completion time in our task (1.96s) was very close to the lower bound admissible by the time tolerance. It suggests that it is perhaps possible to produce faster reaching movements in the conditions of our experiment. However, with shorter T one would expect a slower learning rate as the human control system would get closer to the limits of neuromuscular performance. Obviously, the movement time T cannot be reduced arbitrarily without violating the constraints imposed by the neural and musculoskeletal hardware on the traveling distance, maximal force, and the accuracy of reaching. These constraints are ignored in the optimization problems (54,45,46) and (47,48,55) under the assumption that for comfortable reaching movements the control system operates without undue stress, away from the limits of neuromuscular performance. Taking into account additional considerations would lead to different optimization problems with possibly different criteria of optimality [47].

5 Conclusions

An analysis of reaching movements in the manipulation of multi-mass flexible objects has been undertaken in this paper. Two candidates for the modeling of human-like reaching movements, the lowest-order polynomial model and the minimum hand jerk model, have been analyzed. The models under study have been established in the analytical form for the general case of a linear chain of n masses. Both these models predict qualitatively similar hand velocity profiles featuring at most $n+1$ phases. Qualitatively, however, the velocity profiles are different. The difference becomes sharper with the increase of the number of masses, and in the limiting case of infinite-dimensional object the lowest-order polynomial model does not produce bounded velocity profiles. Finally, theoretical predictions by the models under considerations for a tenmass flexible object have been tested against experimental data obtained using a virtual reality-based setup. It has been demonstrated that the prediction by the minimum hand jerk criterion matches the experimental patterns with reasonable accuracy.

References

1. Morasso P (1981) Spatial control of arm movements. Experimental Brain Research 42:223–227
2. Abend W, Bizzi E, Morasso P (1982) Human arm trajectory formation. Brain 105:331–348
3. Bernstein N (1967) The Coordination and Regulation of Movements. Pergamon, London
4. Arimoto S, Sekimoto M, Hashiguchi H, Ozawa R (2005) Natural resolution of ill-posedness of inverse kinematics for redundant robots: A challenge to Bernstein's degrees-of-freedom problem. Advanced Robotics 19(4):401–434
5. Tsuji T, Tanaka Y, Morasso P, Sanguineti V, Kaneko M (2002) Bio-mimetic trajectory generation of robots via artificial potential field with time base generator. IEEE Transactions on Systems, Man, and Cybernetics—Part C: Applications and Reviews 32(4):426–439
6. Kawato M (1999) Internal models for motor control and trajectory planning. Current Opinion in Neurobiology 9:718–727
7. Flash T, Hogan N, Richardson M (2003) Optimization principles in motor control. In Arbib M, ed.: The Handbook of Brain Theory and Neural Networks. 2nd edn. MIT Press, Cambridge, Massachusetts 827–831
8. Nagasaki H (1989) Asymmetric velocity profiles and acceleration profiles of human arm movements. Experimental Brain Research 74:319–326
9. Milner T, Ijaz M (1990) The effect of accuracy constraints on three-dimensional movement kinematics. Neuroscience 35(2):365–374

10. Wiegner A, Wierbicka M (1992) Kinematic models and elbow flexion movements: Quantitative analysis. Experimental Brain Research 88:665–673
11. Flash T, Hogan N (1985) The coordination of arm movements: An experimentally confirmed mathematical model. The Journal of Neuroscience 5(7):1688–1703
12. Milner T (1992) A model for the generation of movements requiring endpoint precision. Neuroscience 49(2):487–496
13. Berthier N (1996) Learning to reach: A mathematical model. Developmental Psychology 32(5):811–823
14. Arimoto S, Hashiguchi H, Sekimoto M, Ozawa R (2005) Generation of natural motions for redundant multi-joint systems: A differential-geometric approach based upon the principle of least actions. Journal of Robotic Systems 22(11):583–605
15. Arimoto S, Sekimoto M (2006) Human-like movements of robotic arms with redundant dofs: Virtual spring-damper hypothesis to tackle the bernstein problem. In: Proc. IEEE Int. Conf. on Robotics and Automation, Orlando, USA 1860–1866
16. Uno Y, Kawato M, Suzuki R (1989) Formation and control of optimal trajectory in human multijoint arm movement. minimum torque-change model. Biological Cybernetics 61:89–101
17. Aspinwall D (1990) Acceleration profiles for minimizing residual response. ASME Journal of Dynamic Systems, Measurement, and Control 102(1):3–6
18. Meckl P, Seering W (1990) Experimental evaluation of shaped inputs to reduce vibration for a cartesian robot. ASME Journal of Dynamic Systems, Measurement, and Control 112(1):159–165
19. Karnopp B, Fisher F, Yoon B (1992) A strategy for moving a mass from one point to another. Journal of the Franklin Institute 329(5):881–892
20. Meckl P, Kinceler R (1994) Robust motion control of flexible systems using feedforward forcing functions. IEEE Transactions on Control Systems Technology 2(3):245–254
21. Chan T, Stelson K (1995) Point-to-point motion commands that eliminate residual vibration. In: Proc. American Control Conference. Volume 1., Seattle, Washington 909–913
22. Piazzi A, Visioli A (2000) Minimum-time system-inversion-based motion planning for residual vibration reduction. IEEE/ASME Transactions on Mechatronics 5(1):12–22
23. Dingwell J, Mah C, Mussa-Ivaldi F (2004) Experimentally confirmed mathematical model for human control of a non-rigid object. Journal of Neurophysiology 91:1158–1170
24. Flash T (1983) Organizing Principles Underlying The Formation of Hand Trajectories. PhD thesis, Massachusetts Institute of Technology, Cambridge, USA
25. Edelman S, Flash T (1987) A model of handwriting. Biological Cybernetics 57:25–36
26. De Luca A, Oriolo G, Vendittelli M, Iannitti S (2004) Planning motions for robotic systems subject to differential constraints. In Siciliano B, De Luca A, Melchiorri C, Casalino G, eds.: Advances in Control of Articulated and Mobile Robots. Springer Verlag, Berlin 1–38
27. De Luca A (2000) Feedworward/feedback laws for the control of flexible robots. In: Proc. IEEE Int. Conf. on Robotics and Automation. Volume 1., San-Francisco, California 233–240

28. Boerlage M, Tousain R, Steinbuch M (2004) Jerk derivative feedforward control for motion systems. In: Proc. American Control Conference. Volume 3., Boston, Massachusetts 4843–4848
29. Lambrechts P, Boerlage M, Steinbuch M (2005) Trajectory planning and feedforward design for electromechanical systems. Control Engineering Practice 13(2):145–157
30. Todorov E, Jordan M (1998) Smoothness maximization along a predefined path accurately predicts the speed profiles of complex arm movements. Journal of Neurophysiology 80(2):696–714
31. Suzuki K, Uno Y (2000) Brain adopts the criterion of smoothness for most quick reaching movements. Transactions of the Institute of Electronics, Information and Communication Engineers J83-D-II(2):711–722 (In Japanese)
32. Richardson M, Flash T (2002) Comparing smooth arm movements with the two-thirds power law and the related segmented-control hypothesis. The Journal of Neuroscience 22(18):8201–8211
33. Luenberger D (1969) Optimization by Vector Space Methods. John Wiley & Sons, New York
34. Rose N, Bronson R (1969) On optimal terminal control. IEEE Transactions on Automatic Control 14(5):443–449
35. Abramowitz, M. Stegun I, ed. (1972) Handbook of Mathematical Functions With Formulas, Graphs and Mathematical Tables. 10th edn. U.S. Government Printing Office, Washington D.C.
36. Dutka J (1981) The incomplete Beta function—a historical profile. Archive for History of Exact Sciences 24:11–29
37. Samko S, Kilbas A, Marichev Y (1993) Integrals and Derivatives of Fractional Order and Their Applications. Gordon and Breach Science Publications
38. Gel'fand I, Fomin S (1963) Calculus of Variations. Prentice-Hall, Englewood Cliffs, New Jersey
39. Krylow A, Rymer W (1997) Role of intrinsic muscle properties in producing smooth movements. IEEE Trans. on Biomedical Engineering 44(2):165–176
40. Harris C (1998) On the optimal control of behaviour: a stochastic perspective. Journal of Neuroscience Methods 83:73–88
41. Svinin M, Concharenko I, Luo Z, Hosoe S (2006) Reaching movements in dynamic environments: How do we move flexible objects? IEEE Transactions on Robotics 22(4) (in press)
42. Mason J, Handscomb D (2003) Chebyshev Polynomials. Chapman & Hall / CRC Press, Boca Raton
43. Sira-Ramírez H, Agrawal S (2004) Differentially Flat Systems. Volume 17 of Control Engineering. Marcel Dekker, Inc., New York
44. Lévine J, Nguyen D (2003) Flat output characterization for linear systems using polynomial matrices. Systems & Control Letters 48(1):69–75
45. Lee E, Markus L (1967) Foundations of Optimal Control Theory. Wiley, New York
46. Kawato M (1995) Motor learning in the brain. Journal of the Robotics Society of Japan 13(1):11–19
47. Nelson W (1983) Physical principles for economies of skilled movements. Biological Cybernetics 46:135–147

Haptic Feedback Enhancement Through Adaptive Force Scaling: Theory and Experiment

Jaydeep Roy[1]*, Daniel L. Rothbaum[2]†, and Louis L. Whitcomb[3]‡

[1] GE Global Research
Materials Systems Technology
One Research Circle
K1-3A20, Niskayuna, NY 12309, USA
email: jaydeep.roy@ge.com

[2] Atlantic Ear, Nose, and Throat, P.A.
2705 Rebecca Lane, Suite A
Orange City, FL 32763, USA
d_rothbaum@yahoo.com

[3] Johns Hopkins University
Department of Mechanical Engineering
3400 N. Charles Street, 123 Latrobe Hall
Baltimore, MD 21218, USA
email: llw@jhu.edu

Summary. We report the development, implementation, and evaluation of a novel application of robot force control called position based force scaling. Force scaling employs position based force control algorithms to augment human haptic feedback during human-robot co-manipulation tasks. We report the first results in the implementation of asymptotically exact force scaling, as well as the first results in on-line environment compliance adaptation while performing the task of force scaling. One previously reported and two new (one adaptive and one non-adaptive) force scaling algorithms are reviewed. Comparative experiments are reported which quantify the performance of all three algorithms. The data show that the new adaptive force scaling algorithm significantly outperforms both its non-adaptive counterpart as well as

---

* Roy was with the Department of Mechanical Engineering, G.W.C. Whiting School of Engineering, Johns Hopkins University, Baltimore, Maryland.
† Rothbaum was with the Department of Otolaryngology - Head and Neck Surgery, Johns Hopkins School of Medicine, Baltimore, Maryland.
‡ This work was supported by the National Science Foundation under grants BES-9625143, IIS9801684, and EEC9731478. A preliminary version of this chapter was presented at the IEEE/RSJ International Conference on Intelligent Robots and Systems, Lusanne, Oct 2001.

the previously reported set-point regulation based force scaling algorithm over a wide range of user applied input force trajectories and force scale factors.

1 Introduction

Microsurgical tasks often require delicate instrument manipulation and tactile sensing of small interaction forces [1, 2, 3]. Though surgeons possess greater skill than lay persons when performing similar tasks, inherent human limitations of dexterity and tactile sensitivity can complicate and constrain microsurgical procedures [1, 3, 4]. For example, several studies suggest that inexperienced surgeons, particularly residents and fellows, demonstrate low relative success rates in a complex microsurgical procedure called stapedotomy that requires dextrous instrument manipulation in a confined space in the presence of low haptic feedback to the surgeon. [5, 6, 7, 8]. Robot-assist devices of various types, by reducing tremor, directing task execution, and enhancing haptic feedback, promise to overcome some of these limitations [4, 9, 10]. Examples include passive devices that employ motion constraining mechanical designs to guide human task execution [11, 12], and active devices that employ robot control algorithms to assist in the execution of fine manipulation tasks [10, 13, 14]. One such novel application of robot control, previously shown to improve task performance during otologic microsurgery [10], is the subject of the investigation reported in this chapter.

We report the theoretical and experimental development of a robot force control application called position based force scaling, henceforth referred to simply as force scaling. We specifically address the case of human-tool to tool-environment force down-scaling during human-robot co-manipulation tasks. Force down-scaling uses position based force control algorithms to augment human haptic feedback during robot-assisted co-manipulation tasks. We review three previously reported position based force control algorithms [15] and report a theoretical and experimental investigation into their comparative performance when applied to the task of force scaling. To the best of our knowledge, the applications of two of these algorithms to force scaling are the first reported implementations of asymptotically exact force scaling, and the first results in on-line environment compliance estimation while performing the task of force scaling.

Note that the force scaling task in a co-manipulation paradigm, often called force reflection when the scale factor is 1:1, is different from that of force-reflection in master/slave tele-robotic systems. In tele-robotic force reflection, the objective is to provide force feedback during remote manipulation at sites distant from the human operator [16, 13, 14, 17]. Previous work most similar to that presented in this chapter includes that of Kazerooni *et al;* [18, 19, 20], which reports the development of exoskeletons to amplify the strength of a human operator and that of Kumar *et al;* [16], which reports the implementation of one-dimensional position based force scaling on the JHU Steady Hand

Robot. In [18, 19, 20] the authors report a linear systems analysis of the stability and robustness of cooperative human-robot manipulator control systems in which the manipulator scales-up the human operator's force input by a factor of approximately 10. A concise stability analysis of these closed-loop systems, comprising dynamical models of both the robot arm and the human arm, is complicated by the fact that precise mathematical plant models exist for neither the hydraulically actuated robot nor the operator's human arm. In consequence, the authors report a robustness analysis for stable robot force-control laws that accommodate wide variation in both human and robot arm dynamics. In [16] the authors implement one-dimensional force down-scaling using the position based force set point regulator, first reported in [21]. The authors report a stability analysis for one-dimensional position based force scaling and show that the tool tip forces asymptotically track down-scaled user applied forces for the special case of a constant handle force and exact inner loop position control. The experimental results in [16] demonstrate one-dimensional force scaling to work reasonably for several hand generated handle force trajectories. In the present report we extend this previous work and present the complete theoretical and experimental development of three-dimensional position based force scaling.

The specific objectives of this chapter are threefold:

(i) We present a push-poke theory of tool-environment interaction and use this theory, in concert with the position based force trajectory tracking controllers reported in [21, 15, 22], to develop the first complete theory of three-dimensional force scaling.

(ii) We report the first implementation of adaptive position based force scaling. The adaptive force scaling algorithm provides asymptotically exact force reflection and force scaling while simultaneously estimating the environment compliance.

(iii) We report a comprehensive comparative experimental study into the performance of the three position based force control algorithms reported in [21, 15, 22] for the task of force scaling.

The rest of this chapter is organized as follows. Section 2 reviews the co-manipulation paradigm and formulates the force scaling problem for human-robot co-manipulation tasks. Section 3 reviews three previously reported one-dimensional position based force control algorithms. Section 4 presents the push-poke theory of tool-environment interaction and addresses the surface coordinate system representation problem for both these types of tool-environment interactions. Section 5 reports a comparative theoretical analysis of the properties of three one-dimensional force control algorithms, reviewed in Section 3, when applied to the task of three-dimensional force scaling. Section 6 reports the results of an extensive comparative experimental investigation into the performance of the three algorithms for the task of force scaling while pushing and poking. Section 7 summarizes and concludes the main results of this chapter.

2 Position Based 3-D Force Scaling during Human-Robot Co-Manipulation: Problem Statement

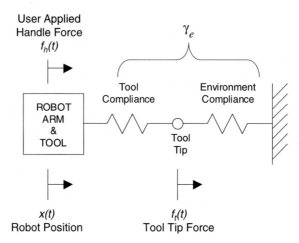

Fig. 1. Simple 1-D position based force scaling in a co-manipulation paradigm. Shown in the figure are the rigid 1-D position controlled robot, compliant tool and environment, and the handle and tip forces.

In the co-manipulation force scaling paradigm, both the robot and the human simultaneously manipulate a single tool that interacts with a compliant environment [16, 13]. The point at which the human user holds the tool is called the handle and the point at which the tool interacts with the environment is called the tool tip. The handle and the tool tip are both instrumented with force sensors called, respectively, the handle sensor and the tip sensor. The robot is assumed to be rigid, non-backdrivable, and under joint level position or velocity control. The task of the force scaling controller then is to synthesize a desired joint position or velocity trajectory based on the sensed handle and tool tip forces, such that the actual tool tip force trajectory asymptotically converges to a scaled version of the user applied handle force trajectory by a user determined scale factor. Figure 1 shows a schematic of the co-manipulation force scaling paradigm for the one-dimensional case.

Figure 2.A shows a schematic block diagram of the system. Figure 2.B defines the notation used in the block diagram and the rest of this chapter. Unless otherwise stated, all forces and positions in Figure 2 and the rest of the chapter are assumed to be expressed in the local surface coordinate system at the point of tool-environment contact. Further, the z-axis of this local coordinate system is chosen to be perpendicular to the surface at the point of contact.

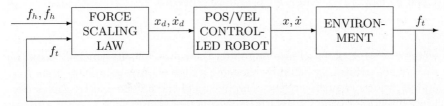

(A) Schematic block diagram

Symbol	Meaning	SI unit
$f_t(t)$	Tip force	N
$\dot{f}_t(t)$	First time derivative of tip force	N/s
$f_h(t)$	Handle force	N
$\dot{f}_h(t)$	First time derivative of handle force	N/s
ν	Force scale factor (Desired tip force/handle force)	-
$\Delta f_t(t) = f_t(t) - \nu f_h(t)$	Tip force tracking error	N
$x(t)$	Actual robot position	m
$\dot{x}(t)$	Actual robot velocity	m/s
$x_d(t)$	Desired robot position	m
$\dot{x}_d(t)$	Desired robot velocity	m/s
$\Delta x(t) = x(t) - x_d(t)$	Robot position tracking error	m
$\Delta \dot{x}(t) = \dot{x}(t) - \dot{x}_d(t)$	Robot velocity tracking error	m/s
x_e	Undeformed environment position (position corresponding to zero force)	m
$k = k_e$	Combined tool, sensor & environment stiffness (henceforth called the environment stiffness)	N/m
$\gamma = \gamma_e = 1/k$	Combined tool, sensor & environment compliance (henceforth called the environment compliance)	m/N
$\hat{\gamma}(t)$	Time varying estimate of γ	m/N
$\dot{\hat{\gamma}}(t)$	First time derivative of $\hat{\gamma}$	m/ N-s

(B) Notation

Fig. 2. 3D Force Scaling: Shown in the figure are (A) the block diagram of the system, and (B) the nomenclature used to describe the control problem. Note that all positions and forces shown are 3-dimensional and expressed in the local surface coordinate frame at the point of tool-environment contact.

2.1 Control Task: Known Environment Compliance

Given a user applied handle force trajectory, $f_h(t) \in R^3$, a positive force scale factor $\nu < 1$, and a known environment compliance γ, the force scaling problem is to synthesize a desired robot velocity trajectory $\dot{x}_d(t)$ such that the tip force $f_t(t)$ satisfies:

Case 1: Asymptotically exact inner loop velocity tracking

$$\lim_{t \to \infty} |f_{tz}(t) - \nu f_{hz}(t)| = 0 \quad \text{if} \quad \lim_{t \to \infty} \Delta \dot{x}(t) = 0. \tag{1}$$

Case 2: Bounded inner loop velocity errors

$$\limsup_{t \to \infty} |f_{tz}(t) - \nu f_{hz}(t)| \leq \epsilon \quad \text{if} \quad \limsup_{t \to \infty} ||\Delta \dot{x}(t)|| \leq M, \quad (2)$$

where $0 < M < \infty$ is a finite bound on the robot velocity tracking error and ϵ can made arbitrarily small by choosing a high enough gain. Note that $f_{tz}(t)$ and $f_{hz}(t)$ are respectively the components of the handle and tip forces along the z-direction of the local surface coordinates at any time t. Also note that the robot position $x(t) \in R^3$ and velocity $\dot{x}(t) \in R^3$ are expressed in the local surface coordinates at any time t.

2.2 Control Task: Unknown Environment Compliance

For the case of an unknown environment compliance, given a handle force trajectory $f_h(t) \in R^3$ and a force scale factor $1 > \nu > 0$, the adaptive force scaling problem is to design a desired robot velocity trajectory $\dot{x}_d(t)$ and an adaptive parameter update law for the compliance estimate $\hat{\gamma}(t)$ such that each of (1) and (2) are satisfied.

3 Three Force Control Algorithms: A Review

In this section we review three one-dimensional position/velocity based force controllers that were previously reported in [21, 15]. The convergence properties of these one-dimensional force control laws, in conjunction with the push-poke theory of Section 4, are used in Section 5 to deduce convergence properties of the three-dimensional force scaling laws based on them. We use the $\tilde{}$ notation to differentiate a one-dimensional variable from the general 3-dimensional variables of Section 2. Let $\tilde{x}(t) \in R$, $\dot{\tilde{x}}(t) \in R$, $\tilde{f}(t) \in R$ be the one-dimensional robot position, velocity and end-effector force respectively.

3.1 Set Point Regulator (SPR)

For the case of a constant desired force $\tilde{f}_d$, in [21] the authors showed that the simple feedback control law:

$$\dot{\tilde{x}}_d(t) = -k_f \Delta \tilde{f}(t), \quad (3)$$

where $\dot{\tilde{x}}_d(t) \in R$ is the desired robot velocity, $\Delta \tilde{f}(t) = \tilde{f}(t) - \tilde{f}_d \in R$ is the force regulation error, and $k_f > 0$ is the force gain guarantees that $\Delta \tilde{f}(t)$ is bounded and that:

$$\lim_{t \to \infty} \Delta \tilde{f}(t) = 0 \text{ if } \lim_{t \to \infty} \Delta \dot{\tilde{x}} = 0 \quad (4)$$

and

$$\limsup_{t \to \infty} \left| \Delta \tilde{f}(t) \right| \leq \frac{M}{k_f} \text{ if } \limsup_{t \to \infty} \left| \Delta \dot{\tilde{x}} \right| \leq M, \quad (5)$$

where $0 \leq M < \infty$ is a finite bound on the steady-state inner loop velocity tracking error $\Delta \dot{\tilde{x}}$.

3.2 Trajectory Tracking Controller (TTC)

When the linear environment compliance γ is known, in [15] the authors showed that the control law:

$$\dot{\tilde{x}}_d(t) = \gamma \dot{\tilde{f}}_d(t) - k_f \Delta \tilde{f}(t), \tag{6}$$

where $\dot{\tilde{x}}_d(t) \in R$ is the desired robot velocity, $\Delta \tilde{f}(t) = \tilde{f}(t) - \tilde{f}_d(t) \in R$ is the force tracking error, and $k_f > 0$ is the force gain guarantees that $\Delta \tilde{f}(t)$ is bounded and satisfies (4) and (5) for all time-varying bounded $\mathcal{C}^2$ desired force trajectories with bounded time derivatives $\dot{\tilde{f}}_d(t), \ddot{\tilde{f}}_d(t) \in R$.

3.3 Adaptive Trajectory Tracking Controller (TTCA)

For the case when the environment compliance is not known, in [15] the authors reported the following adaptive control law:

$$\dot{\tilde{x}}_d(t) = \hat{\gamma}(t) \dot{\tilde{f}}_d(t) - k_f \Delta \tilde{f}(t) \tag{7}$$

$$\dot{\hat{\gamma}} = -\alpha \dot{\tilde{f}}_d(t) \Delta \tilde{f}(t), \tag{8}$$

where $\alpha > 0$ is the adaptive gain and $\hat{\gamma}(t)$ is the time-varying estimate of the environment compliance γ, and showed that the control law guarantees that $\Delta \tilde{f}(t), \hat{\gamma}(t)$ are bounded and that :

$$\lim_{t \to \infty} \Delta \tilde{f}(t) = 0 \text{ if } \lim_{t \to \infty} \Delta \dot{\tilde{x}} = 0 \text{ and } \Delta \dot{\tilde{x}} \in \mathcal{L}_2, \tag{9}$$

for all time-varying bounded $\mathcal{C}^2$ desired force trajectories with bounded time derivatives $\dot{\tilde{f}}_d(t), \ddot{\tilde{f}}_d(t) \in R$.

4 Theory of Pushing and Poking

In order to develop the three-dimensional theory of force scaling, we have identified two types of tool-environment interactions which we call push and poke. Tools that push are called pushers and those that poke are called pokers. The orientation of the local surface coordinate system, which is the coordinate system in which the three-dimensional position based force control and force scaling problem is formulated, depends on type of tool-environment interaction and hence it becomes important to a-priori identify the nature of this interaction. We note here that for tools in contact with the environment, none of three position based controllers reviewed in Section 3 require knowledge of the exact position of the environment and hence we do not concern ourselves with its identification.

In poke interactions, the tool is usually sharp and partially penetrates the surface that it manipulates. The effect is that, locally, the surface takes the

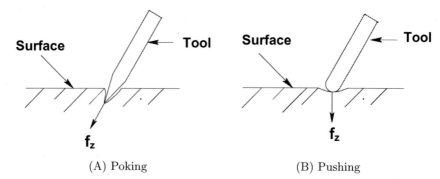

Fig. 3. Poking and pushing: Shown are the tool, environment and the interaction force during (A) a *poke* type tool-environment interaction, and (B) a *push* type tool-environment interaction.

shape of the tool and the predominant interaction force is along the approach axis, defined to be the z - axis, of the tool. As a result, at any given instant of time after contact, the axes of the local surface coordinate system can be chosen to be identical to that of the tool tip coordinate system. Since the tool tip coordinate axes are a known smooth continuous function of the joint variables and kinematic parameters of the robot, the orientation of the surface, at the point of contact, is always known and no additional knowledge is required about the surface. Figure 3.A shows a schematic diagram of a poke-type interaction and the tool-environment interaction forces. Note that a blunt tool that interacts with a very compliant environment and causes the environment to locally take the shape of the tool would be, for force scaling purposes, classified as a poker.

In push interactions, the tool is usually blunt and the surface is semi-rigid. The effect is that the surface does not locally take the exact shape of the tool. In the absence of friction, the tool can only apply forces perpendicular to the surface. The problem statement for force control and force scaling is then identical to the problem statement of hybrid force/position control reported in [22]. To correctly implement force scaling for such a tool-surface interaction, the surface has to satisfy all the properties stated in [22] — specifically, the global surface coordinate system has to be known a-priori. Figure 3.B shows a schematic diagram of a push-type interaction and the the tool-environment interaction forces.

5 Three Force Scaling Algorithms

In this section, we report three force scaling algorithms which are directly derived from the three position/velocity based force controllers reviewed in Section 3.

5.1 Set Point Regulator (SPR) based Force Scaling

Set point regulator (SPR) based force scaling is the most basic position based force scaling approach. The algorithm used to achieve force scaling is:

$$f_{td}(t) = \nu f_h(t), \tag{10}$$

$$\dot{x}_d(t) = -k_f \Delta f_t(t), \tag{11}$$

where $\Delta f_t(t) = f_t(t) - f_{td}(t) = f_t(t) - \nu f_h(t) \in R^3$ is the tip force tracking error expressed in the local surface coordinate system and $k_f > 0$ is the force feedback gain. A one-dimensional version of this algorithm was proposed in [16], where the authors show asymptotically exact force scaling for constant handle forces when the robot is under exact inner loop position/velocity control. However, from Section 3.1, we know that this control algorithm guarantees asymptotically exact force set point regulation for constant desired handle forces both when the robot is under exact inner loop velocity control as well as when it is under asymptotically exact inner loop velocity control, and arbitrarily small force tracking errors under bounded inner loop velocity tracking errors.

This algorithm has two advantages: 1) It requires no knowledge of the environment compliance or orientation. 2) When contact with the environment is lost, the algorithm automatically puts the robot under proportional-velocity control, where the velocity of the tool tip is proportional to the applied handle force, and thus guarantees stable operation of the robot in free space.

The principal disadvantage of this algorithm is that it does not guarantee asymptotically exact force scaling for time varying handle force trajectories even when the robot is under exact inner loop position/velocity control.

5.2 Trajectory Tracking Controller (TTC) based Force Scaling

Trajectory tracking controller (TTC) based force scaling employs the known environment compliance value γ, in conjunction with the trajectory tracking controller of Section 3.2, to provide asymptotically exact force scaling perpendicular to the surface for all smooth time varying handle force trajectories. The control law used to achieve this asymptotically exact force scaling is:

$$f_{td}(t) = \nu f_h(t), \tag{12}$$

$$\dot{x}_{dx}(t) = -k_f \Delta f_{tx}(t), \quad (13)$$
$$\dot{x}_{dy}(t) = -k_f \Delta f_{ty}(t), \quad (14)$$
$$\dot{x}_{dz}(t) = \gamma \dot{f}_{tdz} - k_f \Delta f_{tz}(t), \quad (15)$$

where $\Delta f_{tx}(t) = f_{tx}(t) - f_{tdx}(t) \in R$, $\Delta f_{ty}(t) = f_{ty}(t) - f_{tdy}(t) \in R$, $\Delta f_{tz}(t) = f_{tz}(t) - f_{tdz}(t) \in R$ respectively are the x, y and z components of the tip force tracking errors expressed in the local surface coordinates at the point of tool-environment contact.

This algorithm has three advantages: 1) It provides asymptotically exact force scaling perpendicular to the surface for all smooth time-varying handle force trajectories when the robot is under exact or asymptotically exact inner loop velocity tracking control and arbitrarily small force tracking errors in the presence of bounded inner loop velocity errors. 2) It provides scaled force set point regulation in the remaining directions when the tool is stuck in the environment. 3) It provides proportional-velocity control parallel to the surface when the tool is not stuck in the environment.

The principal disadvantages of this force scaling algorithm are twofold: 1) For its implementation, it requires differentiation of the desired tip force which is a scaled version of a typically noisy sensed handle sensor force. 2) The combined tool environment compliance has to be a-priori known in order to implement this controller.

5.3 Adaptive Trajectory Tracking Controller (TTCA) based Force Scaling

The third force scaling algorithm reported here is based on the adaptive trajectory tracking controller of Section 3.3. Specifically, the algorithm is:

$$f_{td}(t) = \nu f_h(t), \quad (16)$$

$$\dot{x}_{dx}(t) = -k_f \Delta f_{tx}(t), \quad (17)$$
$$\dot{x}_{dy}(t) = -k_f \Delta f_{ty}(t), \quad (18)$$
$$\dot{x}_{dz}(t) = \hat{\gamma} \dot{f}_{tdz} - k_f \Delta f_{tz}(t), \quad (19)$$
$$\dot{\hat{\gamma}}(t) = -\alpha \dot{f}_{tdz}(t) \Delta f_{tz}(t), \quad (20)$$

where $\alpha > 0$ is the adaptive gain.

This algorithm has five main advantages: 1) It does not require knowledge of the environment compliance for its implementation. 2) It estimates the environment compliance on-line and thus provides a powerful tool to do on-line environment (e.g. tissue) property identification. 3) It guarantees asymptotically exact force scaling perpendicular to the surface for all smooth time-varying handle trajectories when the robot is under exact or asymptotically exact inner loop velocity tracking control. 4) It provides scaled force set-point regulation in the remaining surface directions when the tool is stuck in the environment. 5) It provides proportional-velocity control parallel to the surface when the tool is not stuck in the surface.

The main disadvantages of this algorithm are twofold. 1) It requires differentiation of the desired tip force, which is a scaled version of a typically noisy handle sensor signal, for its implementation. 2) When contact is lost with the environment, the compliance estimate adapts to the infinite compliance of free air if the adaptation is active.

6 Experimental Results

This section reports the results of comparative experiments in pushing and poking with the three force scaling algorithms reported in Section 5.

6.1 Experimental Setup

The JHU Steady Hand Robot employed in these experiments is a seven degree of freedom remote-center-of-motion (RCM) robot arm developed at the Johns Hopkins University for medical applications [16, 13]. The arm has a 3-DOF linear base stage, a 2-DOF intermediate RCM stage [23], and a final two degree of freedom z - theta stage. For the experiments reported in this section, all the joints of the arm, except the linear insertion (z) axis of the z - theta stage were active. The ambient noise in the both force sensor readings was of the order of 2-3 mN, leading to a force sensor accuracy of +/- 3 mN.

The tool used for the poking experiments was a 16 cm long thin stainless steel surgical instrument (Storz) with a sharp tip and the tool used for the pushing experiments was a 16 cm long thin aluminum rod having a rounded tip at one end. Both tools were rigidly attached to the tip force sensor at one end and interacted with the environment with the other end. The handle, rigidly attached to the handle force sensor, employed an ergonomic design commonly found in surgical instruments and was manipulated by the user to make the tool move and interact with the environment. Two surfaces provided the environment in the experiments — (1) the compliant surface of a thin plastic box and (2) the more compliant surface of a stretched sheet of paper. Figure 4 shows the JHU Steady Hand robot, the handle, the handle and tip sensors, and the pushing and poking tools used in the experiments.

All six active joints of the robot were under independent joint level PI velocity control. Further, the rotational joints were put under outer loop proportional-velocity control such that the desired angular velocity of the remote center of motion at the tool tip was proportional to the user applied torque at the handle. This allowed the user to orient the the tool to a desired angle of approach to the surface without moving the position of the tool-tip. The desired velocities of the three linear axes of the base were determined

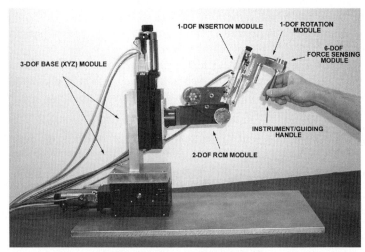

(A) JHU Steady Hand Robot

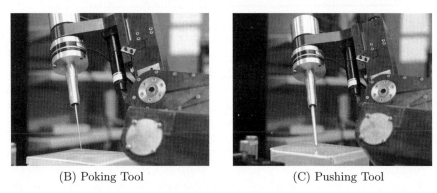

(B) Poking Tool (C) Pushing Tool

Fig. 4. Experimental Setup: Shown in the figure are (A) the JHU Steady Hand Robot, (B) the poking tool, and (C) the pushing tool.

by the user applied linear handle force, the tip-environment interaction force, and the force scaling algorithm being used.

Since we are primarily interested in comparing the relative performance of three force scaling algorithms reported in Section 5 in a fair and unbiased fashion, the force feedback gain, k_f, was held at a constant value of 0.005 N-m/s for all the three algorithms for all the poking and pushing experiments reported in this section. Higher feedback gains were observed, of course, to give lower tracking errors but the relative performance of the three controllers remained unchanged. Further, unless otherwise noted, the adaptive gain, α, was held constant at a value of 0.0005 m/N^3 for all experiments employing the TTCA based force scaling algorithm.

6.2 Algorithm Implementation

This section addresses the practical problems associated with the implementation of the force scaling algorithms reported in Section 5. As noted in Sections 5.2 and 5.3, the TTC and TTCA based force scaling algorithms require feedforward terms that involve the derivative of the measured handle force sensor signal which is typically noisy. Similarly, the adaptation law for the TTCA based force scaling algorithm includes the derivative of the measured handle force sensor signal. To obtain a clean derivative of the measured handle force signal, we filtered the signal using a first order IIR filter having a relatively high time constant of 50 ms. As is well known, filters with high time constants introduce a phase lag between the raw signal and the filtered signal. At a first glance, this would seem to defeat the purpose of the TTC and TTCA based force scaling algorithms, which is to eliminate the phase lag between the desired and the actual force trajectories inherent in the SPR based force scaling algorithm. To overcome this problem, we used a multi-part filter in the implementations of the TTC and TTCA based force scaling controllers. This consisted of using an unfiltered force error feedback term and a filtered feed forward term involving the derivative of the measured handle force. Similarly the implementation of update law (20) for the TTCA based force scaling algorithm used an unfiltered force error term and an IIR filtered handle force derivative term. This resulted in implementations that provided force tracking with almost no phase lag, as will be seen in Sections 6.3 and 6.4.

Another practical issue in the implementation of the TTCA based force scaling algorithm is that the compliance estimate theoretically adapts to ∞ when contact is lost with the environment. To overcome this problem, we disabled the adaptation (that is $\dot{\hat{\gamma}} = 0$), whenever the the tip force fell below the small threshold of 0.05 N, indicating that contact was about to be lost. This resulted in $\hat{\gamma}$ maintaining a constant value when contact was lost and a stable operation of the TTCA based force scaling algorithm in air.

6.3 Poking Experiments

The poking experiments consisted of a human user and the robot co-manipulating the poking tool while the tool-tip interacted with the environment. The user held the tool at the handle and applied a repeatable time varying force for all the experiments reported in this section. The robot sequentially ran the three force scaling algorithms reported in Section 5 at four different force scale factors of 1.0, 0.5, 0.25, and 0.1[4].Experiments were performed with the poking tool both perpendicular to as well at a shallow angle to the surface being manipulated

---

[4] At a scale factor of 0.1, tool tip forces are amplified by a factor of 10 at the handle, which is the haptic interface, thus amplifying the operators haptic sensitivity by a factor of 10

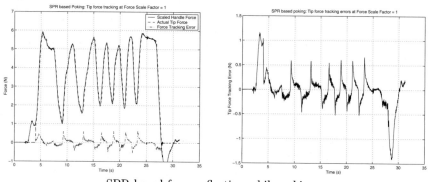

SPR based force reflection while poking -
(Left) Z-component of handle and tool tip forces and (Right) force tracking errors

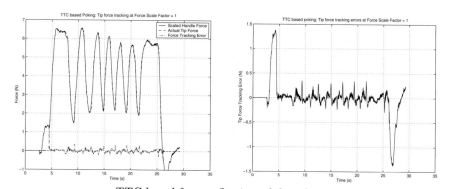

TTC based force reflection while poking -
(Left) Z-component of handle and tool tip forces and (Right) force tracking errors

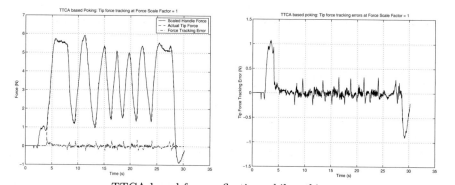

TTCA based force reflection while poking -
(Left) Z-component of handle and tool tip forces and (Right) force tracking errors

Fig. 5. Position based force reflection: Shown (top to bottom) are the tracking performance and errors of the SPR, TTC, and TTCA based force scaling controllers while *poking* at a force scale factor of 1. For the experiments in this figure, the poking tool was perpendicular to a plastic box environment.

Figure 5 shows the tool tip or local surface coordinate z-components of the handle force trajectory, the tool tip trajectory, and the tool-tip force tracking errors for the three algorithms when the poking tool is perpendicular to the plastic box environment and the force scale factor is 1.0. As seen from the figure, the TTC and TTCA based force scaling algorithms, which incorporate a filtered feedforward term, have significantly lower force tracking errors than the SPR based force scaling algorithm. Figure 6 shows the evolution of tool-environment compliance estimate with time for the TTCA based force scaling algorithm. As seen from the figure, the compliance estimate converges to a value of 3×10^{-4} m/N within 0.5 seconds of tool-environment contact. The experimentally measured compliance value, used in the TTC controller, for this tool and environment is 3.1×10^{-4} m/N.

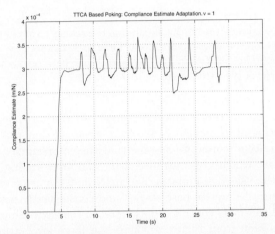

Fig. 6. TTCA based force scaling: Evolution of the tool-environment compliance estimate with time while poking at a force scale factor of 1.0.

Figure 7 shows the average root mean square force tracking errors of the three force scaling algorithms at four different force scale factors for a plastic box environment. For each force scale factor and each algorithm, the average was taken over two experiments — one with the poking tool perpendicular to the surface and the other with the tool at an angle to the surface. The error values have been normalized such that for any given set of experiments (e.g. poking tool interacting with the plastic box environment at an angle and at a force scale factor of 0.1) the root mean square error of the SPR based algorithm is 1.0. The figure shows that the TTC and TTCA based force scaling algorithms significantly outperform the SPR based algorithm at force scale factors of 1.0 and 0.5. At lower force scale factors of 0.25 and 0.1, the TTC based algorithm performs worse than the SPR algorithm.

What is the reason for this change in the order of rankings between the SPR and TTC based algorithms at lower force scale factors? To investigate

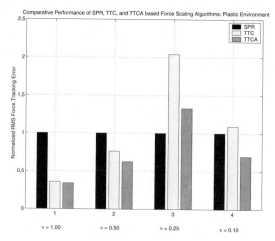

Fig. 7. Poking: Average root mean square force tracking errors of the SPR, TTC, and TTCA based force scaling algorithms at force scale factors of $\nu = 1.0$, 0.5, 0.25, and 0.1 for a plastic box environment. The errors have been normalized so that the average RMS error of the SPR based algorithm is 1.0

this, we revisit one of the assumptions made in the force scaling problem statement formulated in Section 2. We have assumed that the robot is completely rigid. In other words, we assume that the compliance γ_r between the robot base stages, which are controlled using the force scaling laws, and the handle, which the user manipulates, is 0 m/N and the net compliace γ seen by the force controlled active base stages, and consequently used in the feedforward term of the TTC control law, is equal to the environment compliance γ_e. However, as seen from Figure 4, there is a large lever arm between the handle and robot base stages which results in a finite non-zero robot-base to handle compliance γ_r, henceforth referred to simply as the robot compliance. If this compliance is incorporated in the system closed loop model, a simple force balance at the handle and a closed loop error analysis similar to that in [15] yields the following expression for the net effective compliance of the system:

$$\gamma = \gamma_e \left[1 - (n-1)\frac{\gamma_r}{\gamma_e}\right], \tag{21}$$

where $n = 1/\nu$ is the inverse of the force scale factor.

We observe the following from Equation 21:

(i) The net effective system compliance γ equals the environment compliance γ_e when the robot compliance γ_r is 0. This holds for all force scale factors ν, where $\nu = 1/n$.
(ii) The net effective system compliance γ equals the environment compliance γ_e when the force scale factor ν is 1. This holds for all values of the robot compliance γ_r.

(iii) For force scale factors $\nu < 1$ (that is, for $n > 1$) and a non-zero robot compliance, γ_r, the net effective system compliance γ is less than the environment compliance γ_e. As ν decreases (or as n increases), γ becomes progressively smaller than the environment compliance γ_e, until eventually γ equals 0 when $n = 1 + \gamma_e/\gamma_r$.

We note, once again, that the TTC based force scaling algorithm implemented in these experiments uses purely the environment compliance, $\gamma = \gamma_e$, in its feedforward term. As noted in the observations above, this is the correct net effective system compliance only at a force scale factor of 1. Hence the TTC based force scaling algorithm performs best at force scale factors near 1 and its performance progressively degrades relative to the SPR based algorithm at lower force scale factors. However, the TTCA based force scaling algorithm, which requires no initial knowledge of the system compliance, adapts its compliance estimate, $\hat{\gamma}$, to the correct net effective system compliance, γ, at all force scale factors and hence its performance still remains the best amongst the three controllers. Further, we note that at a force scale factor of $\nu = 0.25$, the net effective system compliance is roughly zero for the poking experiments described in this section (see observation iii). Hence the SPR based algorithm now becomes a TTC based algorithm that incorporates the correct effective compliance (0 m/N) of the system in its feedforward term. The compliance estimate of the TTCA based algorithm converges and oscillates around this value, due to force sensor noise, unmodeled friction etc., thus leading to a marginally worse performance of the TTCA controller at this force scale factor. Another interesting observation is that at $\nu = 0.1$, the net effective system compliance is negative. Though the sufficiency proofs used to prove stability and convergence properties of the TTCA algorithm require a non-negative value of γ [15], we observe that in practice, the compliance estimate of the TTCA algorithm correctly converges to the negative value of system compliance γ. Hence the TTCA algorithm outperforms the SPR algorithm even at this very low force scale factor. This theoretically unexplained fact invites further investigation.

Figure 8 shows the normalized average root mean square tracking errors of the SPR and TTCA based force scaling algorithms for the case of the poking tool interacting with a more compliant stretched paper sheet environment. As seen from the figure, the adaptive algorithm outperforms the SPR based algorithm at all the experimentally verified force scale factors.

6.4 Pushing Experiments

Similar to the poking experiments, the pushing experiments consisted of a human user and the robot co-manipulating the pushing tool while the tool-tip interacted with the environment. For all the experiments reported in this section, the human user applied a repeatable time-varying force at the handle while the robot sequentially ran the three force scaling algorithms reported

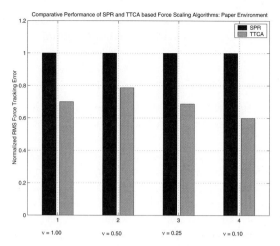

Fig. 8. Poking: Average root mean square force tracking errors of the SPR and TTCA based controllers at force scale factors of $\nu = 1.0, 0.5, 0.25$, and 0.1 for a cantilevered paper sheet environment. The errors have been normalized so that the average RMS error of the SPR based algorithm is 1.0

in Section 5 at four different force scale factors of 1.0, 0.5, 0.25, and 0.1. Experiments were performed with the pushing tool both perpendicular to as well at a shallow angle to the surface being manipulated. In all experiments reported in this section, the environment consisted of the compliant surface of a thin plastic box.

Figure 9 shows the global surface z-components of the handle force trajectory, the tool tip trajectory, and the tool-tip force tracking errors for the three algorithms when the pushing tool is perpendicular to the environment and the force scale factor is 1. As seen from the figure, the TTC and TTCA based force scaling algorithms, which incorporate a filtered feedforward term, have significantly lower force tracking errors than the SPR based force scaling algorithm.

Figure 10 shows the normalized average root mean square force tracking errors of the three force scaling algorithms in global surface z-direction at four different force scale factors. For each force scale factor, the average was taken over two experiments — one with the tool perpendicular to the surface and the other with the tool at an angle to the surface. The error values have been normalized such that for any given set of experiments (e.g. pushing tool interacting with the environment at an angle and at a force scale factor of 0.1) the root mean square error of the SPR based algorithm is 1.0. The figure shows that the TTC and TTCA based force scaling algorithms significantly outperform the SPR based algorithm at force scale factors of 1.0, 0.5, and 0.25. At the force scale factor of 0.1, the SPR based force scaling algorithm outperforms the TTC based algorithm. Again, as explained in Section 6.3, this can be attributed to the fact that at this low force scale factor, the contribution

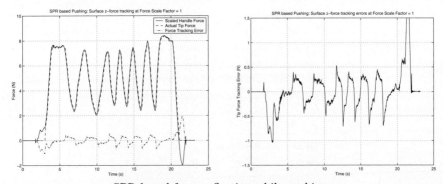

SPR based force reflection while pushing -
(Left) z-component of handle and tool tip forces and (Right) force tracking errors

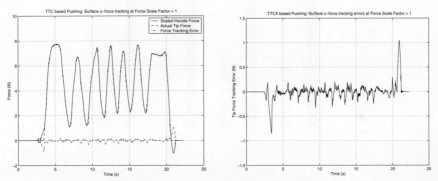

TTC based force reflection while pushing -
(Left) z-component of handle and tool tip forces and (Right) force tracking errors

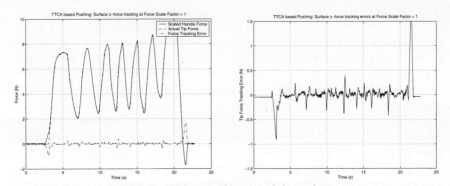

TTCA based force reflection while pushing -
(Left) z-component of handle and tool tip forces and (Right) force tracking errors

Fig. 9. Position based force reflection: Shown (top to bottom) are the tracking performance and errors of the SPR, TTC, and TTCA based force scaling controllers while *pushing* at a force scale factor of 1. For the experiments in this figure, the poking tool was perpendicular to a plastic box environment.

of the unmodeled robot compliance, γ_r, to the net effective system compliance, γ, is significant (see Equation 21). This affects the performance of the TTC based algorithm, which does not take the robot compliance, γ_r, into account and purely uses the environment compliance, γ_e, in its feedforward term. The TTCA based algorithm however adaptively estimates the correct net effective system compliance, γ, and outperforms both the SPR and TTC based algorithms.

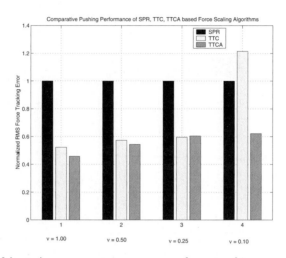

Fig. 10. Pushing: Average root mean square force tracking errors of the SPR, TTC, and TTCA based force scaling algorithms at force scale factors of $\nu = 1.0$, 0.5, 0.25, and 0.1 for pushing on a plastic box environment. The errors have been normalized so that the average RMS error of the SPR based algorithm is 1.0

Figure 11 shows the force reflection ($\nu = 1$) performance of the SPR and TTCA based force scaling algorithms when the user manipulates the handle such that the tool is held at an angle to the surface and pushes perpendicular to the surface while simultaneously sliding along the surface. As seem from the figure, the TTCA based force scaling algorithm, which adaptively estimates the compliance of the surface and uses it in its feedforward term, has significantly lower tip force tracking errors than the SPR based force scaling algorithm.

Figure 12 plots the normalized root mean square force tracking errors of the two controllers for the task of simultaneous pushing and sliding at four different force scale factors. As seen, the TTCA based algorithm outperforms the SPR based algorithm at almost all the force scaling factors studied. Again, the only exception is at a force scale factor of 0.25, where the effective system compliance is roughly zero and hence the SPR based algorithm unintentionally includes this zero feedforward to provide better tracking than the TTCA based algorithm, in which the compliance converges to but oscillates around zero.

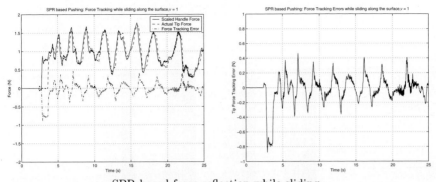

SPR based force reflection while sliding -
(Left) z-component of handle and tool tip forces and (Right) force tracking errors

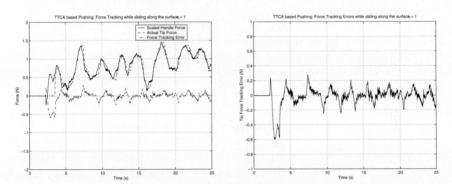

TTCA based force reflection while sliding -
(Left) z-component of handle and tool tip forces and (Right) force tracking errors

Fig. 11. Pushing while Sliding: Shown are the force tracking performance and tip force tracking errors of the SPR (top) and TTCA (bottom) based force scaling algorithms for the task of pushing perpendicular to the surface while simultaneously sliding along the surface. The force scale factor for both experiments is 1.0.

7 Conclusions

This chapter reported the theoretical and experimental development of a novel force control application called position based force scaling. Force scaling uses the position based force control algorithms reported in [21, 15, 22] to augment human haptic feedback during human-robot co-manipulation tasks. Detailed results, presented in Sections 4, 5, and 6 are summarized here:

(i) We present the push-poke theory of tool-environment interaction for co-manipulation tasks and use this theory, in conjunction with the three position based force control algorithms, previously reviewed in [15], to develop the theory of 3-D force scaling.

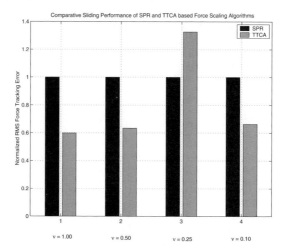

Fig. 12. Pushing while sliding: Normalized root mean square tracking errors of the SPR and TTCA based force scaling algorithms at force scale factors of $\nu = 1.0$, 0.5, 0.25, 0.1. The errors have been normalized so that RMS error of the SPR based algorithm is 1.0 for each force scale factor.

(ii) We report the theoretical and experimental development of adaptive force scaling. The adaptive force scaling algorithm estimates the environment compliance on-line during co-manipulation tasks and uses this estimate to provide asymptotically exact force reflection and force scaling for all smooth time-varying user-applied forces.

(iii) We provide experimental insights into the implementation of asymptotically exact force scaling and force reflection, and directly address the practical issues involved in force sensor signal differentiation. We show that a multi-part filtering approach involving IIR filters provides a good practical way of filtering noisy force sensor signals while ameliorating the known limitations of IIR filters.

(iv) We provide an extensive comparative experimental study into the performance of the three position based force control algorithms reviewed in [15] for the task of force scaling. The adaptive TTCA based force scaling algorithm significantly outperforms both its non-adaptive counterpart and the set point regulator based force scaling algorithms over a wide range of force scale factors for the tasks of poking as well as pushing.

Acknowledgments

We would like to thank Professor Dan Stoianovici for providing the 2 axis RCM stage for use as part of the robotic system used in the experiments. We would also like to thank Dr. Elias Zirhouni for providing laboratory space. Finally, our gratitude is due to Terry Shelley and Jay Burns for giving us excellent machine shop support that was vital in manufacturing some of the tools and other components used in the experimental setup.

References

1. Riviere CN, Khosla PK (1996) Accuracy of positioning in handheld instruments. In: Proc. of the 18th Conf. of the IEEE Engineering in Medicine and Biology Society. 212–213
2. Riviere CN, Rader PS, Khosla PK (1997) Characteristics of hand motion of eye surgeons. In: Proc. of the 19th Conf. of the IEEE Engineering in Medicine and Biology Society, Chicago 1690–1693
3. Gupta PK, Jensen PS, deJuan Jr. E (1999) Surgical forces and tactile perception during retinal microsurgery. In: Medical Image Computing and Computer-Assisted Interventions (MICCAI), Cambridge, England 1218–1225
4. Charles S (1994) Dexterity enhancement for surgery. In: First International Symposium on Medical Robotics and Computer Assisted Surgery. Volume 2. 145–160
5. Hughes GB (1991) The learning curve in stapes surgery. Laryngoscope 11:1280–1284
6. Backous DD, Coker NJ, Jenkins HA (1993) The learning curve revisited: stapedotomy. American Journal of Otology 14:451–454
7. Sargent EW (2002) The learning curve revisited: Stapedotomy. Otolaryngology - Head and Neck Surgery 126:20–25
8. Rothbaum DL, Roy J, Hager GD, Taylor RH, Whitcomb LL, Francis HW, Niparko JK (2003) Task performance in stapedotomy: Comparision between surgeons of different experience levels. Otolaryngology - Head and Neck Surgery 128:71–77
9. Schenker PS (1995) Development of a telemanipulator for dexterity enhanced microsurgery. In: Proceedings of the 2nd International Symposium on Medical Robotics and Computer Assisted Surgery. 81–88
10. Rothbaum DL, Roy J, Stoianovici D, Berkelman P, Hager GD, Taylor RH, Whitcomb LL, Francis HW, Niparko JK (2002) Robot-assisted stapedotomy: Micropick fenestration of the stapes. Otolaryngology - Head and Neck Surgery 127:417–426
11. Ho SD, Hibberd R, Davies B (1995) Robot assisted knee surgery. IEEE EMBS Magazine Special Issue on Robotics in Surgery 292–300
12. Troccaz J, Peshkin M, Davies B (1997) The use of localizers, robots, and synergistic devices in CAS. In: Proc. First Joint Conference of CVRMed and MRCAS, Grenoble 727–736
13. Taylor RH, Jensen P, Whitcomb LL, Barnes A, Kumar R, Stoianovici D, Gupta P, Wang Z, de Juan E, Kavoussi LR (1999) A steady-hand robotic system for microsurgical augmentation. International Journal of Robotics Research 18(12):1201–1210
14. Taylor RH, Funda J, Eldridge B, Gruben K, LaRose D, Gomory S, Talamini M (1996) 45. In: A Telerobotic Assistant for Laparoscopic Surgery. MIT Press 581–592
15. Roy J, Whitcomb LL (2002) Adaptive force control of position/velocity controlled robots: Theory and experiment. IEEE Transactions on Robotics and Automation 18(2):121–137

16. Kumar R, Berkelman P, Gupta P, Barnes A, Jensen P, Whitcomb LL, Taylor RH (2000) Preliminary experiments in cooperative human/robot force control for robot assisted microsurgical manipulation. In: Proceedings of the IEEE International Conference on Robotics and Automation. 610–617
17. Taylor RH, Paul HA, Kazandzides P, Mittelstadt BD, Hanson W, Zuhars JF, Williamson B, Musits BL, Glassman E, Bargar WL (1994) An image-directed robotics system for precise orthopedic surgery. IEEE Transactions on Robotics and Automation 10(3):261–275
18. Kazerooni H (1989) Human/robot interaction via the transfer of power and information signals — Part I: dynamics and control analysis. In: Proceedings of IEEE International Conference on Robotics and Automation. Volume 3. 1632–1640
19. Kazerooni H (1989) Human/robot interaction via the transfer of power and information signals — Part II: an experimental analysis. In: Proceedings of IEEE International Conference on Robotics and Automation. Volume 3. 1641–1649
20. Kazerooni H, Guo J (1993) Human extenders. ASME Journal Dynamic Systems, Measurement, and Control 115:281–290
21. De Schutter J, Van Brussel H (1988) Compliant robot motion II. A control approach based on external control loops. The International Journal of Robotics Research 7(4):18–33
22. Roy J, Whitcomb LL (2002) Adaptive hybrid force/position control of multi-dimensional position controlled robots: Theory and experiment. Manuscript in preparation. Preprint Available at http://robotics.me.jhu.edu/~roy
23. Stoianovici D, Whitcomb LL, Anderson JH, Taylor RH, Kavoussi LR (1998) A modular surgical system for image guided percutaneous procedures. In Wells WM, Colchester A, Delp S, eds.: Lecture Notes in Computer Science 1496: Medical Imaging and Computer-Assisted Intervention - MICCAI'98. Volume 1496. Springer-Verlag, Berlin, Germany 404–410

Learning to Dynamically Manipulate: A Table Tennis Robot Controls a Ball and Rallies with a Human Being

Fumio Miyazaki[1], Michiya Matsushima[2] and Masahiro Takeuchi[3]

[1] Graduate School of Engineering Science, Osaka University, Osaka 560-8531, Japan,miyazaki@me.es.osaka-u.ac.jp
[2] Graduate School of Engineering, Osaka University, Osaka 565-0871, Japan, matsushima@mapse.eng.osaka-u.ac.jp
[3] Akashi National College of Technology, Hyogo 674-8501, Japan, takeuchi@akashi.ac.jp

Summary. We propose a method of controlling a paddle so as to return the ball to a desired point on the table with specified flight duration. The proposed method consists of the following three input-output maps implemented by means of Locally Weighted Regression (LWR): (1) A map for predicting the impact time of the ball hit by the paddle and the ball position and velocity at that moment according to input vectors describing the state of the incoming ball; (2) A map representing a change in ball velocities before and after impact; and (3) A map giving the relation between the ball velocity just after impact and the landing point and time of the returned ball. We also propose a feed-forward control scheme based on iterative learning control to accurately achieve the stroke movement of the paddle as determined by using these maps.

1 Introduction

To perform tasks with intermittent interactions between a robot and its environment, the robot must be able to adjust the strength and timing of interactions during execution of the tasks. These are called hybrid (mixed continuous and discrete) control problems by Burridge et al., who cited hopping, catching, hitting, and juggling as typical examples [1]. These tasks have attracted the attention of researchers in the experimental psychology because they require both receptor anticipation of the environmental situation and effector anticipation of internal movement processes [2]. However, it is difficult to find general approaches that are sufficiently tractable for the robot to perform these tasks. In this paper, we focus on the table tennis task that involves the intermittent nature of the robot-ball interaction in order to explore hybrid control problems.

Andersson constructed a sophisticated robot system that could play table tennis with humans [3]. The expert controller he developed performs task planning and updates the plan as the sensor data change, based on physical models and an exception handling mechanism. Andersson's approach makes full use of human knowledge as explicit models of the task and the environment, and the task performance depends on the system creator's knowledge. In other words, the robot system could not improve its skills through practice or experience.

Researchers in the area of sports science have proposed hypotheses on the nature of the human internal processes to execute complex tasks such as the table tennis stroke. For example, Ramanantsoa proposed simplifying procedures of the table tennis stroke based on Bernstein's hypothesis that expert players limit the degrees of freedom in their movements when planning and performing a shot [4]. The core idea of the procedures he proposed is to identify and reach virtual targets, the point at which the ball should be struck and the paddle velocity just before hitting the ball.

Motivated by Ramanantsoa's idea, we constructed a robot system and let it perform the table tennis task, in which virtual targets were predicted using input-output maps implemented efficiently by means of a k-d tree (k-dimensional tree) [5]. The paddle approached these targets by using a visual feedback control scheme similar to the mirror law proposed by Koditschek [6]. In this control scheme, the feedback gain (mirror gain) is adjusted using an input-output map learned so as to achieve a specified traveling distance of the ball after ball/paddle contact. This method was derived by taking a hint from the idea of "task-level robot learning" proposed by Atkeson [7]. However, this research has revealed the following two problems. One is that the proposed feedback scheme cannot control the height of the ball's return trajectory. This may cause the ball to fail in going over the net. The other is that tracking errors owing to the joint servos hinder the control scheme from functioning properly, because we cannot separate the servo control problem from the trajectory planning problem if tracking errors cannot be ignored.

In this monograph, we propose a method of controlling the paddle so as to return the ball to a desired point on the table with a specified duration of flight. The proposed method consists of the following three input-output maps implemented by means of Locally Weighted Regression (LWR) [8]:

(1) A map for predicting the impact time of the ball hit by the paddle and the ball position and velocity at that moment according to input vectors describing the state of the incoming ball
(2) A map representing changes in ball velocities before and after the impact
(3) A map giving the inverse relation between the ball velocity just after the impact and the bouncing point and time of the returned ball

These maps are employed to predict virtual targets for the above-mentioned paddle control. The third map, which Andersson had not taken into consideration, makes it possible for the robot to aim at the desired point. These maps are implemented by means of Locally Weighted Regression (LWR) [8], though there are alternative approaches such as multi-layer neural networks (NN), radial basis functions (RBF), and so on. The slow convergence rates observed in common NN are impractical in our target problem, though efficient on-line learning algorithms have been proposed [9]. Though RBF is suited for faster learning, applying RBF without removing unreliable sensing data deteriorates the ability of RBF to learn the input-output relation. In comparison with RBF, LWR which is also suited for faster learning can easily incorporate the ability to explicitly identify unreliable training data. In addition, Gorinevsky and Connoly have compared several different approximation schemes (NN, RBF, LWR) on simulated robot inverse kinematics with added noise, and have shown that LWR is more accurate than all other methods [10].

Once a trajectory of the paddle has been planned, it must be executed as accurately as possible. We propose a feed-forward control scheme based on the iterative learning control (ILC) [11] to compensate for servo errors of the robot controller. This scheme synthesizes new inputs given several other inputs that have been correctly learned already.

Contributions of this monograph are summarized as follows:

(i) to verify the effectiveness of the memory based learning approach to the hybrid control problems involved in robotic tasks with intermittent interactions between a robot and its environment,
(ii) to demonstrate that a good combination of forward and inverse maps makes it possible for a robot system to plan appropriate motions in a dynamical environment, and
(iii) to propose a feed-forward control scheme perfectly realizing the robot's action plan using the input-output maps.

An overview of the table tennis robot system is presented in the next section. The third section describes "ball events" in one stroke in connection with tasks the robot has to execute. In the fourth section, we explain the input-output maps to predict virtual targets and how to plan the trajectory of the paddle. The fifth section proposes a feed-forward control scheme based on ILC for precisely tracking the trajectory of the paddle arbitrarily given as a function of time. After that, experimental results including rallies with a human opponent are reported to demonstrate the effectiveness of our approach.

2 Robot Table Tennis

2.1 Table Tennis System

Although Andersson stated that five degrees of freedom are required for the paddle attached to the robot to execute the table tennis task, the minimum number of the degrees of freedom is four, two for its position in a horizontal

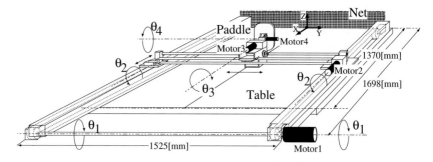

Fig. 1. Table tennis system

plane and two for its attitude because every incoming ball rebounds on a table before being hit during the game. In this paper, we adopt a robot system with the minimum number of degrees of freedom (four) to demonstrate only the effectiveness of our approach to the robot table tennis task without considering the use of redundant degrees of freedom.

Fig. 1 illustrates the table tennis robot system we developed. The robot is driven by four electric motors, motors 1 and 2 for the motion in a horizontal plane and motors 3 and 4 for the paddle attitude. The 155[mm] square paddle moves in parallel with the table at a height of 200[mm]. A stereo vision system (Quick MAG System 3: OKK Inc.), whose two cameras are set behind the paddle at $(x, y, z) = (3200, -1500, 2300)$ and $(2700, 2000, 1800)$ [mm] which are represented with reference to the coordinate frame shown in Fig. 1, extracts the location of the ball's center of gravity from the image every 1/60[s] using the stereo calculation.

2.2 Table Tennis Task

Let us explain the task the table tennis robot must do. The robot plays table tennis according to the same rules as humans. The table's width is 1.5 meters and the net is 0.16 meters high, which is the same as standard table tennis. (Andersson's robot has played against humans according to special rules for robots, scaling down and restricting the area that must be covered by the robots.) We consider that it is basically required for the table tennis robot to have the ability to perform the following task: *returning the ball to a desired point on the table with a specified duration of flight.*

Of course, these terminal conditions of the return trajectory must be chosen freely. In this monograph, we focus on this basic task and describe how to execute it. The quantitative measure of the robot's performance is the resultant accuracy of the terminal conditions (the aim point on the table and the time of flight).

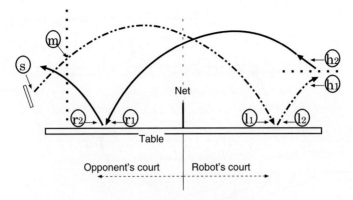

Fig. 2. Definitions of ball events

3 Ball Events in One Stroke

To make the following explanation clear, we define "ball events" by stating the conditions of a ball, which are given below and in Fig. 2.

3.1 Definition of Ball Events

Event- (s) Hit by the opponent
Event- (m) Passing through a virtual plane for the measurement
Event- (l) Bouncing on the robot's court
Event- (h) Hit by the robot
Event- (r) Bouncing on the opponent's court

Numbers 1 and 2 in Fig. 2 mean "before" and "after" the bounce.

The task starts when an opponent player or a pitching machine hits the ball (s) and the ball passes through a virtual plane (m) that is set to measure the motion of an incoming ball (the location of the virtual plane is $x = -1000[\text{mm}]$ as shown in Fig. 7). After the ball flies over the net, it bounces in the robot's court (l_1, l_2) and then hit by the robot (h_1, h_2). The ball is then returned and bounces in the opponent's court again (r_1, r_2).

3.2 Stroke Movement

The table tennis task can be divided into three subtasks as shown in Fig. 3 and summarized below:

TASK A: the planning task in which the robot predicts and plans the hitting trajectory,
TASK B: the hitting and returning task in which the robot achieved the hitting and returning motion planned in TASK A, and

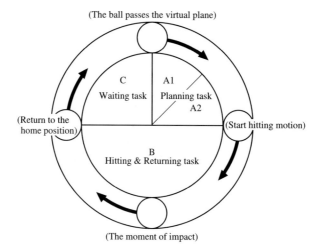

Fig. 3. One stroke movement

TASK C: the waiting task in which the system updates input-output maps (explained in section 4.1) and continues monitoring the ball motion to find that Event m occurs.

We divide TASK A into two parts. In A_1, the system predicts everything required to hit the ball and determines the virtual targets (hitting time, paddle attitude and velocity at the impact, see section 4). In A_2, the system generates the hitting trajectory and motion commands based on the prediction made in A_1 (see section 5).

3.3 Ball's State Estimation

In our approach, robot motions are planned based on the ball's state (position, velocity, acceleration) at Event m. The procedure for estimating the ball's state are given below:

(i) to store a sequence of the ball's location acquired by the stereo vision system until the ball passes through the virtual plane,
(ii) to fit a trajectory described by a first-order polynomial in each direction of X and Y and a second-order polynomial of time in the Z direction, using the least square algorithm,
(iii) to calculate the ball's state at Event m using the trajectory obtained in 2.

The same procedure is employed to estimate the ball's state during other events.

4 Paddle Motion Decision

4.1 Input-Output Maps

Determining the impact time of the ball hit by the paddle and the paddle position/attitude and velocity at that moment is most important for performing the table tennis task. This decision has to be made before the impact occurs. Our approach to making the decision is to use empirically acquired input-output maps instead of following Andersson's approach which is based on explicit models of the task and the environment derived from human knowledge [3].

These input-output maps represent three physical phenomena shown in Figs. 4 to 6 and are defined below.

Map 1 – A map for predicting the impact time of the ball hit by the paddle and the ball's position and velocity at that moment (at Event h_1) according to input vectors describing the state of the incoming ball at Event m.

Map 2 – A map representing a change in ball velocities before (h_1) and after (h_2) the impact of a ball against the paddle with a velocity (V_h) and attitude (θ_3, θ_4).

Map 3 – A map giving the relation between the ball's velocity just after the impact (h_2) and the landing point (p_r) / time (t_r) of the returned ball.

These maps are implemented by means of LWR(see Appendix). We use LWR on memorized data to fit a planar local model at each point where an input-output relation is to be predicted. Each data point is weighted by a function of its distance to the desired point in the regression and the model parameters are determined by the least square algorithm.

Learning Map 1 is implemented by observing the incoming balls hit by a human opponent or a pitching machine and storing the ball's state at the event m and h_1. Once the learning of Map 1 is completed, the robot can hit almost every kind of the incoming ball using Map 1. Learning Map 2 and Map 3 are implemented by observing the ball hit by the paddle with a certain attitude and velocity given at random and storing ball velocities before and after the impact, the paddle's attitude and velocity, and the landing point and time of the returned ball. We explain these maps in more detail hereinafter.

4.2 Prediction of the Ball's State at the Impact Using Map 1

The aerodynamics of a ball and the physics of a ball's bounce off of a table govern a change in the ball's state between Events m and h_1. Andersson formulated the ball's equations of motion:

$$a = -C_d|\boldsymbol{V}|\boldsymbol{V} + C_m|\boldsymbol{V}|\boldsymbol{W} \times \boldsymbol{V} - g \qquad (1)$$

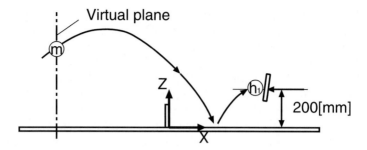

Fig. 4. [Map 1] – A map for predicting the impact time of the ball hit by the paddle and the ball position and velocity at that moment (h_1) according to input vectors describing the state of the incoming ball (m)

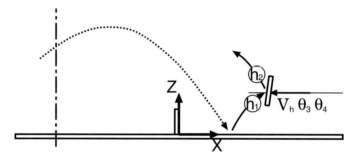

Fig. 5. [Map 2] – A map representing the relation between paddle state ($[V_h, \theta_3, \theta_4]$) and a change in ball velocities before and after the impact ($h_1 \rightarrow h_2$)

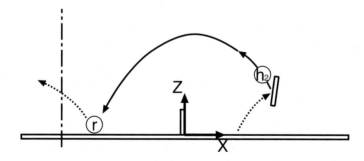

Fig. 6. [Map 3] – A map giving the relation between the ball velocity just after the impact (h_2) and the landing point and time of the returned ball (r)

where V is the velocity vector, W is the spin vector, a is the acceleration vector, g is the acceleration of gravity, and C_d and C_m are the drag coefficient & the coefficient for the Magnus effect. He also derived algebraic equations of a ball's bounce. We do not use these explicit models to predict the ball's state at the event h_1. Instead, we take notice that the ball's state at Event h_1 depends on the ball's state at Event m, that is, the ball's position vector $(p_{bmx}, p_{bmy}, p_{bmz})$, the velocity vector $(v_{bmx}, v_{bmy}, v_{bmz})$, and the spin vector $(w_{bmx}, w_{bmy}, w_{bmz})$. However, we cannot directly observe the spin vector. We substitute the acceleration vector for the spin vector as components of the ball's state at Event m because Eq.(1) holds at any point in time. Moreover, considering that most of the spin is along the Y axis (w_{bmy}), we use a_{bmz} in place of w_{bmy}. Additionally considering the condition that Event h_1 occurs, that is, $p_{bhz} = 200$[mm], we can represent a change in the ball's state between the Events m and h_1 as a nonlinear input/output relation of the form:

$$[p_{bmz}, v_{bmx}, v_{bmy}, v_{bmz}, a_{bmz}] \to [dt, dx, dy], V_{bh1} \qquad (2)$$

where $dt = t_h - t_m$, $dx = p_{bhx} - p_{bmx}$, $dy = p_{bhy} - p_{bmy}$ and $V_{bh1} = [v_{bh1x}, v_{bh1y}, v_{bh1z}]$ (See Fig. 7). We call these maps Map 1.

Learning Map 1

In the learning phase of Map 1, we store the ball's state at Events m and h_1 as a set of input/output pairs. The procedure of constructing Map 1 is as follows:

(i) Measuring the trajectories of the incoming ball without hitting.
(ii) The ball's state at Event m (B_m) is obtained using the measured position data around the virtual plane.
(iii) The ball's state at Event h (B_h) is obtained using the measured position data around the hitting plane.
(iv) Each input vector is standardized independently.
(v) We use the cross-validation error check [8] to store only reliable data sets.

Utilization of Map 1

In the lookup phase, the impact time $^*t_h$ and the ball position $(^*p_{bhx}, ^*p_{bhy})$ and velocity $(^*v_{bh1x}, ^*v_{bh1y}, ^*v_{bh1z})$ at that moment are predicted by Map 1 with the aid of LWR and the following relations:

$$^*t_h = t_m + {}^*dt \qquad (3)$$
$$^*p_{bhx} = p_{bmx} + {}^*dx \qquad (4)$$
$$^*p_{bhy} = p_{bmy} + {}^*dy \qquad (5)$$

where $^*dt, ^*dx$, and $^*dy$ are interpolated outputs of Map 1. The ball velocities $(^*v_{bh1x}, ^*v_{bh1y}, ^*v_{bh1z})$ are also predicted by Map 1 with the LWR interpolation.

4.3 Decision of the Hitting State of the Paddle Using Map 2 and Map 3

The remains of virtual targets, the paddle attitude and velocity at the impact, are determined using Map 2 and Map 3.

Learning Map 2 and Map 3

The collision dynamics between the ball and the paddle govern the transition of the ball's state from Event h_1 to Event h_2. The flight dynamics after hitting govern the transition of the ball's state from Event h_2 to Event r. We consider that these transitions are expressed as input-output maps of the form

$$[V_h, \theta_3, \theta_4] \rightarrow \boldsymbol{V}_{bh12}(= [v_{bh12x}, v_{bh12y}, v_{bh12z}]) \tag{6}$$

$$\boldsymbol{V}_{bh2} \rightarrow [dt_{hr}, d\boldsymbol{p}_{bhr}(= [dp_{bhrx}, dp_{bhry}])] \tag{7}$$

where $\boldsymbol{V}_{bh12}$ means the difference between the velocities just before hitting $\boldsymbol{V}_{bh1}$ and just after hitting $\boldsymbol{V}_{bh2}$, that is, $\boldsymbol{V}_{bh12} = \boldsymbol{V}_{bh2} - \boldsymbol{V}_{bh1}$. dt_{hr} is the flight duration of the returned ball. $d\boldsymbol{p}_{bhr}$ is the flight distance, that is, $d\boldsymbol{p}_{bhr} = \boldsymbol{p}_{br} - \boldsymbol{p}_{bh}$. Since the acquired data in the learning phase should be distributed uniformly in the input space, we randomly choose input variables V_h and (θ_3, θ_4). The measurement is executed by hitting a ball under these paddle conditions at the hitting point determined by using Map 1.

Utilization of Map 2 and Map 3

If we regard the input-output maps of Eq. (6) and Eq. (7) as forward maps involved in the return shot, we need their inverse maps to determine appropriate control variables V_h and (θ_3, θ_4). Andersson modeled the collision dynamics between the ball and the paddle for the special case of a spinless ball with the paddle normal parallel to the paddle velocity [3]. According to his model, the paddle attitude (θ_3, θ_4) is uniquely determined by $\boldsymbol{V}_{bh12}$ and the paddle velocity $V_h(=[v_{hx}, 0, 0])$ is uniquely determined by $\boldsymbol{V}_{bh12}$ and $\boldsymbol{V}_{bh1} \cdot \boldsymbol{V}_{bh12}$ ("·"

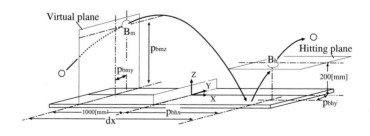

Fig. 7. Prediction of the Ball Trajectory

means the inner product). Referring to his model, we consider the following inverse maps to determine the control variables.

$$[dt_{hr}, d\boldsymbol{p}_{bhr}] \to \boldsymbol{V}_{bh2} \tag{8}$$

$$[\boldsymbol{V}_{bh12}, \boldsymbol{V}_{bh1} \cdot \boldsymbol{V}_{bh12}] \to [V_h, \theta_3, \theta_4] \tag{9}$$

which are constructed using the same stored data that constitute the forward maps of Eq. (6) and (7).

Once the desired $[dt_{hr}, d\boldsymbol{p}_{bhr}]$ is given, the required ball velocity just after hitting $\boldsymbol{V}_{bh2}$ is determined using the inverse map of Eq. (8). A change in velocities at the impact point $\boldsymbol{V}_{bh12}$ is then calculated by $\boldsymbol{V}_{bh2}$ and $^*\boldsymbol{V}_{bh1}$ (predicted by Map 1), and the control variables V_h and (θ_3, θ_4) are determined using the inverse map of Eq. (9).

4.4 Paddle Trajectory

Given virtual targets using the above-mentioned input-output maps, we can define a trajectory of the paddle which attains the designated final position (p), velocity (v) and acceleration (a) corresponding to the virtual targets. We use a fifth-order polynomial as the position trajectory of each axis and set the velocity trajectory of the form

$$v(t) = c_1 t^4 + c_2 t^3 + c_3 t^2 \tag{10}$$

considering the boundary conditions:

$$\begin{aligned} p(0) &= p_i, \ v(0) = 0, \ a(0) = 0 \\ p(t_h) &= p_h, \ v(t_h) = v_h, \ a(t_h) = 0 \end{aligned} \tag{11}$$

where t_h is the time for the entire trajectory which is predetermined by taking account of the robot's torque limits. As may be seen from these conditions, the paddle motion begins at rest and ends with zero acceleration. The coefficients c_1, c_2, and c_3 are obtained by applying these boundary conditions. After completing the paddle motion to hit a ball, the paddle has to return to the waiting position as soon as possible. We use the same trajectory as Eq. (10) reversing the boundary conditions of Eq. (11).

5 Generation of Paddle Movement

Once a trajectory of the paddle has been planned, it must be executed as accurately as possible. The iterative learning control [11] based on the iterative operations of a robot is an effective method to accurately track a desired motion pattern without modeling the dynamics of the controlled object. Unfortunately, this control method is not directly applicable to the table tennis task because the desired motion patterns of the paddle vary depending on the motion of the incoming ball. In this section, we propose a method of synthesizing new control inputs without the use of iterative operations in learning control given several other inputs that have been correctly learned already.

5.1 Synthesizing a New Trajectory

Let us assume that the following three desired motion (velocity) trajectories of the paddle are given:

$$v_a(t)(0 \leq t \leq T_a), v_b(t)(0 \leq t \leq T_b), v_c(t)(0 \leq t \leq T_c) \quad (12)$$

where each trajectory is represented as a fourth-order polynomial given by Eq. (10) and its coefficients are expressed by $\mathbf{c}_a = [c_{a1}, c_{a2}, c_{a3}]^T$, $\mathbf{c}_b = [c_{b1}, c_{b2}, c_{b3}]^T$ and $\mathbf{c}_c = [c_{c1}, c_{c2}, c_{c3}]^T$ respectively. Then, an arbitrary trajectory $v_d(t)(0 \leq t \leq T_d)$ expressed by the same fourth-order polynomial as Eq. (10) with coefficients $\mathbf{c}_d = [c_{d1}, c_{d2}, c_{d3}]^T$ can be represented as a linear combination of the trajectories $v_a(t), v_b(t)$, and $v_c(t)$, that is,

$$v_d(t) = k_a v_a(t) + k_b v_b(t) + k_c v_c(t) \quad (0 \leq t \leq T_d) \quad (13)$$

where $T_d \leq \min(T_a, T_b, T_c)$ and the coefficients $[k_a, k_b, k_c]^T$ are determined by solving

$$\begin{bmatrix} c_{a1} & c_{b1} & c_{c1} \\ c_{a2} & c_{b2} & c_{c2} \\ c_{a3} & c_{b3} & c_{c3} \end{bmatrix} \begin{bmatrix} k_a \\ k_b \\ k_c \end{bmatrix} = \begin{bmatrix} c_{d1} \\ c_{d2} \\ c_{d3} \end{bmatrix} \quad (14)$$

The 3×3 matrix on the left side of Eq. (14) has to be nonsingular. It should be noted that the relation of Eq. (13) holds even if each duration of the motion trajectory is different from the rest. Concerning the number of trajectories provided to synthesize a new trajectory, it depends on the degree of the polynomial function and its boundary conditions and is equal to the number of coefficients representing the polynomial.

5.2 Synthesizing New Input Commands

In the following, we assume that the controlled object can be represented as a linear system with an input u and an output v to simplify the explanation. If a new desired trajectory $v_d(t)$ is given by Eq. (13), the approximate ideal input $u_d(t)$ for the desired trajectory $v_d(t)$ can be obtained by

$$\mathbf{u}_d = k_a \mathbf{u}_a + k_b \mathbf{u}_b + k_c \mathbf{u}_c \quad (15)$$

where the coefficients $[k_a, k_b, k_c]^T$ are the same as those in Eq. (13). Although Kawamura et al. also proposed an interpolation method based on a linear relation of the controlled object, they did not mention the details of interpolation [12].

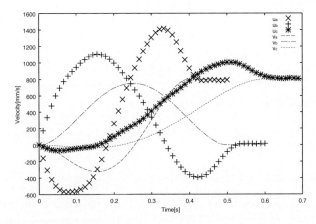

Fig. 8. Patterns A, B and C (ui: Learned input. vd$i$: Desired trajectory.)

5.3 Experimental Verification

To demonstrate the effectiveness of the proposed method, we now present experimental results. In the following, we focus on the motion of the actuator on the X axis. Fig. 8 shows the approximate ideal inputs for the three desired trajectories learned by using ILC. In this figure, we omit the actual trajectory of each motion pattern because it coincides almost perfectly with the desired one.

Figure 9 shows an experimental result obtained by applying the proposed method. A new trajectory is accurately achieved by the input $u_d(t)$ synthesized by combining the learned inputs for the three trajectories given in Fig. 8. We can also see that the synthesized input $u_d(t)$ coincides almost perfectly with the input $u_{ILC}(t)$ obtained by using ILC.

Next we explain how to achieve an accurate stroke movement and demonstrate an experimental result. This is an application of the proposed method to the table tennis task. As mentioned in Section 3, one stroke motion of the paddle consists of two parts. One is the "hitting motion" that starts from a home position (at $t = 0$) and ends at a predicted hit point p_h with a desired velocity v_h and attitude of the paddle (at $t = t_h$). Another is "returning motion" that starts from the final state of the "hitting motion" (at $t = t_h$) and ends at the home position (at $t = t_f$). Fig. 10 illustrates an experimental result of the stroke motion achieved by using the proposed method. In this stroke motion, the desired velocity at the hit point $p_h = 500[\text{mm}]$ is set $v_h = 200[\text{mm/s}]$. Hitting motion ends at $t_h = 0.4[\text{s}]$ and returning motion ends at $t_f = 0.8[\text{s}]$. From this figure, we can see that a given stroke movement of the paddle is accurately achieved by the input determined using the proposed method.

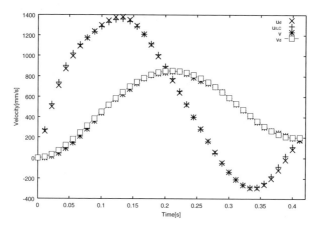

Fig. 9. A new input and output trajectory generated by patterns A, B, and C (u_d: Input calculated by the proposed method, u_{ILC}: Input obtained by ILC, v: Actual trajectory, v_d: Desired trajectory)

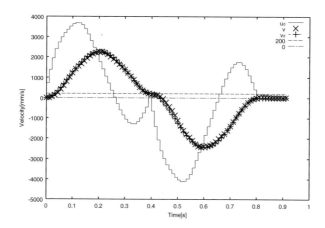

Fig. 10. One stroke movement (u_d: Input calculated by the proposed method, v: Actual trajectory, v_d: Desired trajectory)

6 Experimental Results

In this section, we demonstrate the effectiveness of our approach by presenting experimental results achieved by the table tennis robot. Short movies of the robot's learning, returning the ball, and rallying with a human can be found at http://robotics.me.es.osaka-u.ac.jp/MiyazakiLab/Research/pingpong/2003/movie-e.html.

6.1 Experimental results of the "ball controlling task"

Ball Prediction with the Map 1

We evaluated the prediction accuracy of our method through the following experiments carried out under the following three spin conditions:

(a) knuckle balls (without fast spin) pitched by a pitching machine
(b) three kinds of ball spins (top spin, back spin and knuckle) pitched by a pitching machine
(c) slow spin ball hit by a human

The ball states were extracted offline from 300 data of trajectories measured in advance. Two hundred ball states were used to construct the database and others were used to evaluate the prediction accuracy. The results of evaluation are shown below.

There are two major sources of prediction error. One is the propagation of measurement noises. Another is a shortage of the data that constitute the learned maps. The former is an inevitable source which remains even if the learned maps are perfect.

The standard deviation of the ball position data acquired by the vision system is 5.0 [mm]. In the process of state estimation using the least square algorithm, the standard deviation is reduced (2.5 [mm] in position and 95 [mm/s] in velocity). We can estimate the prediction error using the error analysis, provided that the learned maps are perfect. For example, if we assume that the ball flies in 0.4 [s] from the virtual plane at $x = -1000$[mm] to the hitting position (Fig. 7), the prediction error of the ball's state at the impact in X direction is about 40 [mm] (standard deviation) in position and 95 [mm/s] in velocity.

Prediction of the Ball Trajectory Pitched by the Machine

The ball trajectories were given by the pitching machine, which is set to pitch knuckle balls with various angles, positions and velocities. In the process of prediction with Map 1, we chose the weight matrix of the distance function to be a unit matrix I, and the band width h is to be 0.8 in the LWR (see Appendix). Table 1 presents the averages and standard deviations of the prediction errors in time, position, and ball velocities at the impact.

The error in the hitting time is a few milliseconds, and the position errors are within a few centimeters. This means that if the paddle reaches the predicted position at the predicted time, it can hit the ball around the center of the paddle. The errors in the ball positions and velocities are close to the result of the error analysis mentioned above, which is enough to determine the paddle conditions at the impact.

Table 1. Prediction error (knuckle)

	Average	Standard deviation
dt [s]	0.003641	0.003364
dx [mm]	29.7799	20.4901
dy [mm]	18.0082	14.8297
vx [mm/s]	64.6510	51.9007
vy [mm/s]	46.6689	70.5712
vz [mm/s]	42.5393	55.5754

Prediction of the Ball Trajectory with Different Spins

We evaluated the prediction accuracy of the ball motion with different spins. The map acquired in the learning phase consists of three data groups, each of which has one hundred data elements and is characterized by the ball spin such as knuckle (no spin), top spin, and back spin. The pitching machine with various positions and various angles pitched the balls at various velocities and spins. The prediction error was evaluated using the data not included in the map.

Tables 2 and 3 depict the results for balls with top spin and back spin. Though the prediction errors are larger than those for the knuckle ball only (as compared with Table 1), they are acceptable for hitting a ball. The results also indicate the ability to recognize the spin of the ball. In other words, the system can properly predict a ball's different spins.

Table 2. Prediction error (top spin)

	Average	Standard deviation
dt [s]	0.004534	0.009044
dx [mm]	16.9239	26.5432
dy [mm]	8.8132	14.0514
vx [mm/s]	117.0519	229.9383
vy [mm/s]	33.0024	55.7955
vz [mm/s]	41.2663	65.5895

Table 3. Prediction error (back spin)

	Average	Standard deviation
dt [s]	0.006952	0.010768
dx [mm]	18.9122	25.9342
dy [mm]	10.7785	16.0699
vx [mm/s]	66.1913	102.2234
vy [mm/s]	33.1297	44.9824
vz [mm/s]	67.0640	91.1075

Prediction of the Ball Trajectory hit by a chuman

We evaluated the prediction accuracy of the ball motion hit by a human. The pitching machine set behind the robot's court pitched the ball to the human. The human returned the ball to the robot's court and the ball trajectory was measured. In this experiment we requested the human to hit the ball normally.

Figures 11 and 12 depict the predicted hitting time and position with prediction errors. The prediction accuracy was almost the same as in the knuckle case.

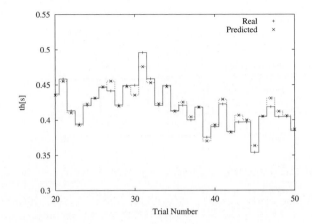

Fig. 11. Prediction accuracy of the hitting time

Map for the Decision of the Hitting Motion

We experimentally acquired Map 2 and Map 3 with which the robot decides the hitting motion.

In the learning phase, the robot hit the balls pitched by a pitching machine with a constant angle, velocity, and spin using a paddle driven by the commanded velocities and angles in the range of

$$800 \leq V_h \leq 2000 [\text{mm/s}] \quad (16)$$
$$-50 \leq \theta_3 \leq 50 [\text{deg}] \quad (17)$$
$$-20 \leq \theta_4 \leq 20 [\text{deg}] \quad (18)$$

Fig. 13 illustrates a map representing the relation between $\boldsymbol{V}_{bh12}$ and V_h acquired by learning 120 trajectories of the returned ball, as an example of the acquired maps. Three surfaces corresponds to the cases of $v_{bh12z} = 0, -500$ and 500 [mm/s] respectively. Using this map, the robot can determine the paddle velocity that gives the required ball velocity.

Next we evaluated the capability of controlling the flight duration of the returned ball with the acquired maps. These maps were well learned in 140 trials. We used these well-learned maps in the following experiments. We fixed the desired landing point as $x = -1100 [\text{mm}], y = 300 [\text{mm}]$. We also set the desired duration of flight as 0.5 [s] and 0.7 [s] alternatively. In this lookup phase, the robot continues to acquire the data of the ball and paddle movements for Map 2 and Map 3 to improve the hitting motion, that is, the robot continues learning in parallel. Figs. 14 and 15 show experimental results of the duration of flight and errors of landing point for the last 40 trials performed by a robot.

The influence of measurement noises on the landing point can be estimated similarly to the prediction error. Assuming that the ball bounds on the oppo-

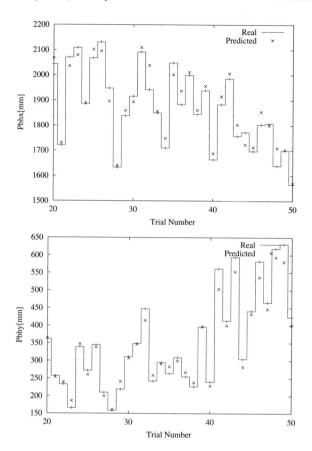

Fig. 12. Prediction accuracy of the hitting position

nent's table in 0.8 [s] after passing through the virtual plane, we can estimate the returning position error to be about 80 [mm] (standard deviation) through the error analysis. The returning position errors shown in Fig. 15 are close to this value, which supports the effectiveness of Map 2 and Map 3.

We also measured the accuracy of the landing point by making a comparison between the robot and a human player with ten years experience playing table tennis. Fig. 16 is the errors in the landing point for the last 40 trials performed by the human. We can see that the robot achieves almost the same accuracy as the human. Both the human and a robot tend to return balls over the destination.

Figs. 17 and 18 present the ball and paddle trajectories over the 179th and 180th trial. We can see that the robot achieves different returning trajectories with the same landing position by controlling the duration of flight.

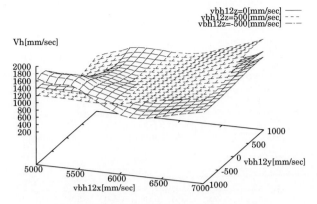

Fig. 13. Maps between V_{bh12} and V_h

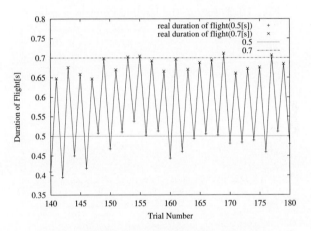

Fig. 14. Duration of flight after hitting

6.2 Experimental Results of the Rally Task with a Human

Rally Task Experiment

We demonstrate the robot rally with a human as an application of our approach (Fig. 19). The "rally task" means the table tennis rally that people generally play. We consider it as the repetition of "ball controlling task" described in the previous subsection.

Experimental Results

We demonstrate that the robot can perform the rally task with a human using the proposed method described previously. In the experiment, a human hit a ball toward the robot at random and the robot returned the ball with a fixed duration of flight (dt_{hr}=0.55 [s]) to a desired landing point ($p_{rx} = 1550$ [mm],

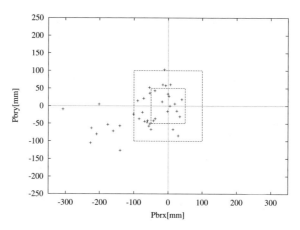

Fig. 15. Errors in the landing point by a robot

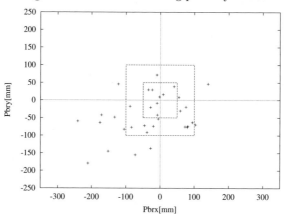

Fig. 16. Errors in the landing point by a human

$p_{ry}=0.3 \times {}^*p_{bhy}$) for the opponent's easy hitting, where ${}^*p_{bhy}$ is a predicted impact point.

Figures 20 to 22 present some part of the data acquired in the rally. Figure 20 shows the time history of the ball motion in X direction. Figures 21 and 22 show the ball and paddle trajectories on plane where the waiting position of the paddle is $x=900$ [mm], $y=0$ [mm]. It should be noted that the rally task comprises three subtasks (Task A~C) described in section 3.2. We can see that the robot returns the ball to the point the opponent can hit easily by changing the impact point back and forward, right and left.

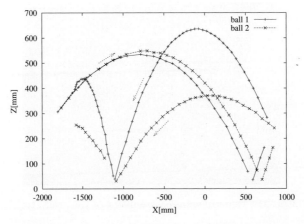

Fig. 17. Ball trajectory in X-Z plane (179th and 180th trial)

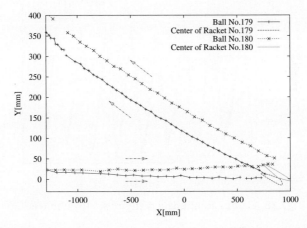

Fig. 18. Ball and paddle trajectory in X-Y plane (179th and 180th trial)

7 Conclusion

We have described an approach for a robot to play the table tennis task on the premise that no explicit models concerning the task are employed and determining the impact time of the ball hit by the paddle and the paddle position/attitude and velocity at that moment is most important for performing the table tennis task. Our approach to making the decision is to use empirically acquired input-output maps which are implemented by means of LWR. These maps represent three physical phenomena, the aerodynamics of a ball, physics of a ball's bounce off of a table, and physics of a ball's bounce off of a paddle. We have also proposed a feed-forward control scheme based on ILC to accurately achieve the stroke movement of the paddle as determined by using

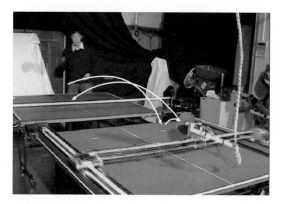

Fig. 19. Experimental environment of the "rally task"

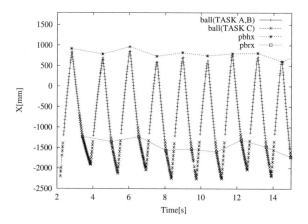

Fig. 20. Ball motion in the "rally task" (in X direction)

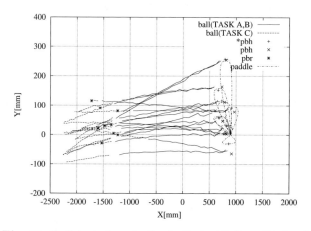

Fig. 21. Ball trajectory in the "rally task" (in X-Y plane)

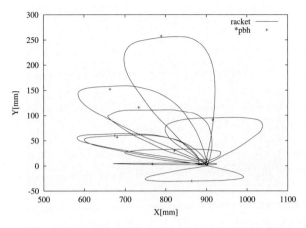

Fig. 22. Paddle trajectory x in the "rally task"

these maps. Experimental results including rallies with a human opponent also have been reported.

Appendix: Locally Weighted Regression(LWR)[8]

Let us consider approximating the n training values $\{y_1, y_2, ..., y_i, ..., y_n\}$ taken under different n conditions $\{\boldsymbol{x_1}, \boldsymbol{x_2}, ..., \boldsymbol{x_i}, ..., \boldsymbol{x_n}\}$ to the linear model

$$y = \beta_0 + \beta_1 x_1 + ... + \beta_j x_j + ... + \beta_m x_m = \boldsymbol{x}^T \boldsymbol{\beta} \tag{19}$$

where $x_j (j = 1, ..., m)$ is the jth components of the input vector $\boldsymbol{x}$.

In Locally Weighted Regression (LWR), the linear local model can be specialized to the query by emphasizing nearby points. Weighting the criterion is done in the following way

$$C(\boldsymbol{q}) = \sum_{i=1}^{n}[(\boldsymbol{x}_i^T \boldsymbol{\beta} - y_i)^2 K(d(\boldsymbol{x}_i, \boldsymbol{q}))] \tag{20}$$

where $K()$ is the weighting or kernel function and $d(\boldsymbol{x}_i, \boldsymbol{q})$ is a distance between the stored data point $\boldsymbol{x}_i$ and the query point $\boldsymbol{q}$.

We can predict $\hat{y}$

$$\hat{y}(\boldsymbol{q}) = \boldsymbol{q}^T \hat{\boldsymbol{\beta}} \tag{21}$$

using the estimated $\hat{\boldsymbol{\beta}}$ to minimize the criterion (20).

There are many different approaches to defining a distance function. We use the following model in which the distance function between the query point $\boldsymbol{q}$ and ith data $\boldsymbol{x}_i$ and the weighting function $K(d_i)$ are

$$d_i = \sqrt{\sum_{j=1}^{p} m_j^2 (x_{ij} - q_j)^2} \qquad (22)$$

$$K(d_i) = w_i^2 = exp(-\frac{d_i^2}{h^2}) \qquad (23)$$

where the m_j is the feature scaling factor for the jth dimension and a smoothing or bandwidth parameter h defines the scale or range over which generalization is performed.

Then, introducing a matrix $\boldsymbol{X}$ whose ith row is $\boldsymbol{x}_i$ and a vector $\boldsymbol{y}$ whose ith element is y_i, the estimated $\hat{\boldsymbol{\beta}}$ is given by

$$\hat{\boldsymbol{\beta}} = (\boldsymbol{Z}^T \boldsymbol{Z})^{-1} \boldsymbol{Z}^T \boldsymbol{v} \qquad (24)$$

where $\boldsymbol{Z} = \boldsymbol{W}\boldsymbol{X}$, $\boldsymbol{v} = \boldsymbol{W}\boldsymbol{y}$, and $\boldsymbol{W}$ is a $n \times n$ matrix with diagonal elements w_i.

References

1. Burridge R, Rizzi A, Koditschek D (1999) Sequential composition of dynamically dextrous robot behaviors. The International Journal of Robotics Research 18(6):534–555
2. Schmidt R, Lee T (1999) Motor Control and Learning: A Behavioral Emphasis. Human Kinetics
3. Andersson R (1988) A Robot Ping-Pong Player: Experiment in Real-Time Intelligent Control. The MIT Press
4. Ramanantsoa M, Duray A (1994) Towards a Stroke Construction Model. The International Journal of Table Tennis Sciences 2(2):97–114
5. Miyazaki F, Takeuchi M, Matsushima M, Kusano T, Hashimoto T (2002) Realization of table tennis task based on virtual targets. In: Proc. IEEE Int. Conf. on Robotics and Automation. Washington D.C. 3844–3849
6. Bühler M, Koditschek D, Kindlmann P (1994) Planning and control of a juggling robot. The International Journal of Robotics Research 13(2):101–118
7. Aboaf E, Drucker S, Atkeson C (1989) Task-level robot Learning: Juggling a tennis ball more accurately. In: Proc. IEEE Int. Conf. on Robotics and Automation. Scottsdale, Arizona 1290–1295
8. Atkeson C, Moore A, Schaal S (1997) Locally weighted learning. Artificial Intelligence Review 11(1-5):11–73
9. MacDorman K (1999) Partition nets: An efficient on-line learning algorithm. In: Proc. 9th Int. Conf. on Advanced Robotics, ICAR'99. Tokyo 529–535
10. Gorinevsky D, Connolly T (1994) Comparison of some neural network and scattered data approximations: The inverse manipulator kinematics example. Neural Computation 6:521–542
11. Arimoto S, Kawamura S, Miyazaki F (1984) Bettering Operation of Robots by Learning. Journal of Robotic Systems 1(2):123–140
12. Kawamura S, Fukao N (1994) Interpolations for Input Torque Patterns obtained through learning control. In: Proc. Int. Conf. on Control, Automation, Robotics, and Vision. Singapore 2194–2198

13. Myers R (1990) Classical and Modern Regression with Applications. Duxbury, New York
14. Press W, Teukolsky S, Vetterling W, Flannery B (1993) Numerical Recipes in C. Gijutsu Hyouronsha (Japanese Ed.)
15. Cross R (2002) Measurements of the horizontal coefficient restitution for a super ball and a tennis ball. The American Journal of Physics 70(5):482–489

Printing: Krips bv, Meppel
Binding: Stürtz, Würzburg

REF
TJ 211.35 .A38 2006
Kawamura, Sadao.
Advances in robot control : from ev(

DATE DUE

GAYLORD			PRINTED IN U.S.A.